Springer Series in Advanced Microelectronics

Volume 40

For further volumes:
http://www.springer.com/series/4076

The Springer Series in Advanced Microelectronics provides systematic information on all the topics relevant for the design, processing, and manufacturing of microelectronic devices. The books, each prepared by leading researchers or engineers in their fields, cover the basic and advanced aspects of topics such as wafer processing, materials, device design, device technologies, circuit design, VLSI implementation, and subsystem technology. The series forms a bridge between physics and engineering and the volumes will appeal to practicing engineers as well as research scientists.

Detlev Richter

Flash Memories

Economic Principles of Performance, Cost and Reliability Optimization

Springer

Detlev Richter
Institute of Technical Electronics
Technical University of Munich
Munich
Germany

ISSN 1437-0387
ISBN 978-94-007-6081-3 ISBN 978-94-007-6082-0 (eBook)
DOI 10.1007/978-94-007-6082-0
Springer Dordrecht Heidelberg New York London

Library of Congress Control Number: 2013942008

Printed on acid-free paper

Springer is part of Springer Science+Business Media (www.springer.com)

To Svea and Annett

Preface

- System development and optimization is an iterative process of technical and commercial decisions between hardware and software solutions based on different architectures and concepts. Decisions are based on technical specification of the current product generation while often they do not reflect the long-term product roadmap. The long-term development roadmap of a memory product family has a significant influence on system cost and system performance and on the commercial success of a new memory-centric system family.
- Non–volatile solid-state memories changed the way how to interact with our world. NAND flash became the enabling technology for portable storage. The availability of solid-state non–volatile memories large enough and fast enough for a certain application space is enabling year by year new commercial products in this application area.
- The NAND flash product roadmap doubles year-by-year the memory density and outperforms the conservative predictions of the ITRS roadmap. In the past, the read access of a memory cell through a NAND string of 32 cells was commented as a challenging and less reliable memory concept. NAND flash was becoming the driver of the whole semiconductor memory industry. This book will systematically introduce cell, design, and algorithm aspects of NAND flash, compare them to alternative solutions and make conclusions valid for flash memories.
- A complexity figure is introduced to illustrate the challenges along a nonvolatile memory development. Performance indicator values are calculated based on the introduced memory array model to reduce this complexity and derive quantitative figures to guide performance, cost, and durability optimization of nonvolatile memories and memory-centric systems.
- A performance indicator methodology is developed and the main graphical representation is based on performance and application trend lines. The strength of this method is the quantitative judgment of different cell and array architectures. The methodology enables the opportunity to compensate insufficient

margins in the durability parameters on system level by increasing other values like memory density and program performance. The introduced model-based performance indicator analysis is a tool set supporting iterative decision processes for complex systems.

Important Notice Regarding the Use of this Thesis

The information contained in this work may describe technical innovations, which are subject of patents held by third parties. Any information given in this work is in no form whatsoever an inducement to infringe any patent.

April 2013 Detlev Richter

Acknowledgments

The idea to describe all the different aspects—optimizing a complex nonvolatile semiconductor-based systems—and to make an assessment of each method was born during my employment time at Qimonda Flash GmbH.

During my time working as an independent technology consultant, I missed a systematic analysis of all effects influencing a nonvolatile system. The systematic definition of each nonvolatile effect was the starting point to develop the nonvolatile memory lectures given at Technical University Munich.

My studies are focused to work out the dependencies between nonvolatile memory key parameters within complex systems and translating them into structured models. These performance models are part of nonvolatile memory lectures to demonstrate the importance of certain hidden parameters for a commercially successful product and system development roadmap.

A model-based approach is introduced and used to analyze the evolutionary development of complex systems and to apply the key performance parameter methodology to predict disruptive innovations for nonvolatile memories as the most complex semiconductor devices with a regular structure and cutting-edge lithography requirements.

I would like to thank Prof. Dr. rer. nat. Doris Schmitt-Landsiedel for supporting, accompanying, and accepting this work.

Further thanks are due to:

Dr. Andreas Kux for support, suggestions, and discussions for almost 12 years;

Dr. Christoph Friederich for support, sharing new ideas especially on nonvolatile memories;

Dr. Gert Koebernik for programming, testing, and designing flash algorithm on highest level;

Andreas Täuber for a productive time working together on embedded DRAM's, on charge trapping, and on floating gate flash memory designs.

Dr. Thomas Kern, Dr. Josef Willer, Dror Avni, Shai Eisen, Asaf Shappir, and Boaz Eitan for an intensive and innovative working time improving charge trapping flash memories. Doris Keitel-Schulz, Dr. Markus Foerste, Dr. Jan Hayek, Dr. Michael Specht, Johann Heitzer, and Rino Micheloni for the great support during my product innovation times; Further thanks to all people working on flash memories during my times at Infineon Flash GmbH and Qimonda Flash GmbH.

Contents

Abbreviations and Symbols

Abbreviations

ABL	All Bit Line
BBM	Bad Block Management
BFR	Bit Failure Rate
BL	Bit Line
BOM	Bill of Material
BPD	Background Pattern Dependency
BTBHH	Band-To-Band Hot Hole
CD	Coupling Dielectric
CG	Control Gate
CHE	Channel Hot Electron
CTF	Charge Trapping Flash
DIMM	Dual Inline Memory Module
DRAM	Dynamic Random Access Memory
ECC	Error Correcting Code
EDC	Error Detection Code
EEPROM	Electrical Erasable Programmable ROM
EOT	Equivalent Oxide Thickness
EPROM	Erasable Programmable ROM
ERS	Erase
ETOX	Erase Through OXide
F	Minimal feature size
FG	Floating Gate
FLOTOX	FLOating-gate Tunneling Oxide
FTL	File Translation Layer
FN	Fowler Nordheim
GIDL	Gate Induced Drain Leakage
GSL	Ground Select Line
GWL	Global Word Lines
HDD	Hard Disk Drives
HPC	High-Performance Computing

IPD	Inter Poly Dielectric
ISPP	Incremental Step Pulse Programming
ITRS	International Technology Roadmap for Semiconductors
JEDEC	Joint Electron Devices Engineering Council
LSB	Least Significant Bit
MCP	Multi-Chip Package
MLC	Multi-Level Cell (applied for 2 bit/cell–4 level per cell)
MNOS	Metal Nitride Oxide Semiconductor
MOSFET	Metal Oxide Semiconductor Field Effect Transistor
MSB	Most Significant Bit
NROM	Nitride ROM
NVM	Non-Volatile Memory
ONFI	Open NAND Flash Interface Working Group
OTP	One Time Programmable
PAE	Program After Erase
PBE	Program Before Erase
PGM	Program
PROM	Programmable ROM
PV	Program Verify
RAM	Random Access Memory
RBER	Raw Bit Error Rate
ROM	Read Only Memory
RTN	Random Telegraph noise
SADP	Self-Aligned Double Patterning
SAQP	Self-Aligned Quadruple Patterning
SA-STI	Self-Aligned-Shallow Trench Isolation
SBPI	Self Boosted Program Inhibit
SL	Source Line
SLC	Single Level Cell (applied for 1 bit/cell–2 level per cell)
SLN	Source Line Noise
SONOS	Silicon-Oxide-Nitride-Oxide-Silicon
SRAM	Static RAM
SSD	Solid-State Disk
SSG	String Select Gate
SSL	String Select Line
STI	Shallow Trench Isolation
TCO	Total Cost of Ownership
TD	Tunneling Dielectric
TOX	Tunnel Oxide
TLC	Triple Level Cell (applied for 3 bit/cell–8 level per cell)
TSV	Through Silicon Via
WL	Word Line
XLC	e**X**tended Level per Cell (applied for 3 bit/cell and 4 bit/cell up to 16 level per cell)

Symbols

α_G	Gate coupling ratio
α_D	Drain coupling ratio
BW_{RD_DRAM}	Read data bandwidth (DRAM). [MB/s]
BW_{WR_DRAM}	Write data bandwidth (DRAM). [MB/s]
$C_{[x, y, xy]}$	Coupling capacitance between adjacent cells in x, y and xy directions. [F]
C_{BD}	Capacitance between bulk and nitride layer (bottom dielectric) [F]
C_B	Capacitance between floating gate and bulk [F]
C_D	Capacitance between floating gate and drain [F]
C_{FC}	Capacitance between floating gate and control gate [F]
C_S	Capacitance between floating gate and source [F]
C_{TD}	Capacitance between nitride layer and control gate (top dielectric) [F]
d_{BD}	Thickness of bottom dielectric
d_{TOX}	Thickness of Tunnel Oxide
d_{TD}	Thickness of top dielectric
DR	Data Rate for sequential data bursts [MB/s]
DR_{RD_Burst}	Data Rate for read burst access to an open row [MB/s]
DT_{RD}	Read Data Throughput [MB/s]
DT_{PGM}	Program Data Throughput [MB/s]
DT_{WR}	Write Data Throughput [MB/s]
f_{IF}	Clock frequency of the data interface [Hz]
g_m	Transconductance of the MOS transistor (flash cell) [V/decade]
I_{cell}	Cell current for a flash cell transistor [A]
I_{DS}	Drain current for MOS transistor (flash cell) [A]
$I_{leakage}$	Leakage current flow in a cell array [A]
I_{sense}	Sense current [A]
I_{string}	Current flow through the NAND string [A]
ϕ_B	Energy height of a potential barrier in the device's band structure [eV]
Q_{FG}	Overall charge on the floating gate capacitor [C]
R	Resistance [Ω]
σ	Standard deviation of the distribution
t	Time [s]
t_{BERS}	Block Erase time for NAND flash [s]
t_{CAS}	Column Access Strobe [s]
t_{CL}	CAS Latency == Column Access Strobe [s]
t_{DI}	Data in time [s]
t_{IF}	Data interface cycle time [s]
$t_{NV\_storage}$	Nonvolatile physical storage time [s]
t_{PROG}	Page program cycle time for NAND flash [s]
t_R	Read time for NAND flash—data transfer from flash array to data register [s]

t_{RACT}	Read Access Cycle Time [s]
t_{RAS}	Row Access Strobe [s]
t_{RC}	Read data interface cycle time (asynchronous NAND data interface) [s]
t_{RCD}	Row address to Column address Delay [s]
t_{RP}	Row Pre-charge cycle time [s]
t_{WC}	Write data interface cycle time (asynchronous NAND data interface) [s]
t_{WCT}	Write Cycle Time [s]
t_{WR}	Write Recovery time [s]
τ	Tau is characterizing the rise time [s]
V	Applied voltage [V]
V_{cc}	Voltage of the common collector [V]
V_{CG}	Voltage of the Control Gate [V]
V_{DS}	Voltage applied between Drain and Source of the cell transistor [V]
V_{FG}	Voltage of the Floating Gate [V]
V_{GS}	Voltage applied between Gate and Source of the cell transistor [V]
V_{SE}	Voltage of the Storage Element [V]
V_{SS}	Ground potential [V]
V_{PP}	Voltage applied to selected word line during program (program pulse) [V]
ΔV_{PP}	Program pulse voltage increment [V]
V_{RD_PASS}	Voltage applied to unselected word lines during read operation [V]
V_{th}	Threshold voltage of the MOS transistor [V]
V_{th_LL}	Left side of the erased V_{th} cell distribution [V]
V_{th_LH}	Right side of the erased V_{th} cell distribution [V]
V_{th_HL}	Left side of the programmed V_{th} cell distribution [V]
V_{th_HH}	Right side of the programmed V_{th} cell distribution [V]

Recent Publications by the Author

[FHK+08] C. Friederich, J. Hayek, A. Kux, T. Müller, N. Chan, G. Köbernik, M. Specht, D. Richter, and D. Schmitt-Landsiedel, "Novel model for cell - system interaction (MCSI) in NAND Flash," *International Electron Devices Meeting, Tech. Digest*, Dec. 2008.

[R06] D. Richter, "Multi-bit per cell Flash Memories–Interactions between digital and analogue solutions," *U.R.S.I.Kleinheubacher Tagung 2006.*

[FSL+06] C. Friederich, M. Specht, T. Lutz, F. Hofmann, L. Dreeskornfeld, W. Weber, J. Kretz, T. Melde, W. Rösner, E. Landgraf, J. Hartwich, M. Stadele, L. Risch, and D. Richter, "Multi-level p+ tri-gate SONOS NAND string arrays,"*International Electron Devices Meeting, Tech. Digest*, pp. 1–4, Dec. 2006.

[AGB+03] Z. Al-Ars, A.J. van de Goor, J. Braun and D. Richter, "Optimizing Stresses for Testing DRAM Cell Defects Using Electrical Simulation," in Proc. Design, Automation and Test in Europe (6th DATE'03), Munich, Germany, March 3-7, 2003, pp. 484-489.

[AGB+01] Z. Al-Ars, A.J. van de Goor, J. Braun and D. Richter, "Simulation based Analysis of Temperature Effect on the Faulty Behavior of Embedded DRAMs," in Proc. IEEE International Test Conference (32nd IEEE ITC'01), Baltimore, Maryland, October 28-November 2, 2001, pp. 783–792.

[AGB+01a] Z. Al-Ars, A.J. van de Goor, J. Braun, B. Gauch, D. Richter and W. Spirkl, "Development of a DRAM Simulation Model for Fault Analysis Purposes,'' in Proc. 13th Workshop on Testmethods and Reliability of Circuits and Systems, Miesbach, Germany, February 18–20, 2001.

[FR00] T. Falter, D. Richter, "Overview of Status and Challenges of System Testing on Chip with Embedded DRAMs," in Solid-State Electronics, no. 44, 2000, pp. 761–766.

[RK99] D. Richter, V. Kilian, "The embedded DRAM test dilemma". 1999 3rd IEEE International Workshop on Testing Embed-ded Core-based Systems Chips, CA.

[MMR98] R. McConnell, U. Möller and D. Richter, "How we test Siemens' Embedded DRAM cores". 1998 In-ternational Test Conference, Washington, DC.

[MMR97] R. McConnell, U. Möller and D. Richter, "A Test Concept for Embedded DRAM Cores". 1997 1st IEEE International Workshop on Testing Embed-ded Core-based Systems, Washington, DC.

[KGFR10] G. Koebernik, J. Gutsche, C. Friederich, and D. Richter, "Integrated Circuits and Methods for Operating the Same Using a Plurality of Buffer Circuits in an Access Operation," Application US Patent 20 100 002 503, Jan. 7, 2010.

[RS09] D. Richter, K. Seidel, "Semiconductor memory and method for operating the same," CN Patent CN000100524523C, Aug. 05, 2009.

[RRS+08a] E. Ruttkowski, D. Richter, M. Specht, J. Willer, D. Manger, K. Oisin, S. Meyer, K. Knobloch, H. Moeller, D. Keitel-Schulz, J. Gutsche, G. Koebernik, and C. Friederich, "Integrierte Schaltkreise, Verfahren zum Herstellen eines integrierten Schaltkreises, Speichermodule, Computersysteme," Application DE Patent DE102 007 033 017A1, Nov. 20, 2008.

[RRS+08a] W. von Emden, G. Tempel, D. Richter, A. Kux, " Halbleiterbauelement und Verfahren zum Programmieren und Löschen von Speicherzellen," Application DE Patent DE102007003534A1, Aug. 07, 2008.

[SR07] K. Seidel, D. Richter, "Semiconductor memory and method for operating a semiconductor memory," CN Patent CN000001941 202A, Apr. 04, 2007.

[RS05] D. Richter, W. Spirkl, " Verfahren zum Testen der Refresheinrichtung eines Informationsspeichers," EP Patent EP0000011227 42A3, Aug. 17, 2005.

[MR05] R. McConnell, D. Richter, " Schaltungsanordnung mit temperaturabhängiger Halbleiterbauelement-Test- und Reparaturlogik," EP Patent EP000001008858B1, Jul. 14, 2004.

Abstract

The work presents the outcome of investigations of cost optimized volatile and nonvolatile memory product optimizations along the semiconductor shrink roadmap. A model-based quantitative performance indicator methodology is introduced applicable for performance, cost, and reliability optimization. The methodology has been developed based on industrial 2-bit to 4-bit per cell flash product development projects. A graphical representation based on trend lines is introduced to support a requirement-based product development process.

Cell, array, performance, and reliability effects of flash memories are introduced and analyzed. Key performance parameters of flash memories are systematically derived to handle the flash complexity. A performance and array memory model is developed, and a set of performance indicators characterizing architecture, cost, and durability is defined. This higher abstraction level is introduced to guide engineers responsible to develop system solutions to overcome the complexity and weaknesses of flash memories.

The performance indicator methodology is applied to demonstrate the importance of hidden memory parameters for a cost and performance optimized product and system development roadmap.

The unique features of the introduced methodology for a semiconductor product development process are:

- Reliability optimization of flash memories is all about threshold voltage margin understanding and definition;
- Product performance parameters are analyzed in depth in all aspects in relation to the threshold voltage operation window;
- Technical characteristics are translated into quantitative performance indicators;
- Performance indicators are applied to identify and quantify product and technology innovations within adjacent areas to fulfill the application requirements with an overall cost optimized solution;
- Cost, density, performance, and durability values are combined into a common factor—performance indicator—which fulfills the application requirements.

This dissertation demonstrates the feasibility to quantify technology innovations based on the introduced model-based performance indicator set. The complexity figure consists of semiconductor manufacturing, memory design, embedded control of internal flash operations, and the system hardware, and software architecture is selected to apply the methodology and guide design decisions by quantitative performance indicator values derived from application requirements.

November 2012

Chapter 1
Introduction

1.1 Scope of Work

Economic rules have a strong impact on technology and design and define or impact the architecture trends in the semiconductor industry. The scope of this work is to develop a model-based performance indicator methodology applicable for performance, cost and reliability optimization of non-volatile memories.

Semiconductor memory devices have a long history and are selected to analyze the development process and identify successful optimization strategies. This work is focused on the interaction between new application requirements and the evolution of memory device architectures offering solutions to fulfill a subset of the specified requirements. The array architecture selection and the memory optimization strategy focused on density, performance, energy and reliability parameters is described in detail to identify key parameters within these areas. A methodology is developed to limit the number of iterations during the development process of complex electronic systems.

The growing DRAM market has generated a lot of new memory architectures mid of the nineties. One decade later only a limited number of these architectural concepts have gained significant market share and commercial success. A similar selection process of memory architectures is analyzed for non-volatile memories to gain understand of the rules behind balancing between cost, performance and reliability.

The non-volatile memory market is becoming the dominant part of the semiconductor market due to the fact that NAND flash was becoming the driver of semiconductor lithography. Solid-state based storage solutions are the enabler for a lot of mobile applications. Over a long period of time the cost per bit position for NAND flash was reduced faster than predicted by the ITRS roadmap. The density of a memory is only one important parameter. Other product parameters have to be improved with the same momentum generation by generation along the shrink roadmap.

Memories are regular structured semiconductor devices. A memory is defined by an array with columns—called bit lines—and rows—called word lines—to access

D. Richter, *Flash Memories*, Springer Series in Advanced Microelectronics 40,
DOI: 10.1007/978-94-007-6082-0_1, © Springer Science+Business Media Dordrecht 2014

a memory cell located at each cross point. Flash memories are becoming complex devices along the shrink roadmap due to bit to bit interferences and multi-bit per cell concepts. The non-volatile memory development for data storage applications combines the most aggressive usage of technology (lithography), analog design complexity in terms of sensing concepts and accuracy requirements to voltages for all cell operations as well as digital design complexity in terms of embedded algorithm and adaptive techniques. The product development effort has to be limited to meet the small time window to successfully enter the commodity memory market. Design and technology of a non-volatile memory has to ensure the data integrity over life time—clearly the most important parameter for the end customer.

The non-volatile device complexity is combined with Multi-Bit and Multi-Level Cell concept to increase the memory density and reduce cost per bit faster than supported by the shrink roadmap. Multi-Level Cell flash impacts all design and technology elements and influences the application specific parameter. The quantitative analysis of the interference between the memory technology and the application space is one focus of this work.

A memory-centric system based on flash memories can overcome all weaknesses of a single memory device and boost the system performance by utilizing the massive parallel accessible memories.

A Performance Indicator Methodology is developed to analyze and predict the dependency of all parameters impacting performance, cost and reliability optimization of non-volatile memories. Flash memories are selected to apply the methodology to quantify design and technology innovations and to support a requirement based memory development process.

The complex example of flash memories is used to derive key performance parameters as the basis to introduce an abstraction level defined as performance indicator to handle the complexity of non-volatile systems. Different abstraction layers—a concept of thinking—are introduced to enable innovations on application level applicable to complex non-volatile semiconductor memories:

- The number of people is growing slowly compared to the bit growth rate in the semiconductor world. The generated and stored information is doubled year by year. A first rule can be derived for solid-state based non-volatile memories:
- **Read disturbance will become less important due to the fact that the probability reading certain information over a time scale is becoming every year less probable by a factor of X.**
    - ***"Theorem 1" Reliability for Solid State Storage - 15-Sep-2009 23:09:09 by Detlev Richter***

Reliability and Durability parameter are becoming a part of the performance indicator set guiding the product development and are part of the complete system optimization setup.

1.2 Overview About the Structure

The book "Flash Memories—Economic Principles of Performance, Cost and Reliability Optimization" is structured into eight chapters. A short abstract for each chapter is given to guide the readers, which subjects can be used independently and which are in a consecutive order to define and introduce step by step the model-based performance indicator methodology.

- Chapter 1—Introductory to the subject of this work and the structure of the book.
- Chapter 2—Introductory to the fundamentals of non-volatile memories. The physics of non-volatile storage elements are introduced. Electron and non-electron based cells are described and compared. A subset of flash array architectures are analysed and memory performance parameter are derived. Memory building blocks, sensing of flash and embedded flash program and erase algorithm are introduced. The chapter ends with a detailed discussion of flash threshold voltage window margin and multi-level cell concepts.

The performance indicator methodology is developed on the basis of memory array architecture figures like performance, cost and durability parameters. Chapter three to five introduce the corresponding topics performance, durability and efficiency and define and discuss the required memory performance parameters in detail for each subject.

- Chapter 3—Introduction to the performance definition for volatile and non-volatile memories. Performance parameters for flash memories are derived as basis for the key performance indicator definition in Chap. 6. The link between data throughput and durability is introduced and counter measures are derived.
- Chapter 4—introduces a comprehensive overview about the reliability figures and the critical corner cases which could force additional development effort and system cost.
- Chapter 5—Focus on the product development process and the hidden complexity of a memory-centric system development. Application based system requirements are in conflict with rules given by memory product and technology roadmaps. Efficiency parameters to guide this development process are defined in detail and are the basis for the memory array model development in Chap. 6. The dilemma to optimize in parallel performance, cost and reliability and to meet the application requirements is described.

The history of memories shows a couple of disruptive design and technology innovations important for the commercial success of the specific memory architecture. Design and technology decisions have to be made in a high volume commodity memory market in time. A systematic methodology is developed to compare design concepts proven in volume with design innovations in this work.

- Chapter 6—develops a model-based performance indicator methodology to support a quantitative judgement of design, technology and algorithm options.

A performance and array model is developed and a set of performance indicators is defined. Flash memories are selected to apply the performance indicator methodology to quantify design and technology innovations and to support a requirement based system development process.
- Chapter 7—System optimization based on an exemplar application of the performance indicator methodology. NOR and NAND flash is used to apply the access and storage oriented memory concept. A memory-centric system development is used to discuss the predicted need for 3D-NAND on the basis of performance indicator trend lines.
- Chapter 8—is used to summarize the key elements of the methodology and to give an outlook.

This work is based on experiences of development decisions, concrete project challenges developing volatile and non-volatile memories based on floating gate and charge trapping cells in both NAND and NOR flash memory array architectures over the last 10 years.

The methodology is developed to guide the reader how to find for each challenge the cost-optimized solution for the investigated application case. Weaknesses of non-volatile memories have to be classified and the performance indicator methodology helps to find a solution which translates the weakness of the memory into strength of the system based on this memory type.

The basic rules of the performance indicator methodology can be applied to any complex system in the electrical engineering world in case the core product or technology defines a major part of the key system parameters.

The model-based performance indicator analysis was applied to quantify the target architecture and make visible the capability of the selected flash architecture to outperform competing concepts.

Chapter 2
Fundamentals of Non-Volatile Memories

The subject of this chapter is to introduce the fundamentals of non-volatile memories. An overview about electron and non-electron based cells is given followed by a cell assessment for high density non-volatile memories. The link between memory cell and memory array performance parameters is introduced and in depth analysed for NAND and NOR array architectures. The design specific aspects of sensing and program and erase algorithm techniques are introduced for floating gate and charge trapping cell based flash memories.

The threshold voltage window is introduced as the major tool to design and optimize flash memories. A detailed threshold voltage window margin analysis is applied to explain the sensitivity of different array and algorithm concepts for NOR and NAND flash memories. Multi-level and multi-bit per cell flash architectures are introduced as the main development direction to increase the bit density per cell and to accelerate the memory cost reduction in parallel to the shrink roadmap.

A performance assessment is made on cell, on array and on flash memory level including algorithm aspects to setup the performance parameter selection process.

2.1 Moore's Law: The Impact of Exponential Growth

2.1.1 History of Non-Volatile Floating-Gate Memories

The non-volatile storage of information on a semiconductor device was invented by Kahng and Sze in 1967 [1]. The deep understanding of the device physics and an incremental improvement of cell and technology took place during the next 15 years.

The erasable programmable read-only memory (EPROM) was developed as a ROM replacement during program code development phase. The application of microcontrollers required an erasable programmable ROM to accelerate the system development. The debugged program code could be programmed electrically to the EPROM. The erasure of the stored information was done with UV light. The next

D. Richter, *Flash Memories*, Springer Series in Advanced Microelectronics 40,
DOI: 10.1007/978-94-007-6082-0_2, © Springer Science+Business Media Dordrecht 2014

innovation was the development of an electrically erasable programmable read-only memory (EEPROM).

The application complexity was growing and the required amount of code memory increased. The storage element had to become smaller in cell size and the single floating gate transistor was developed. The floating gate (FG) cell could be combined with different memory array architectures. NOR- and NAND-type Flash EEPROM was invented and published [2]. The name "flash" was suggested by Dr. Masuoka's colleague, Mr. Shoji Ariizumi, because the erasure process of the memory contents reminded him of the flash of a camera.

At that time—mid of the eighties—the semiconductor memory market was dominated by the two main memory architectures Dynamic RAM and Static RAM. Cellular phones were becoming the first volume application requiring a non-volatile memory for fast code execution. The selected memory had to be large enough to store the complete program code of the cell phone and the user and application data. At that time NOR-type flash was becoming large enough, had a very short access time and could be designed with a low standby current. The application requirements, which were mandatory for the cell phone market in the time from 1990 to 2000, were fulfilled by NOR-type flash memories.

The functionality of mobile cellular phones increased continuously year by year and the demand increased for larger flash memories. Multi-level NOR flash [3] was introduced storing two bit per cell which reduces the cost per bit of NOR flash. The usage of flash memories for the main memory instead of DRAM reduces the power consumption of cell phones during standby significantly. DRAM manufacturer adapted their products to the low power requirements of cell phones. Low power DRAM (LPDRAM) architectures were developed. The DRAM manufacturers accelerate the low power DRAM development for the growing mobile market and **M**ulti-**C**hip **P**ackage (MCP) products appeared on the market combining NOR and LPDRAM memories.

The NAND-type flash memory was developed as storage oriented non-volatile memory to replace the disc and the hard disc drives by block based storage of data similar to the magnetic competitor devices. NAND flash was the most dense memory architecture and 100 % cost and die size optimized. The page size was specified with additional spare area at the end of each page to enable error detection and correction coding. NAND introduced a so called spare block handling over lifetime. NAND flash covers defective erasable sectors by the methodology to replace the bad ones with good sectors. Failing bits and failing operations were becoming first time part of a memory specification. The first applications of NAND were limited to the storage of digital data (pictures and music). Again the demand of mobile applications (smart phones) for large non-volatile memories storing all kind of data made the paradigm shift possible NAND flash was becoming the dominant non-volatile memory.

The increase of memory density is based on a continuously reduction of feature size year by year. The increase of process deviation is the price to be paid for smaller cell sizes. Random bit failures are becoming statistically more relevant for floating gate cell based memories in the GBit density range. The effort for NOR flash memories increases dramatically to improve technology and test to guarantee a fault free lifetime. The strong benefit of NOR flash the fast access time is becoming a design challenge for an effective EDC and ECC implementation. NAND flash is compensating an increase in single bit failure rates simply by an increased spare area for advanced ECC concepts.

NAND flash performance and energy efficiency was improved year by year. Every bit line belonging to a cell which is not intended to be programmed requires an inhibit scheme. Historically the supply voltage (5.0 V) was applied to these bit lines and this inhibit scheme was a limit for the NAND architecture. A "self-boosted" inhibit scheme [4] was invented reducing the current and energy consumptions dramatically. A robust and stable NAND program inhibit scheme was a prerequisite for a successful multi-level cell NAND development roadmap.

The multi-level cell NAND flash architecture based on floating gate cells accelerates the memory density increase to more than two bits per cell. 3D cells (FinFET) have shown their improvement potential for charge trapping based flash architectures [5, 6]. The next development step is to build a three dimensional NAND flash array in which the NAND string is oriented into the Z-direction. This 3D-memory array is utilizing automatically all benefits of 3D flash cells and will enable the continuously increase of flash memory density over the next decade.

2.1.2 *Moore's and Hwang's Law*

Memories have two strong benefits to be the development driver of the semiconductor industry. A regular array structure combined with algorithmic test patterns is enabling fast technology learning and focused failure analysis. The semiconductor industry started with SRAM test arrays and then DRAM's became the technology driver. Over decades DRAM memories were driving lithography and process improvements. In sub 100 nm technologies development of DRAM capacitor based on high-k dielectrics requires more time and NAND flash memories take over the leadership.

The cell and array structure of NAND flash is simpler compared to DRAM based on a capacitor and a select transistor. Applications based on portable data storage create a huge and increasing market demand for NAND flash to support solid-state storage based devices. The combination out of these two aspects was pushing NAND flash into the leading position for technology and lithography development.

1965 Gordon Moore published his famous paper [7]—"Cramming more components onto integrated circuits". Moore's law was later modified—the number of transistors on a chip doubles every two years. The combination out of device design, process development and application as well as marketing by Intel was the dominant factor for the microcomputer revolution.

In this century mobile applications were becoming the most growing markets. This growth is linked to the density increase of a non-volatile solid-state storage medium. Approximately 40 years after Moore's law the semiconductor branch of Samsung had defined an aggressive roadmap for NAND flash—the bit density for

NAND flash memories has to double every year. The flash community called this "Hwang's law". The former head of Samsung Electronics Hwang Chang-gyu set this target making NAND technology more competitive to replace HDD storage solutions by Solid-State Disc products.

NAND flash is an enabling technology for mobile applications like digital cameras, MP3 music players and smart phones. The first product innovation wave was portable storage, next wave is solid-state disc are entering the mobile and server computing and the last wave will influence the computer architecture fundamentally. Multi-core microprocessors combined with next generation non-volatile solid state memories will become the performance, cost and energy optimized new computing architecture.

Forty years after floating gate cell devices were invented and twenty years after NAND flash memories were announced the combination out of floating gate cell and NAND array is changing the quality of our life. All kind of data are becoming mobile and access and usage of digital information will influence mostly all areas of our life. Moore's and Wang's law strongly impact our life.

2.1.3 History of Solid-State Storage: Cell and Application Overview

A short history of non-volatile cell concepts—see Table 2.1—is compiled as introductory to this work. The cell concepts are linked to the dominating application if possible. The main application market is added if figures are available.

The detailed analysis of the history of solid-state storage is the basis for this work. This work develops answers for two questions:

Table 2.1 History of cell concepts

Year of invention	Cell concept	Product/Memory device	Application / market share
1967	Floating gate		
1971	EPROM-FAMOS		EPROM
1976	EEPROM-SAMOS		EEPROM
1978	NOVRAM		
1984	Flash memory		EEPROM
1988	ETOX flash memory	FG cell & NOR array	Mobile phone
1995	Multi-level cell NOR	FG NOR	Mobile phone
1995	2-bit multi-level NAND	FG NAND	Mobile camera
1999	NROM virtual ground NOR	Charge trap & VG NOR	Mobile storage
2005	4-bit per cell—NROM	Charge trap & VG NOR	OTP data storage
2006	4-bit per cell—MLC NAND	FG cell & NAND	All mobile devices
2009	2-Bit per cell PCM		
2007–11	3D-NAND	3D charge trap NAND	

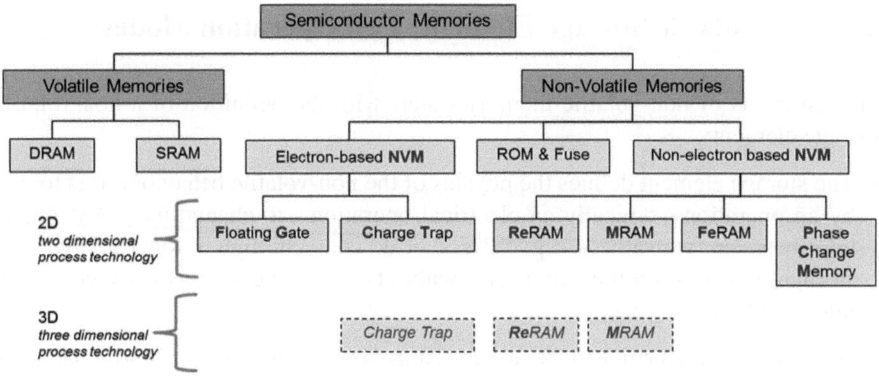

Fig. 2.1 Semiconductor memory cell overview and classification of non-volatile memories

- What are the reasons for the success of specific cell concepts?
- Which methodologies could predict the cell and architecture development?

2.1.4 Memory Cell Classification

The differentiation between electron-based and non-electron-based non-volatile cells is introduced to classify non-volatile memory cells. Beside the physical differences the electron-based non-volatile memories are the dominant commercially successful memory types compared to the non-electron based types. The non-electron based non-volatile memories are also called emerging memories.

The memory cell overview and classification is shown in Fig. 2.1.

The analysis of memory data over technology nodes requires commercially successful memory cell concepts with a market share greater 5 %. Statistically significant technology data are available for electron-based non-volatile memories. The commercial success of product innovations added to these memory and cell concepts are analysed in this work over the last two decades.

Expected performance and reliability values based on model-based calculations are compared with characterization data and are finally validated on high-volume flash memory production data.

Electron-based non-volatile memories - especially flash memories based on floating gate cells—are responsible for more than 95 % market share. The next development step of commercially successful non-volatile cells will be based on 3D memory architectures. Potential non-volatile cell candidates are marked with a dotted line in Fig. 2.1.

A first performance indicator analysis of 3D memory array architecture is started in chapter seven based on electron-based non-volatile cells.

2.2 Non-Volatile Storage Element: Cell Operation Modes

Fundamentals of non-volatile memories start with the definition of a non-volatile storage element:

- The storage element defines the physics of the non-volatile behaviour. It is forced by an operation—typically an electrical operation—to change the behaviour at least between two values—e.g. charged or de-charged; high or low resistive. The storage element saves the written data without any voltage supply over a predefined retention time (10 years).

The storage element can be a transistor—in which the threshold voltage is changed—or another resistive or ferroelectric structure—in which for example the resistance is changed.

Theoretically most of the storage elements can store more than one bit. The application potential within a real memory product of this enhanced storage density of multiple bits depends on the non-volatile cell architecture, the memory array structure and the sensing concept.

The next level is the non-volatile cell, which is defined by a direct connection to bit- and word lines. It is important to highlight this difference between a non-volatile storage element and a non-volatile cell shown in Fig. 2.2. The flash transistor is a non-volatile cell with an integrated storage element. This can be a floating gate or a charge trapping layer required to influence the threshold voltage of the transistor.

A group of memories are based on a non-volatile cell consisting out of a storage element and a select device. An example for the combination out of storage element and select device is thc EEPROM cell.

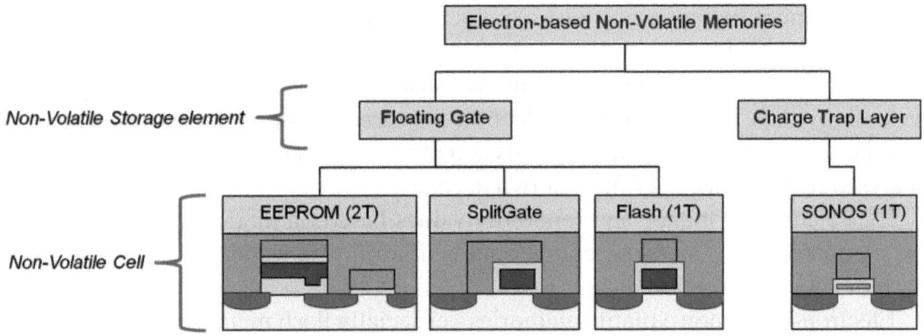

Fig. 2.2 Example for the link between non-volatile storage element and non-volatile cells

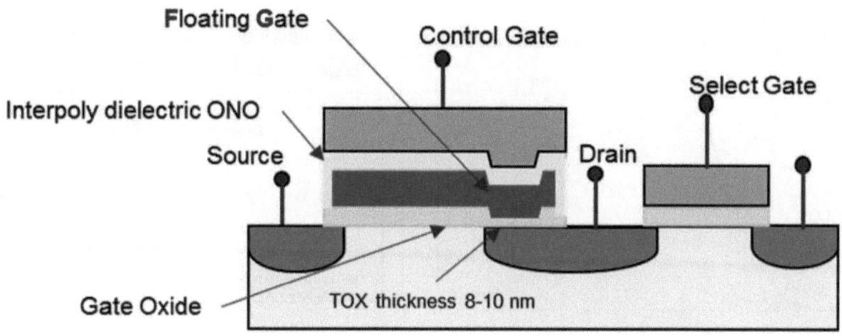

Fig. 2.3 Schematic of a FLOTOX cell including the select transistor

2.2.1 EEPROM: The Classical Two Transistor Memory Cell

The EEPROM [8] is used to introduce physical basics and main operation modes. The **E**lectrically **E**rasable **P**rogrammable **R**ead **O**nly **M**emory is a classical two transistor cell to allow bit-wise program and erase operation.

- EEPROM—bit-wise programming and bit-wise selective erase

The basic EEPROM cell is shown in Fig. 2.3 called the ***FLO***ating gate ***T***hin ***OX***ide (FLOTOX) memory cell [9].

Programming of this cell is achieved by applying a high voltage to the control gate, the drain is low biased. The voltage on the floating gate is increased by capacitive coupling so that a FN tunneling current from the drain into the floating gate through the tunnel oxide—TOX—at the 8–10 nm thin position starts to inject electrons.

Erasing of this cell is achieved applying a high voltage to the drain and the control gate is grounded. The floating gate is again capacitive coupled to the low voltage and FN tunneling enforces electrons to flow from the floating gate into the drain. The select gate is used to control the drain bias.

The storage element combined with a select transistor enables the capability for a bit wise change of the EEPROM as known from random access memories. The limiting factor of such a concept is the large cell size enforced by the additional select transistor per storage element.

2.2.2 Single MOS Transistor: Electron Based Storage Principle

A standard MOS transistor controls the current flow between source and drain supplying a gate voltage value to create a channel below the gate oxide. A gate voltage higher than the threshold voltage turns the MOS transistor on and a drain source current starts to flow if there is a drain voltage applied. The positive gate voltage creates a negative charge within the channel resulting in the $I_{DS}=f(V_{GS})$ curve shown in Fig. 2.4.

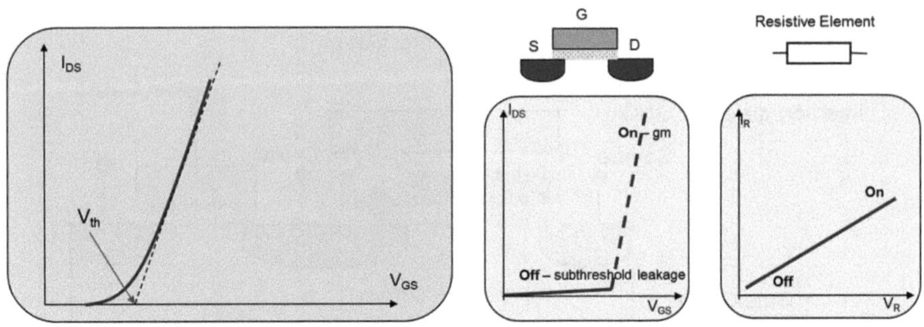

Fig. 2.4 MOS transistor: I_{DS} current dependency from gate voltage, threshold voltage and on/off ratio

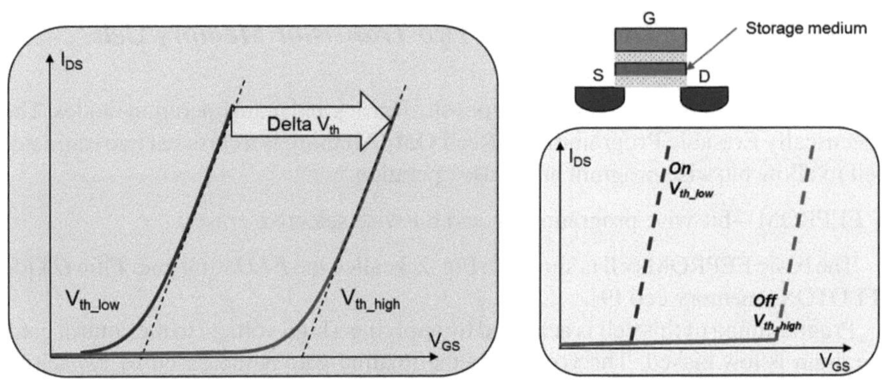

Fig. 2.5 Floating gate transistor: I_{DS} current dependency on gate voltage and on/off ratio

A MOS transistor has an excellent on/off ratio and is therefore a preferred storage element. The difference to a linear resistive element is shown in the following figure.

A MOS transistor with a storage medium above the gate oxide creates automatically a non-volatile storage element. The storage medium—a floating gate or a charge trapping nitride layer—can be charged with electrons. This additional charge on top on the gate oxide influences the threshold voltage V_{th} of the device characteristic.

Figure 2.5 shows the two $I_{DS}(V_{GS})$ curves for the same function—$I_{DS} = f(V_{GS})$—for a non-volatile cell transistor with an uncharged and a charged storage medium. The stored charge (e.g. Q_{FG}) in the storage medium is translated into a shift of the threshold voltage of the storage cell transistor.

The delta of the threshold voltage can be calculated with the charge stored in the storage medium over the capacitance:

$$\Delta V_{T,CG} = -\frac{Q_{FG}}{C_{FC}}$$

where C_{FC} is the capacitance between floating gate and control gate.

A MOS transistor with an additional storage layer for electrons (storage element) is an excellent non-volatile cell, which could be charged and discharged via the bottom oxide and the read operation could be controlled by the threshold voltage itself. The charging and discharging of the storage layer depends in detail on the specific construction of the non-volatile cell transistor.

2.2.3 Operation Modes of Floating Gate MOSFET

The read, program and erase sub chapters introduce the flash cell operation principles. The introduction is focusing on the logical behaviour. The physical effects linked to flash cell operations are described in the literature [10] in more detail.

2.2.3.1 Read Operation

The transistor characteristic of the flash cell defines the read parameter. The transconductance curve [V_{GS} versus I_{DS}]—the slope of the curve is defined as g_m—is an important parameter defining the read operating point. The accuracy of **all** voltage levels applied to the cell terminals influence the read accuracy. The applied gate voltage defines the read threshold level. During a read operation the V_{th} of the cell is compared versus the applied read reference V_{th} level (see Fig. 2.6):

- A V_{th} ***level higher than the read reference level*** defines the cell as programmed.
- A V_{th} ***level lower than the read reference level*** defines the cell as erased.

The threshold voltage of a cell within a memory array cannot be measured directly. The read current at the border of the array is used to evaluate the status of the cell.

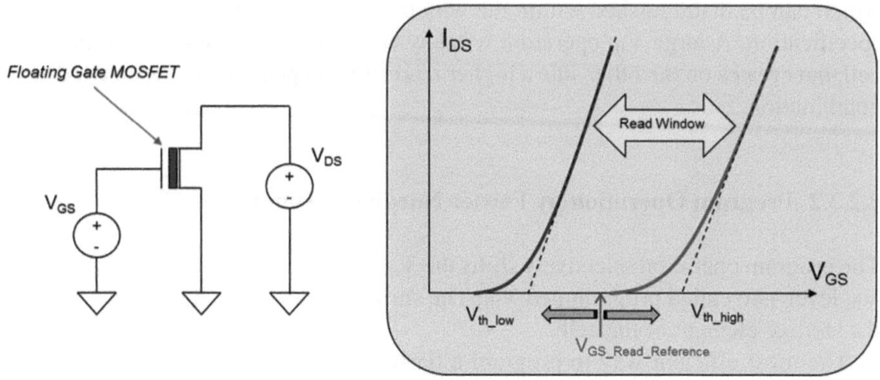

Fig. 2.6 Read operation—gate voltage classifies between erased and programmed cells

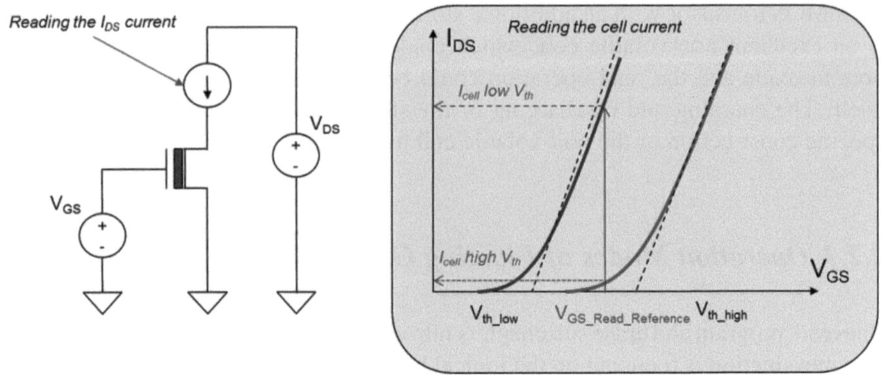

Fig. 2.7 Read operating point—defined by Cell V_{th} and read operation cell current

- ***A V_{th} level higher than the read reference level*** results into a leakage/sub-threshold current for the defined reference gate voltage and identifies the cell as programmed.
- ***A V_{th} level lower than the read reference level*** forces a measurable high cell current which allows the sense amplifier to identify the cell as erased.

Figure 2.7 shows a read—current sensing on the drain side—of the cell transistor current I_{DS}:

Sense amplifiers circuits measure the cell current at the border of the memory array. Sense amplifier principles are introduced in Sect. 2.5. A fast sensing can be achieved applying the classical differential sensing technique used in other memories [11, 12].

Read operation requirements define the basic cell V_{th} operation window which is a key parameter for every flash memory product. The read operating voltage conditions (V_{GS} and V_{DS}) define the achievable operating window—the number of V_{th} levels which can be distinguished within the window—and the achievable read durability specification. A large V_{th} operation window ensures a robust read operation of the cell, but creates on the other side a higher read stress depending on the cell and array combination.

2.2.3.2 Program Operation by Fowler Nordheim Tunneling

The program operation selectively shifts the V_{th} level of an erased cell up to the target V_{th} level—so called programmed V_{th}. The shift is achieved by injecting charge into the storage element of the cell.

The most efficient way to program a flash cell in terms of energy is the tunnel process. A high electric field across the bottom oxide ensures a direct tunnelling of electrons into the storage element (floating gate) of the flash cell. This is achieved by **F**owler-**N**ordheim (FN) tunneling [13].

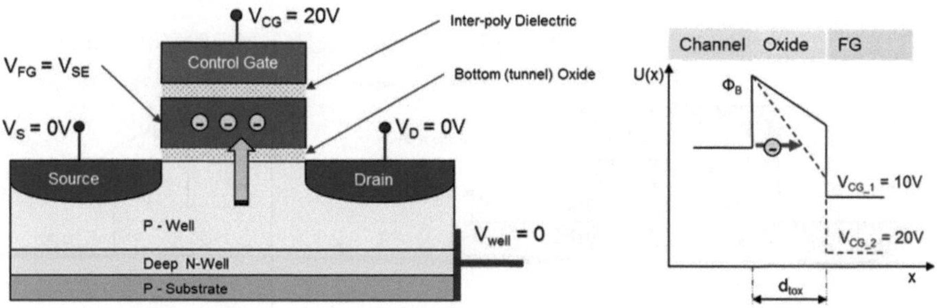

Fig. 2.8 Cell program operation—Fowler Nordheim tunneling through a potential barrier

The FN programming is controlled by the gate voltage shown in Fig. 2.8. If the electrical field is high enough the FN tunnelling starts and the voltage of the storage element—V_{SE}—follows the gate voltage. The program time depends on the gate voltage value applied and the thickness of the bottom oxide—also called tunnel oxide.

FN programming requires that the gate has to be accessed. The potential of the other terminals depends on the selected cell/array combination.

- A high electric field is required so that the barrier becomes small enough (Fig. 2.8).
- A high voltage is applied to the gate so that electrons start to tunnel through the bottom oxide—also called tunnel oxide—into the floating gate.
- The program efficiency is excellent (=1); all electrons are injected into the storage element.
- No significant program current flow.

During the program time charge is injected into the floating gate—V_{FG} (V_{SE}) increases—and the electric field over the bottom oxide is reduced. A reduced field decreases the electron flow. This dependency is linear and the V_{SE} potential follows exactly the change of the applied gate voltage [14].

The quantum mechanical tunnelling process through a barrier is a slow operation and the low tunnelling current depends exponentially on the barrier height ϕ_B to the 3/2 power. The tunnelling process is reversible, so that the opposite voltages will erase the storage element.

2.2.3.3 Program Operation: Channel Hot Electron Injection

A faster way to inject charge into the storage element of a flash cell uses the energy of hot electrons. This physical effect enables the electrons to overcome the potential barrier of the bottom oxide. This is achieved by **C**hannel **H**ot **E**lectron—**CHE**—programming.

The CHE programming is a fast operation and the amount of charge or the efficiency can be controlled by both the gate and the drain voltage shown in Fig. 2.9.

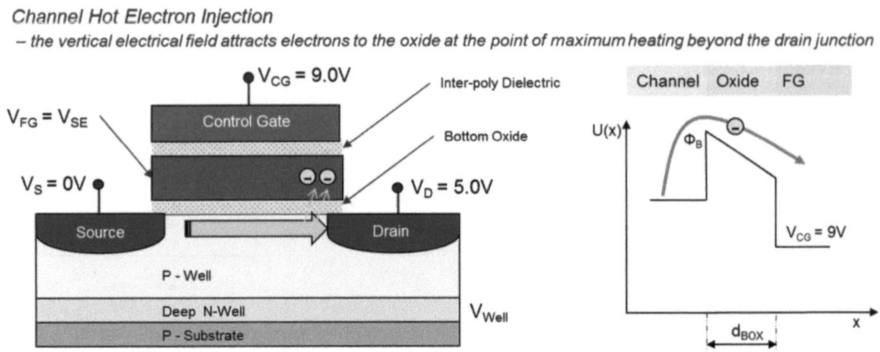

Fig. 2.9 Cell program operation—channel hot electron injection electrons crossing the barrier

Both voltages combined have to be above the level to generate hot electrons; the drain voltage has to be at least higher than the barrier height (3.2 V for Si) to generate hot electrons. The program time depends on the values of the voltage levels applied to all terminals of the cell transistor and the gate length of the flash cell.

CHE programming is a fast operation, in which all cell terminals have to be accessed.

- A current flow between source and drain is required, which generates hot electrons. The gate voltage has to be higher than the drain voltage level and generates an electrical field which accelerate the generated hot electrons into the floating gate.
- The program efficiency is based on statistical electron energy distribution functions, only those electrons with significant high kinetic energy cross the barrier → 10^{-5}.
- Medium high drain and high gate voltages are required.
- A current flow between the cell terminals source and drain is the basis of the physical effect.

At the beginning of the CHE programming a fast and rapid change of the cell V_{th} occurs, slowing down, because V_{SE} becomes lower than V_D and the electron injection saturates [15].

2.2.3.4 Erase Operation

The step from an EPROM—an electrically programmable ROM—to an EEPROM—an electrically erasable programmable ROM—replaces the UV erase of the ***complete memory*** with an electrical controlled erase procedure of ***pre-define erase blocks***.

The erase operation reverses the threshold voltage shift of the program operation. The tunnelling current flows into the substrate, which is common for all cells within an erase block. The erase operation shifts a large quantity of cells—programmed and erased ones—below the erase V_{th} level.

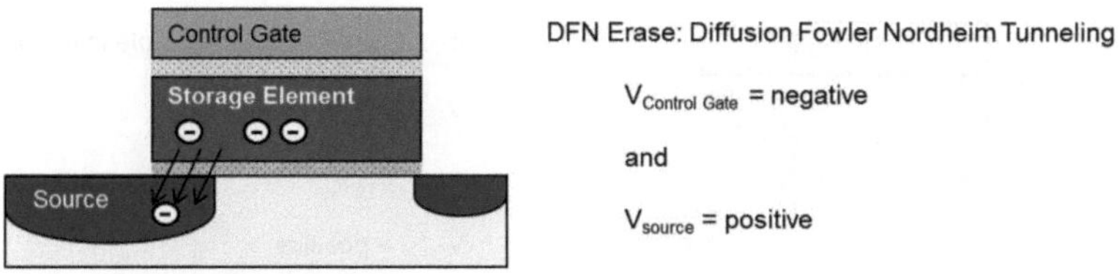

Fig. 2.10 Diffusion FN tunneling Erase—DFN

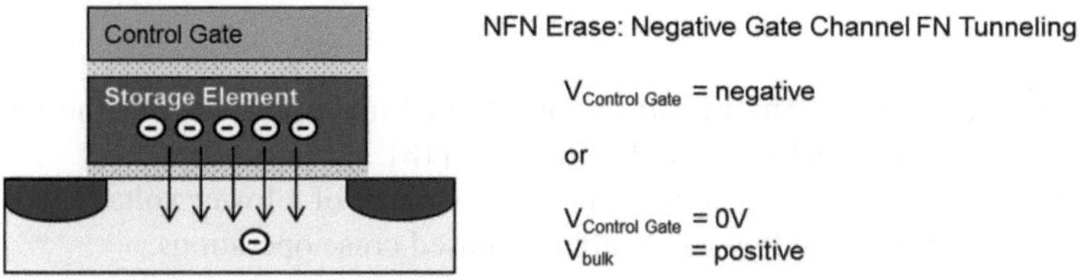

Fig. 2.11 Negative Gate Channel FN tunneling Erase—NFN

The erase operation can be based on a tunnel process described in Figs. 2.10 and 2.11.

In a FLOTOX cell the so called **D**iffusion**F**owler-**N**ordheim Tunneling—**DFN**—was used to erase all cells. The required electrical field could be achieved by applying a positive voltage to the respective source or drain area and a negative voltage to the control gate. This voltage has to be applied a certain time interval in the range of milliseconds. The erase time depends on the voltage difference between gate and source and the thickness of the tunnel oxide. As described in Fig. 2.3 this could be different than for the rest of the channel region.

Along the shrink roadmap a large overlap between the diffusion and the gate limits the capability to reduce the cell size. Consequentially the **N**egative Gate Channel **FN** Tunneling—**NFN** —was introduced to erase the cell via the bulk.

A negative high voltage has to be applied to the selected cells for a certain time—hundreds of µs to ms—to ensure the electron flow from the floating gate into the bulk region.

High electric fields applied to the tunnel oxide result into a degradation of the tunnel oxide quality. The tunnel oxide degradation accelerates over lifetime and leakage paths are created which affects the quality of the dielectric barrier and therefore the data retention parameter of the flash cell.

Another physical erase operation principle was applied erasing a flash cell using the band-to-band **H**ot **H**ole **I**njection (**HHI**) erase mode [16, 17].

A negative high voltage pulse is applied to the gate which has only half the value compared to the channel FN tunnel erase. The cell is forced into a snap-back condition. A number of hot holes are generated by impact ionization near the drain region and are accelerated back into the floating gate, which is erased quite fast by a flow of holes recombining with the stored electrons shown in Fig. 2.12. A significant current flow is required to generate the conditions to allow the **H**ot **H**ole **I**njection

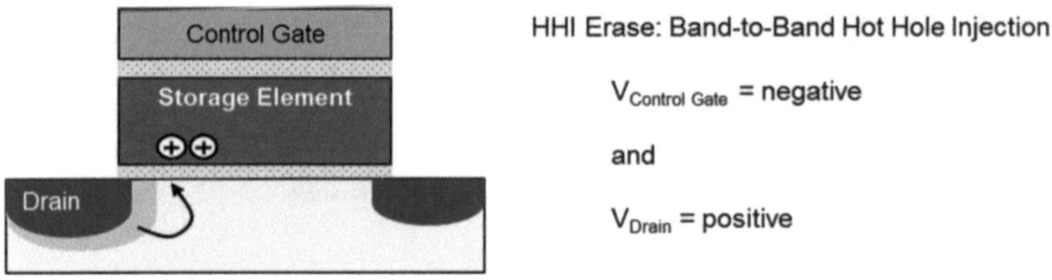

Fig. 2.12 Band-to-Band Hot Hole Injection—HHI

mechanism to work. "(This) band-to-band tunnelling contributes to the so called Gate Induced Drain Leakage (GIDL) current." [18].

Band-to-band hot hole injection has the advantage of a lower voltage bias and a faster erasing speed than Fowler–Nordheim based erase operations.

The importance of the erase operation for performance and reliability of each flash memory is discussed in the corresponding chapters including the erase algorithm complexity.

2.2.4 *Flash Cell Operation Summary*

An EEPROM and one transistor flash cell are selected to describe the relationship between the non-volatile storage element and the non-volatile cell. The basic operation modes and the physical principles to charge and discharge an electron-based non-volatile cell are introduced and summarized in Table 2.2. The physics to charge and discharge flash cells are discussed in detail in the literature [1, 19].

The introduced physical operation principles can be now combined on cell level to achieve the specified program and erase performance. The energy efficiency of the cell operation is considered as one of the most important parameter in combination with the applied memory array architecture. Slow cell operations are characterized with an excellent energy efficiency and fast operation with a medium one. The challenge selecting the right combination of cell architecture and physical operation modes is visible in Table 2.2.

The cell operation modes can be only determined in combination with a memory array. The following two chapters focus on the combination out of cell, physical principles and array architecture. It will be shown that the read operating point and the achievable V_{th} operation window are impacted by program and erase in combination with different array architectures.

The focus is put on selecting the most cost efficient memory architecture. The two dominant storage elements of flash cells—floating gate and charge trapping layer—will be investigated in more detail.

Table 2.2 Physical operation modes to charge and discharge

Cell operation principle	Charge / program	Discharge / erase	Performance / efficiency
Channel Hot Electron Injection (CHE)	Hot electron injection (~100 μA/cell)		Fast operation: 1 μs Efficiency low: 10^{-5}
Channel FN tunnelling (CFN)	Fowler Nordheim Tunneling		Slow operation: ms Efficiency excellent: 1
Source Side Injection (SSI)	Hot electron injection (~10 μA/cell)		Fast operation: 10 μs Efficiency low: 10^{-5}
Hot Hole Injection (HHI)		Hot hole injection induced by GIDL	Slow operation: 1 ms Efficiency middle: 10^{-3}
Diffusion FN tunnelling (DFN)		Fowler Nordheim Tunneling	Slow operation: 1 ms Efficiency middle: 10^{-4}
Negative Channel FN tunneling (NFN)		Fowler Nordheim Tunneling	Slow operation: 1 ms Efficiency excellent: 1

2.3 Non-Volatile Cell: Electron and Non-Electron Based

A memory is defined as a matrix of memory cells. At the intersection of each row—word line—and each column - bit line - a memory cell is located. The non-volatile memory is mainly defined by the selected type of the array organization, how the cells are connected to word and bit lines (NOR, AND, NAND). The design implementation details of the array impact significantly the performance and reliability values of the memory product (local or global bit lines, twisted or not twisted lines, ground connection or plates).

The combination of cell and array define most key parameters of the non-volatile memory:

- the access to the cell: direct within a NOR-Array or indirect within a NAND-Array;
- the number of lines used to read, program and erase the selected cells.

An overview of electron-based non-volatile memories derived from one transistor flash cells is shown in Fig. 2.13 combining different array architectures with two different storage elements.

A detailed investigation of the one transistor flash cell is done specifically for the two shown storage elements combined with the best matching array architecture. Floating gate cells are applied to a NAND array which requires two times FN tunneling—the most efficient way to program and erase. This combination enables

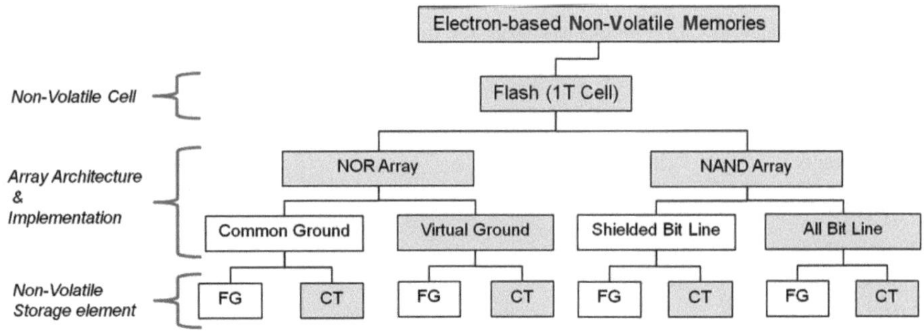

Fig. 2.13 Non-volatile memory—cell and array overview

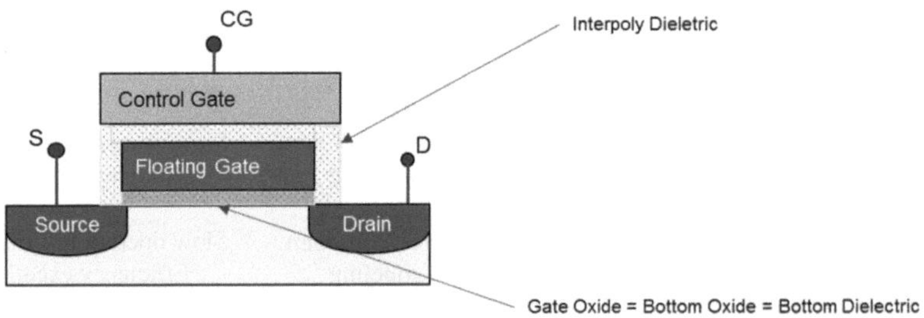

Fig. 2.14 Floating gate flash cell

a cost and energy optimized NAND flash memory product. NAND is the basis for the performance indicator methodology developed in this work.

Charge trapping cells applied to a NAND array are used to assess the differences between FG and CT and derive key performance parameters to characterize cell and array combination. Charge trapping cells applied to a VG-NOR array are capable to store physically two bits per cell and enable a fast read and program access—CHE programming—and a dense non-volatile memory architecture.

2.3.1 Flash Cell: Floating Gate Technology

The most commercially successful non-volatile memory technology over the last thirty years is based on the floating gate cell architecture and is widely used for NOR and NAND memories as well as for a lot of special embedded flash architectures [2].

Figure 2.14 shows a schematic cross section of a floating gate cell transistor, the floating gate is deposited above the gate oxide and completely isolated by the surrounded inter-poly dielectric.

The NAND floating gate program operation based on Fowler Nordheim tunnelling requires a transfer of electrons from the bulk to the floating gate through the thin-gate oxide layer – also called Bottom Oxide. The voltage at the Control Gate is increased

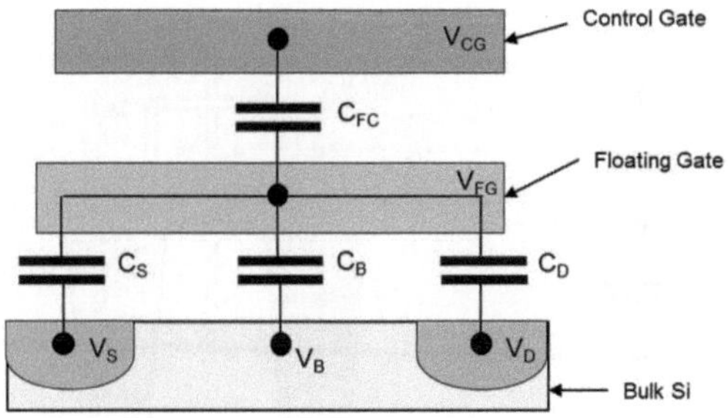

$$gate\ coupling:\ \alpha_G = \frac{C_{FC}}{C_T}$$

$$drain\ coupling:\ \alpha_D = \frac{C_D}{C_T}$$

$$total\ cap.:\ C_T = C_{FC} + C_D + C_S + C_B$$

Fig. 2.15 Equivalent circuit for a FG charge coupling calculation

to a level that the effective electric field across the Bottom Oxide is high enough and the tunnel process starts. As the charge tunnelled into the floating gate is increased, the electric field is reduced, which slows down the electron injection.

The FN tunnelling process is reversible, which results into an erase process with inverted polarity. During the history of the floating gate cells a couple of different erase procedures were introduced. In today's floating gate devices channel FN tunnelling is used as the default erase operation principle.

The floating gate transistor can be described as a network of capacitors connected to the floating gate, shown in Fig. 2.15.

Based on the above equivalent circuit the floating gate voltage V_{FG} can be calculated as:

$$V_{FG} = \alpha_G(V_{CG} + fV_D)$$
$$f = \frac{\alpha_D}{\alpha_G} = \frac{C_D}{C_{FC}}$$

Where C_{FC} is the poly to poly capacitance, C_D is the floating gate to drain capacitance, C_S is the floating gate to source capacitance and C_B is the floating gate to bulk capacitance.

A key design parameter for floating gate cells using FN tunneling is the program coupling ratio. The coupling ratio ensures an asymmetry of the electrical field across the tunnel oxide (dielectric) and the inter-poly dielectric. The electric field across the tunnel oxide is higher and therefore the injected charges stay on the floating gate and are not able to overcome the inter-poly dielectric barrier into the control gate.

An effective coupling of the programming voltage can be achieved by construction of the cell, where a high poly to poly capacitor C_{FC} is obtained by a control gate surrounding the floating gate, shown in Fig. 2.16. A cell with a good coupling ratio is a tradeoff between bottom oxide thickness and effective cell height.

The ratio between the poly to poly capacitor C_{FC} and the bottom oxide capacitor C_B to the bulk defines the coupling ratio which impacts directly the size of the V_{th} operation window of the cell. This construction principle based on the different capacitor sizes is one of the major success factors of floating gate flash cells and memories.

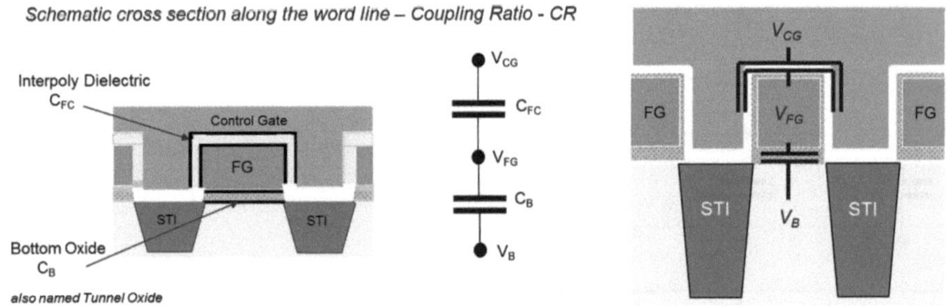

Fig. 2.16 Schematic cross section along the WL and NAND FG to illustrate the capacitive coupling ratio

The floating gate cell combined with FN tunneling operations offers strong benefits:

- The coupling ratio and therefore the effectiveness of the program operation and the size of the V_{th} operation window are well defined by the construction of the cell. This deterministic and predictable behavior is the key benefit of the NAND FG cell combination. Maintaining this excellent coupling ratio is a serious challenge for the scalability of floating gate cells.
- The floating gate is a conductor and the programmed charge has the same voltage potential at each position of the floating gate. This ensures threshold voltage stability for the cell transistor as long as the encapsulation of the floating gate is defect free.

A consequence of a conducting storage element is that each single defect causing leakage can cause discharge of the floating gate to the neutral V_{th} level. The process challenge of the floating gate technology is the encapsulation of the floating gate, leakage of charge stored in the floating gate has to be suppressed to a very small pre-defined level over life time.

2.3.2 Flash Cell: Charge Trapping Technology

This non-volatile memory technology is based on charge storage in localized traps within a nitride layer. Non-volatile cell devices which have the nitride storage layer above the oxide are called MNOS—Metal Nitride Oxide Silicon—and are introduced in 1967 [20] and used as programmable ROM. The **S**ilicon **O**xide **N**itride **O**xide **S**ilicon (SONOS) cell which adds another oxide layer between the poly silicon gate and the nitride storage layer was introduced in 1977 [21] and is the basic cell architecture for today's charge trapping flash cells.

The robustness against physical defects of the storage element is one benefit of the SONOS cell concept compared to floating gate cells. A defect in one of the oxide

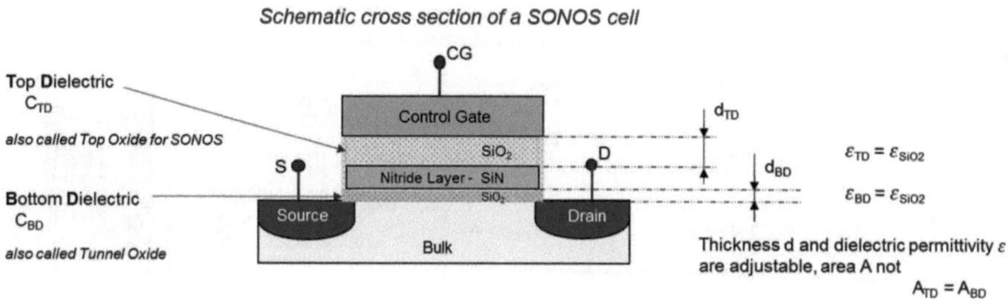

Fig. 2.17 SONOS flash cell based on charge trapping nitride layer

layers forces only the loss of a limited number of electrons linked to the traps near by the defect.

A complicated encapsulation of a floating gate is not required. The start of volume production can be managed much faster compared to other memory technologies and the achievable yield during ramp and in high volume production is outstanding. The production requirements are less challenging and the principle scalability is excellent. These are the two major benefits for charge trapping based non-volatile technologies.

The following Fig. 2.17 shows a typical SONOS cell to explain the functionality.

The FN tunnelling of the SONOS flash cell is performed as known from the FG flash cell. The voltage at the storage layer increased so that the effective electric field across the tunnel oxide is high enough to start the tunnel process of electrons from bulk into the nitride layer. As charge tunnels and get trapped into the nitride, the electric field over the tunnel oxide is reduced, which decreases the electron injection process.

The coupling ratio of planar SONOS cells cannot be improved by mechanical cell construction, because the size of the top dielectric is the same as that of the bottom dielectric for 2D cells as shown in Fig. 2.17. The cell operation has to be optimized by varying the thickness of tunnel oxide or exchange the top oxide layer by a material combination with a higher dielectric permittivity (ε).

- For the same material—silicon oxide—the ratio between top dielectric capacitance C_{TD} and bottom dielectric capacitance C_{BD} is defined by different thicknesses d_{TD} and d_{BD}.
  - $C = \varepsilon * \frac{A}{d}$ (the size A and ε are the same)

The material combination of the top dielectric defines the cell construction and the name SANOS—**S**ilicon-**A**luminum oxide-**N**itride-**O**xide-**S**emiconductor [22]. The TANOS cell has a dielectric composite of $TaN/Al_2O_3/SiN/SiO_2$. TaN suppresses the unwanted backward Fowler-Nordheim tunneling current of electrons through the top dielectric significantly due to its higher work function [23]. We use still the terminology of a coupling ratio for SONOS cells in this work to describe the amount of asymmetry of the electrical field applied. Different solutions are published in the literature. Charge trap flash technology development directions are described to set up the performance indicator analysis for the corresponding non-volatile memories.

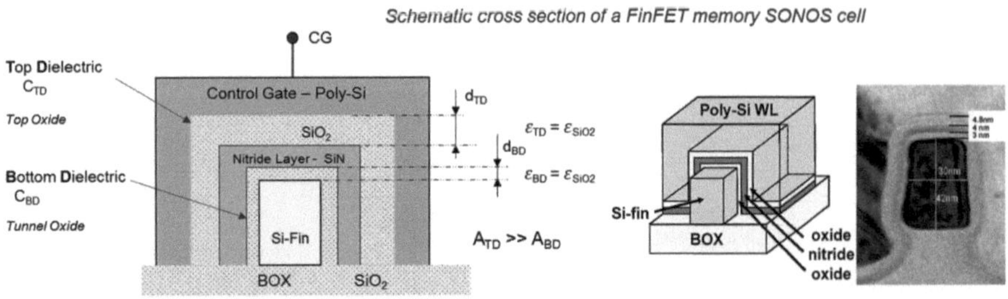

Fig. 2.18 FinFET SONOS cell and a p+ tri-gate SONOS NAND cell example

- Memory developments based on the SONOS cell principle are listed below to accept or to overcome the retention and performance limitations compared to floating gate cell based memories:
  - Optimized SONOS thicknesses for embedded flash cells with limited retention requirements based on FN tunneling program and erase.
  - The combination of SONOS cells with CHE programming overcomes the program performance conflict with the retention requirements. CHE enables a fast programming even with higher bottom oxide thicknesses (>4–5 nm).
  - FN tunneling operation replacement with CHE programming and Hot Hole Injection erase improves reliability figures of charge trapping cells for the same bottom oxide thickness.
    - This SONOS cell can be improved to locally store the charge in separate areas of the storage element—the so called two bit per cell Nitride ROM flash cell is discussed in more detail in Sect. 2.3.3.
- Flash memory development strategies based on SANOS and TANOS cells with application specific adapted program performance and reliability requirements.
  - Different material combinations and cell encapsulations are published for the material improved SANOS and TANOS cell based NAND flash memories.
  - A dedicated focus has to be put on a specific charge trap reliability issue, the non-stability of charge position within the trapping layer [24], causing a shift in the cell's threshold voltage. The threshold voltage function is $V_{th} = f(Q, x)$ (Q is the charge stored in the layer, x is the position of the charge). The surrounding electric field can move the charge within the storage element.
- 3D based non-volatile cell concepts based on charge trapping material, in which the 3D construction (tri-gate) overcomes most of the limits of the SONOS principle.
  - A coupling ratio based on the different sizes (A_{TD} and A_{BD}) ensures a large V_{th} cell operation window and an excellent erase efficiency [25].

Figure 2.18 shows a FinFET structure as an example for 3D cell constructions [25, 26].

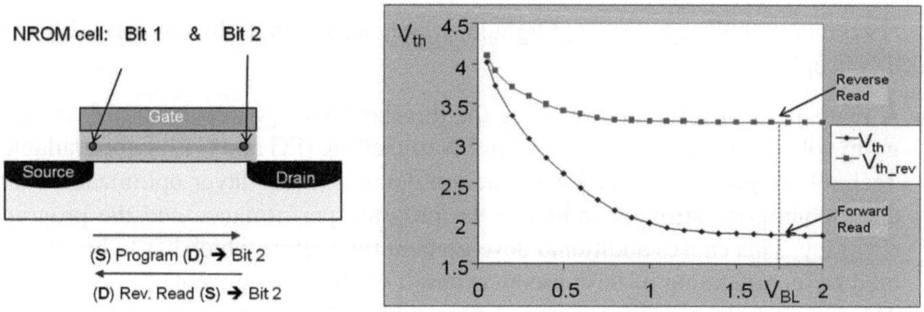

Fig. 2.19 Two bit per cell Nitride ROM [29] and forward and reverse read threshold voltage dependencies

2.3.3 Two Bit per Cell Nitride ROM: Charge Trapping Technology

A SONOS cell can be programmed by channel hot electron injection. The charge is locally injected above the drain junction of the cell transistor in case the correct high voltage values are applied at gate and drain. The interchange of source and drain potential enables to inject charge above each of the two junctions of the cell transistor as shown in Fig. 2.19.

The CHE programming principle operates with lower voltages compared to FN tunneling. The stress to the bottom oxide is reduced and thicker bottom oxides can be used to improve the retention. Hot Hole Erase (band-to-band tunneling hot holes) allows a further optimization and overcomes the discussed FN tunneling—coupling ratio—construction principle trade-off of the planar SONOS cell.

The electron storage based on traps in the nitride layer of both sides of the transistor combined with a reverse read concept creates a physically separated two-bit flash cell [27]. The two bit per cell Nitride ROM (NROM) is described in the literature [28, 29]. The accuracy of the charge injection exactly above the junction is a key performance parameter for program, read and erase. The threshold voltage value and the stability depend on the injected charge Q, the position of the traps and the quality (energy depth) of the traps.

$$V_{th} = f(Q, x, d)$$

The Hot Hole Injection erase principle can selectively erase one side of the cell and is applied for the two bit per cell NROM.

This multiple-bit charge trap cell concept offers two benefits:

- Two independent bits based on locally injected and trapped charge within the nitride storage layer result into a high physical robustness (comparable to one bit per cell).
- The density increase by a factor of two does not impact the cell operation performance—read and program operations are as fast as for Single-Level Cell flash concepts.

Two major challenges are highlighted, which are not that obvious assessing this cell concept:

- A mathematical formula describing the threshold voltage dependency from program voltage conditions as known for FN tunnelling (FG and CT) is not available. Technology parameters, cell construction details, nitride layer optimization and erase conditions strongly influence the program performance and the program accuracy. This causes additional development time effort which has to be considered for qualification of flash products based on this cell concept.
- The injected charge can move over time within the charge trapping layer (also called charge migration) and influence the threshold voltage of the cell [30], which requires design concepts to detect and overcome moving distributions.

The two bit per cell Nitride ROM flash cells are an excellent cell concept to analyse performance and reliability issues of charge trapping based electron storage and develop countermeasures to overcome this behaviour on technology [31], design, algorithm and application level.

High volume production of charge trapping Nitride ROM based flash products requires a complete and deep understanding of all specific charge trap cell effects.

2.3.4 Non-Electron Based Cells: PCRAM, MRAM, FeRAM

Flash memories have a time consuming erase operation which is applied to a large quantity of cells organized in an erase block. This third additional memory operation is unknown for all other memories.

The emerging non-volatile memories (NVM)—PCRAM, MRAM and FeRAM—have two fundamental benefits:

- They behave like a RAM—no dedicated erase operation on block level.
- The physical storage principle is not based on electrons. Therefore all of them claim excellent scalability along the shrink roadmap.
- In principle all of them can be integrated in real 3D next generation memory architectures.

The non-electron based non-volatile memories are on the market either as standalone memories with a smaller density or as embedded non-volatile solution. Emerging memories are claiming significant improvement over flash memories, but still cannot compete on the cost per bit figures.

The basic cell principles are introduced to generate a reference of available non-volatile cell types and operation modes. In the system optimization chapter a DRAM replacement will be analysed and emerging memories have serious benefits based on short read access times.

2.3.4.1 FeRAM

Research and development activities are reported for ferroelectric RAM's since 1982. Today a wide application range is targeted with FeRAM cells from NV FeNAND flash page buffer for solid-state disc (SSD), low current leakage architecture for TAG RAM (a TAG RAM is a specialized RAM architecture to hold address tags) up to SRAM and DRAM replacements.

The one transistor one capacitor (1T1C) FeRAM cell uses design principles known from DRAM designs and offers the feature to store information after power-off. The 1T1C FeRAM is widely used and shown in Fig. 2.20 together with the physical storage principle.

A ferroelectric capacitor is the non-volatile storage element. The storage principle is based on ferroelectric polarization which enables fast and robust memory operations:

- Orientation of the spontaneous polarization with remanent charges is reversible by an applied electrical field. Low power read and write operation are ensured due to this principle
  - Binary State 0: positive electric field results into positive polarization
  - Binary State 1: negative electric field changes into a negative polarization

The comparison between the capacitive storage elements utilized within FeRAM and DRAM are showing the similarity and the major difference being the non-linearity of the charge voltage function.

DRAM	FeRAM
Linear Q–V with constant capacitance	Non-linear Q–V with hysteresis
C has the same value for storage of 1 and 0	C has a higher value for storage of 1 (C1 > C0)

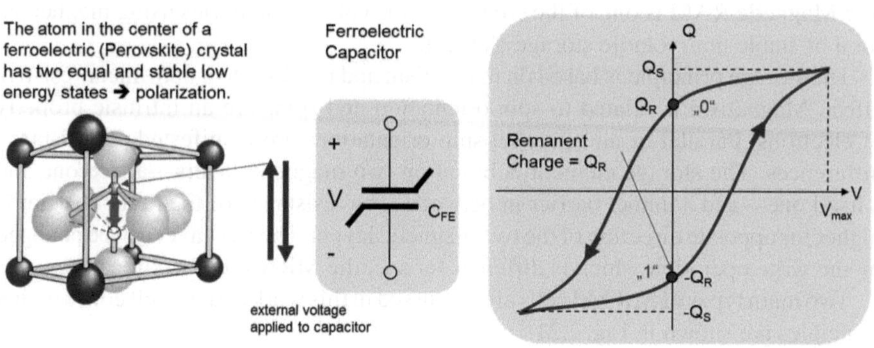

Fig. 2.20 FeRAM: ferroelectric polarization, capacitor and counter-clockwise hysteresis operation principle

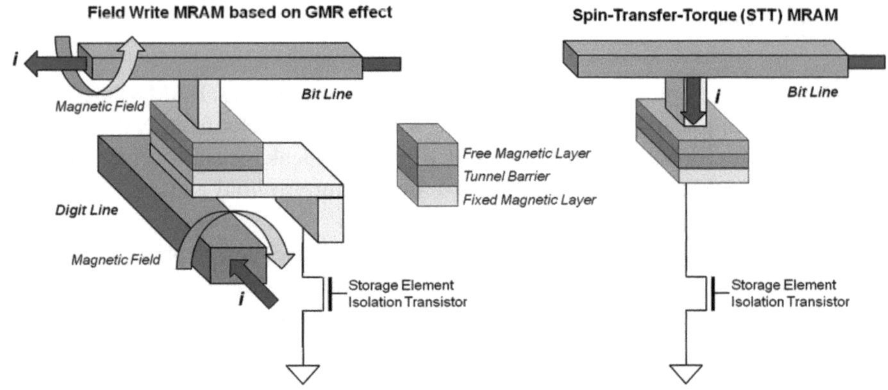

Fig. 2.21 MRAM cell construction principles based on GMR [34] and on STT

Research activities are focusing on reduction of cell size to ensure scalability and on addressing the reliability issues due to fatigue. Especially the reference cells fatigue related issues limit the theoretically good endurance values of FeRAM memories, due to the fact that the reference cells have to withstand orders of magnitude more destructive read operations.

A very competitive new ferroelectric cell concept is the **MFIS**—**M**etal-**F**erroelectric-**I**nsulator-**S**ilicon—**FET**. A one transistor cell with a ferroelectric gate insulator is based on a low voltage controlled physical operation principle. The MFIS is becoming a very competitive emerging non-volatile cell concept after research for several decades on ferroelectric memories. An overview about the research status is given in the literature [32, 33].

2.3.4.2 MRAM

The Magnetic RAM is one of the emerging non-volatile memories using magnetism for a bi-stable non-volatile storage element.

The storage principle is based on magnetism and results into a non-linear resistive effect. Magnetism is related to spin orientation and spins are an intrinsic property of electrons. Parallel or anti-parallel spin orientations are manifested as resistance differences. The storage element is based on two magnetic layers—a free one and a fixed one—and a tunnel barrier in between. The resistance of the tunnel barrier is higher for opposite direction of the two magnetic layers. The free layer can be changed by the write operation which is different for specific MRAM cell architectures.

Two main types of MRAM cells are discussed in this work and the cell construction principles are shown in Fig. 2.21:

Field Write MRAM cells:

- The ***magnetic field induced switching MRAM cell*** is using the field around a current line to flip the polarization of the free magnetization layer. This type is in production for high reliability applications and for embedded memories. The required high write current in both the bit line and the digit line is a serious issue for a scaling roadmap of this cell.

- The cell is also known as toggle MRAM, because the current to write the cell can be toggled step by step to reduce the disturbance to the cell and to the neighbor cells.

Spin Transfer Torque (STT) MRAM cells:

- The ***spin transfer torque cell*** requires a lower current to polarize the electron spin and has an excellent scalability compared with the magnetic field induced switching MRAM.

MRAM cells are characterized by a fast read and write cycle time and an excellent durability behavior. The excellent read disturbance is only achieved for the Field Write MRAM, the Spin-Transfer-Torque cell has the same read disturbance issues as other NVM cells.

The construction principle of the STT MRAM makes it a preferred architecture for 3D memory array solutions. The current STT-MRAM development and research status is summarized in [35].

2.3.4.3 PCRAM

The basic non-volatile storage principle in phase change memories is based on the usage of a material which can exist in two different structural states in a stable fashion. The structural state can only be changed if an energy barrier is overcome. This energy can be supplied to the material in various ways—laser impulses are used for recording of an optical memory (CD, DVD) and electrical current pulses are used for Phase-Change RAM (PCRAM).

A Phase-Change RAM (PCRAM) or Ovonic Universal Memory (OUM) is one type of non-volatile memory using a thin film of chalcogenide alloy like GST (GeSbTe = germanium, antimony and tellurium) that can be switched between crystalline and amorphous states as a bi-stable memory device.

The operation principle of a PCRAM is the reversible phase change between amorphous and crystalline by heating the storage element layer shown in Fig. 2.22.

- Elements in a cell: a ***chalcogenide film***, a ***heater*** and an ***isolation device*** (transistor or diode)

The PCRAM has the potential for unlimited read cycles and write cycles in the order of 10^{12}. The scalability is already demonstrated down to 3 nm feature size [36, 37]. Phase-Change Memory cells offer enough window space for the storage of more than one bit per cell [38].

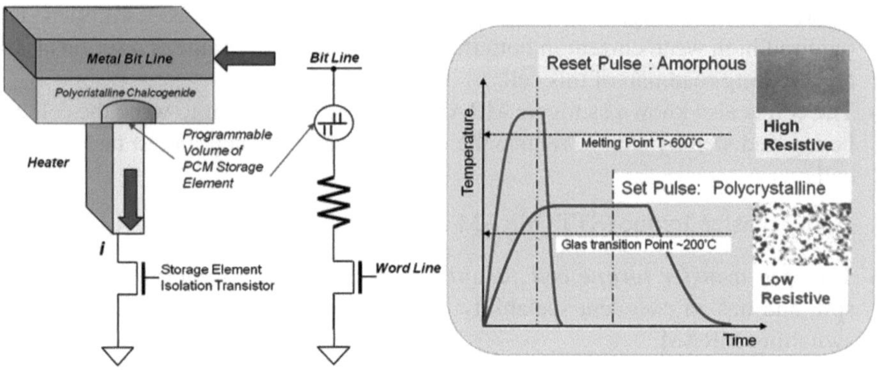

Fig. 2.22 PCRAM operation principle—set and reset

2.3.5 Summary: Non-Volatile Cell

The non-volatile cell or the non-volatile memory element defines the reliability parameters of the non-volatile memory product. Design and Algorithm can influence the level of degradation, but they can't substantially improve the reliability values.

The performance of the memory product is based on the combination of cell and memory array. The cell operation window will impact the capability to store more than one bit and influences product performance parameters too.

Part one of the cell summary compares major characteristics given in Table 2.3:

Part two of the summary derives features based on the analysis of the cell operations for an ideal non-volatile cell/element:

- Program or Write
  - Short program time
  - Program and Erase operation are voltage controlled operation
    Impacts the array requirements strongly
  - Linear dependency of the "V_{th}" of the cell from the program voltage
  - Required program voltage level is reduced automatically with the shrink of the cell
    As lower the max rating of the program voltage as better
  - A large usable operation window to support multiple bit storage per cell
- Read
  - Non-destructive read operation
  - A large distance between read and program voltage conditions to suppress read disturbances
  - High sensitivity of cell current to stored state to enable a simple sensing of the stored information on the same time with a high accuracy

Table 2.3 Non-volatile cell summary table

Topics	NOR flash	NAND flash	FeRAM	MRAM	PCM
Material	SiOx/Poly-Si	SiOx/Poly-Si High-k, metal gate, metal FG	PZT, SBT	CoFeB etc.	GST
Cell size (F^2)	8–10	4	15–40	8–25	6–20
Read op.	Non-destructive	Non-destructive	Destructive	Non-destructive	Non-destructive
Read latency	20 ns	40 μs	50 ns	<10 ns	20 ns
Write latency	1 μs	>500 μs	50 ns	<10 ns	100 ns
Endurance	10^5	10^4	10^8	10^{16}	10^{6-12}
Power in Write pW sec/bit	10^5	0.1	30	40 (field) / 20(spin)	20–1000
Advantage	Conventional material Low latency	Cost per bit Low energy	Fast write Low Latency	Fast write Low latency	Scalability
Challenge	Endurance Failure rate Write energy	Endurance 3-bit per cell	Scalability Fatigue	Magnetic field disturbance	Thermal disturbance Write energy

The third part of the cell assessment is to judge the capability to support innovative multi-bit or multi-level cell operations. The overall size of the memory window and the stability of the programmed level are two important parameters.

- The ideal non-volatile cell/element would store the "PGM energy of each PGM pulse in different levels".
- A bit selective write operation combined with an innovative selective read operation is one of the major research directions.
  - The major role behind this is cost driven: The option to increase the bit density per cell by such a kind of write/read selectivity is more important than other parameters.

As long as the proposed innovative multi-bit selectivity is not available to reduce cost per bit and increase density and performance a certain flexibility of the selected cell concept is required. To fulfill these needs the selected non-volatile cell has to be capable to support or enable 3D integration.

The industrial success of the floating gate cell is the proof of concept, that a combination

- out of a transistor—excellent on/off ratio-
- including a conductive storage element—the floating gate–
- with a coupling ratio to program the FG well defined by the cell geometry
- and a linear dependency of cell V_{th} from the programming voltage

is one of the most robust and predictable non-volatile cell architectures.

The replacement of the floating gate by a charge trapping layer enables a robust technology and the opportunity for a real 3D integration. The cell size has to be small enough to enable 3D cell transistor structures—the benefits were demonstrated based on FinFET SONOS devices [39]—to compensate known issues of the charge trapping layer compared to the floating gate based storage layer.

2.4 Flash Memory Array

Cell type and cell behaviour are discussed often as key features defining the non-volatile memory product performance and specification. This is true for specific non-volatile cell parameters, but product specification and performance parameters are defined by array architecture as well as by cell architecture and cell physics.

The **K**ey **P**erformance **P**arameters of a non-volatile memory product are defined by

- the selected cell in combination with the best fitting array or in most cases
- the selected memory array architecture in combination with the best fitting cell.

This section introduces the next level of memory parameter derived from the array architecture. Dependencies between cell and array especially for the applicable memory operations, interferences between cells within the selected array, physical layout considerations and types of array disturbance are introduced and briefly described for NOR, VG-NOR and NAND array types.

2.4.1 Array of Cells: Threshold Voltage Distributions

A non-volatile flash cell is characterized by a specific threshold voltage which can be shifted by program and erase operations. The target of a memory array is the dense fabrication of billions of cells addressable in columns and rows. The memory array is surrounded by column—bit line—decoder and row—word line—decoder and buffer circuits. Figure 2.23 shows a typical view of a memory array. A small snapshot—nine cells—are zoomed out of the upper left corner of the memory array to illustrate the distance to the row and column decoder and the sensing circuit.

Billions of memory cells produced by hundreds of semiconductor process steps have significant statistical deviations. The strong order of the memory cell array achieves a high homogeneity, but still every cell is individually connected by bit lines, which vary in length and resistance. The select and decoder circuits are introducing inhomogeneities for most of the memory array types. Both effects have an impact on the cell V_{th} of each individual cell—either by technology deviation to the cell itself—or by the so called array effects due to the different position of each cell within the memory array.

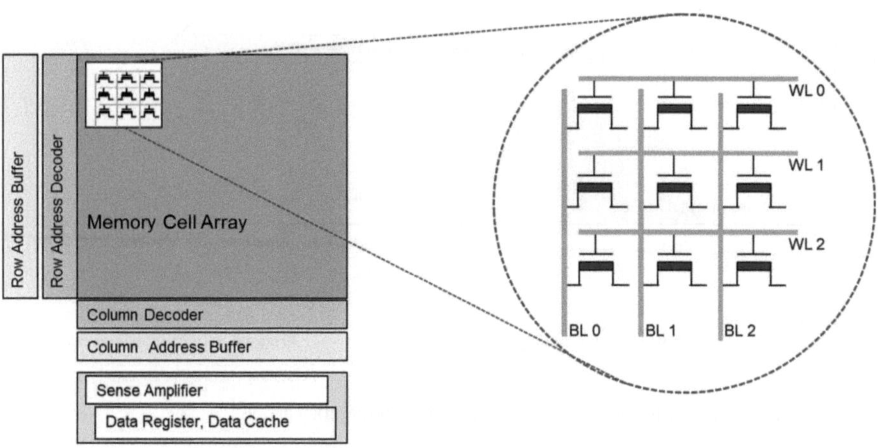

Fig. 2.23 Memory cell array and peripheral circuits

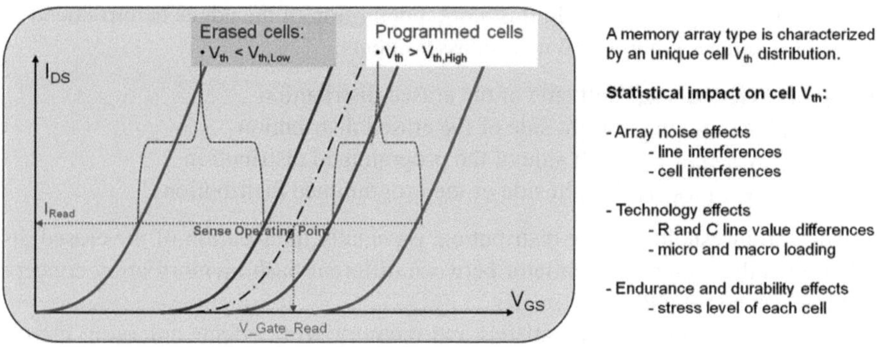

Fig. 2.24 Cell threshold voltage range within a memory array

Figure 2.24 shows the two above mentioned effects and their impact on the cell threshold voltage.

The programmed and erased cells are distributed over a certain threshold voltage range. The threshold voltage of all cells within a memory array is measured and the individual values are plotted resulting into a distribution curve. Figure 2.25 shows the V_{th} distribution of a NOR flash memory array.

The V_{th} distribution of all programmed cells is above the program verify level (PV) and the width is a result of the program algorithm introduced in Sect. 2.6. The V_{th} distribution of all erased cells is below the erase verify level (EV), but all erased cells have a positive threshold voltage.

The statistical deviations of all cells combined with array effects require a defined read window to read fault free the cell information out of the memory array. Each flash cell has its own read or sense trip point, on memory array level the sense operation requires a read window.

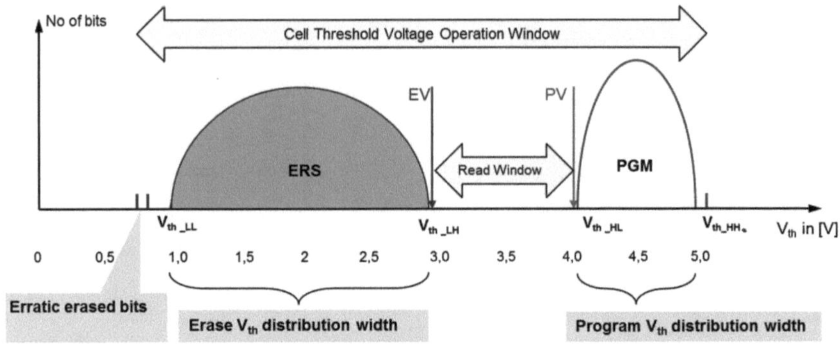

Fig. 2.25 Cell V_{th} distribution for a programmed flash memory array

The edges of each threshold voltage distribution have a strong influence on memory performance parameters. In this work a definition of the edges is introduced for threshold voltage window and margin assessments:

- V_{th_LL} "$V_{th\_Low\_Low}$" left side of the erased distribution
- V_{th_LH} "$V_{th\_Low\_High}$" right side of the erased distribution
- V_{th_HL} "$V_{th\_High\_Low}$" left side of the programmed distribution
- V_{th_HH} "$V_{th\_High\_High}$" right side of the programmed distribution

The cell threshold voltage distribution, especially the location of the erased distribution is the major differentiator between different flash memory array concepts and impacts all reliability parameter.

High volume production statistics and memory array effects enlarging the cell threshold voltage operation window for a memory array. This behavior and the control of reliability effects is introduced and discussed in the algorithm and reliability chapters.

2.4.2 NOR Array: Direct Cell Access

The NOR flash memory was introduced as a memory array architecture by Intel in 1988. The specific characteristics of the NOR memory array is the direct access to each cell terminal.

NOR is defined as a logical operator that consists of a **logical** *OR* followed by a **logical** *NOT* and returns a true value **only if both** operands are **false**.

Figure 2.26 shows the typical logical NOR array schematic, the so called ETOX® cell developed by Intel—**E**rase **T**hrough **Ox**ide—and an array cross section showing the shared bit line contact. The optimized array structure is discussed in more detail in Sect. 2.4.2.4. The NOR cross section in Fig. 2.26 shows two NOR cells followed by one shared bit line contact.

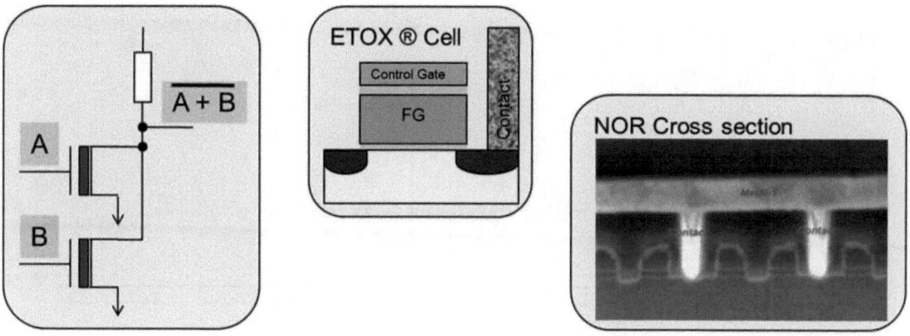

Fig. 2.26 NOR array principle based on ETOX® cell and NOR silicon cross section [40]

The NOR array architecture allows single bit, row and column operation and is the perfect memory array architecture for an electrically erasable and programmable Read Only Memory.

The NOR flash memories are also called Code Flash driven by the application usage for program code storage. The NOR array offers fast random access to the complete memory, and its capability to change data words or single bits generates a perfect code memory. The program code of a microcontroller could be executed directly out of the NOR (CODE) flash memory.

Flash performance parameters are now discussed for a NOR flash memory array.

2.4.2.1 NOR Read Operation and Sensing Principles

The NOR array is characterized by a direct access to each memory cell. The direct cell access allows a fast sense operation of the cell current, which guarantees a fast random read access in the range of 25–90 ns.

The accuracy of the sense operation is a key parameter for read. The sense accuracy influences also every program operation applying the same sensing principle during each program verify operation. The read accuracy is a key parameter for the V_{th} window margin definition which defines the performance and the reliability capability of a flash memory product.

A sense operation is always a current voltage conversion. The so called trip point is the sensing point for a selected read gate voltage of the cell to be read. This sensing trip point should be exactly the same for program verify, erase verify and read to compensate all effects and ensure the accuracy.

Physically this could not be achieved. Therefore two basic sensing principles are applied:

- Constant gate voltage—different sensing currents represent different V_{th} position of the cell;
- Constant sense current—different gate voltages represent different V_{th} position of the cell;

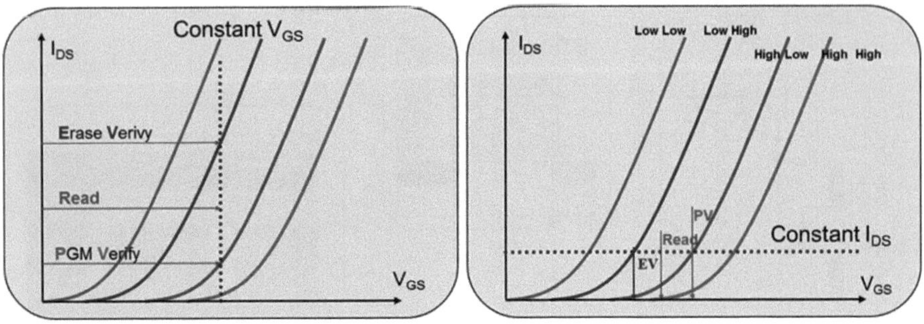

Fig. 2.27 constant gate voltage and constant sense current sensing scheme

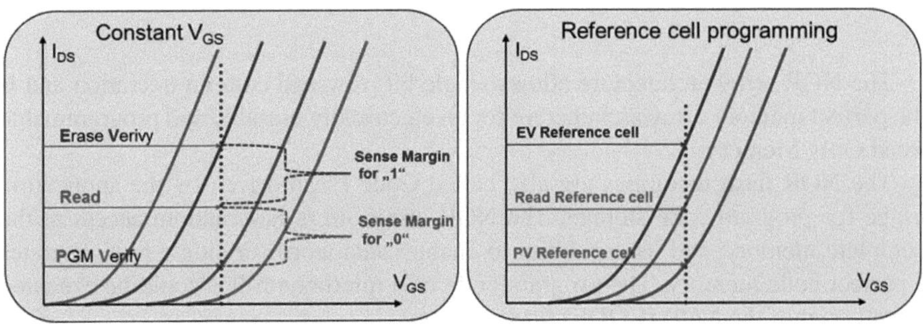

Fig. 2.28 Sense margin and reference cell position for erase verify (EV) and program verify (PV) reference cells

Starting from the principles shown in Fig. 2.27, different sensing implementations are possible, resulting into different sense concepts or better sense architectures.

Different sense amplifier architectures could be used for a NOR based memory array.

- Direct integration of the sense current and comparison with a reference voltage.
- Direct integration of the sense current and comparison with a reference cell or structure.

Especially for the constant gate voltage sensing a reference cell or a reference cell array has strong benefits, because the reference cell has normally the same temperature coefficient and will follow the array cell and compensate any shift of the target cell during every read operation.

Figure 2.28 shows the reference cells which are required for NOR flash sensing. The reference cells have to be programmed during the wafer testing with a very high accuracy, which is time consuming and adds significant test costs to the NOR flash product. The fact that the array cells will degrade over time for every non-volatile memory and the reference cells are seeing a different stress – the read disturbance is orders of magnitude higher – is one weakness of a reference cell based sensing concept.

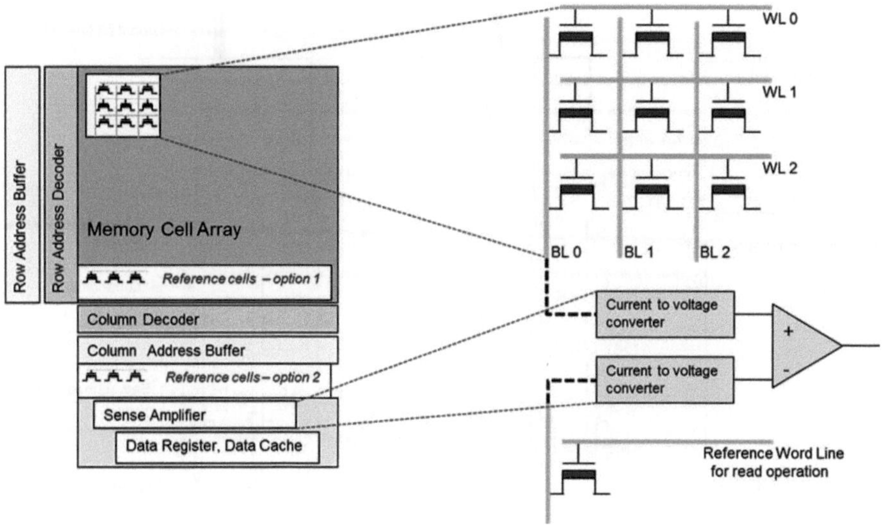

Fig. 2.29 Simplified sense concept and layout options to place the reference cells

Figure 2.29 shows a simplified NOR flash sensing implementation using a separate reference cell array. Two different options are shown in this figure suitable for the reference cell array placement.

NOR Sensing circuits for a fast current to voltage conversion are discussed in detail in the literature including such important design topics like offset compensation [41].

The read performance within a NOR array is defined by the sensing concept, circuit and layout details and the reference cells. An accurate sensing requires a stable gate voltage and a stable bit line voltage applied with a very high accuracy.

Low resistive and a low capacitive bit lines are the target for a memory array. For NOR memories a local and global bit line architecture is widely used to achieve the required low bit line rise time values. The requirement to the word line rise time is in the same range.

The segmentation of the NOR memory—applying a local and global bit line scheme—ensures the necessary values for a fast sensing.

Figure 2.30 illustrates the principles of a local bit line decoding (local y-select). The design and the layout of the y-select have to made carefully to ensure the same electrical behaviour of each path. The remaining array effects will create an imprint during each program and erase cycle. This effect will influence the sensing and the reliability behaviour of flash cells within the NOR array. A more detailed discussion of these effects will be done within the VG-NOR array chapter.

The read operation accuracy is impacted by effects enforced by the local Y-select structure and by global dependencies shown in Fig. 2.31. Each NOR array block has a different distance to the sense amplifier at the end of the global bit lines.

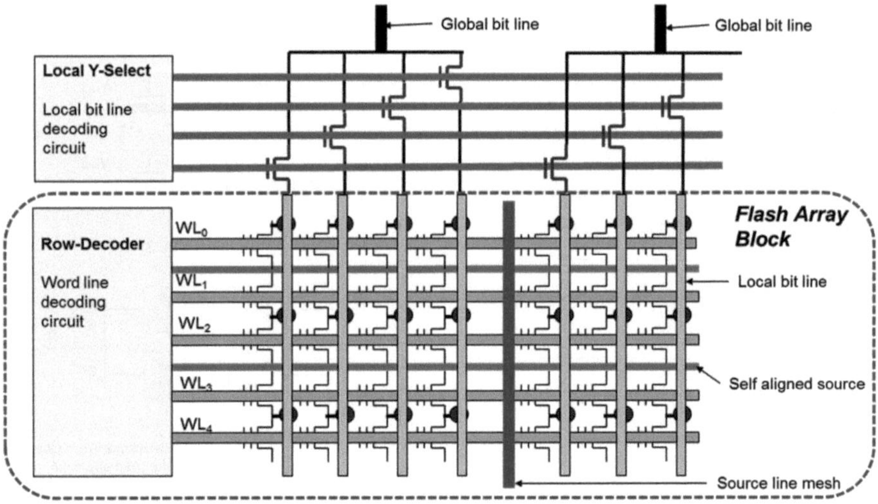

Fig. 2.30 NOR flash array block including local and global bit line architecture

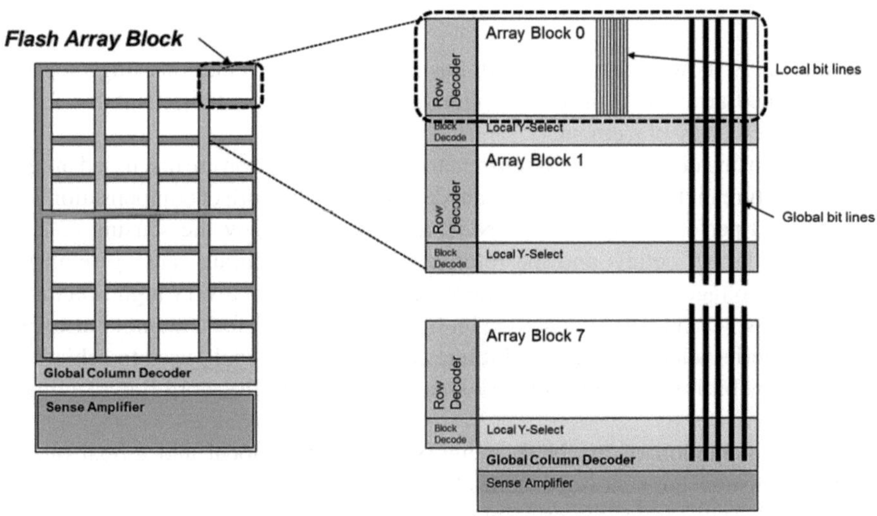

Fig. 2.31 Distance between Sense amps and array blocks—local and global bit line decoding

The NOR sense architecture and the reference cell scheme have to compensate the local Y-select effects, the distance between different blocks and the reference cell block. Temperature changes, cell alteration and reliability effects have to be considered on top for the V_{th} read margin window.

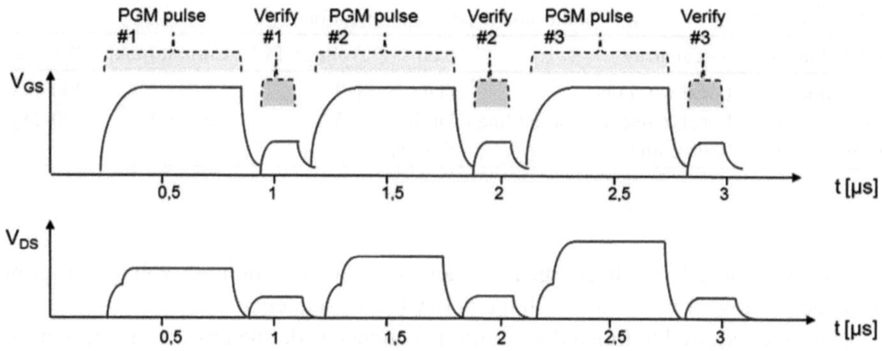

Fig. 2.32 NOR flash CHE program pulse operation with drain voltage stepping

2.4.2.2 NOR Write Operation and Program Principles

The program operation within a NOR array is executed at one cell within a certain sub array.

→Single Cell Operation Principle;

The NOR array offers the access to all cell terminals and enables the usage of the fast channel hot electron (CHE) program principle. Applying CHE programming has two benefits:

- lower voltages than FN tunnelling results into less damage in the gate oxide;
- fast program operation in the range of µs;

The program operation is based an incremental step pulse program algorithm which is introduced in detail in Sect. 2.6. A number of program pulses is applied, each pulse followed by a verify operation to compensate statistical cell and array effects described in Fig. 2.24. Each cell V_{th} has to pass the program verify threshold voltage level.

The NOR array offers two options to increase step by step the threshold voltage of the cell to be programmed. An increase of the gate voltage as well as the drain voltage increases the efficiency of the channel hot electron programming operation.

- Gate voltage stepping → gate voltage is increased step by step, drain voltage is constant;
- Drain voltage stepping → gate voltage is constant, drain voltage is increased step by step;

Figure 2.32 shows a typical timing sequence to program one cell by CHE within a NOR array belonging to the corresponding sense amplifier at the end of the global bit line.

The program operation—either gate or drain stepping—is followed by a verify operation and only afterwards the next cell could be programmed. Bit and word lines

Table 2.4 NOR array voltage conditions for program operation

Global array	Local array	Cell	Voltage [V]	Current [μA]	Time [ns]
Word line	Local word line	Gate	8–9		300–800
Global bit line	Local Y-select, local bit line	Drain	4–5	100–200	200–500
Source mesh	Source line	Source	0 V		

have to be charged and discharged frequently to execute the high voltage program sequence followed by the low voltage verify (read) sequence.

The time required to finish the program sequence is defined by the program pulse plateau time and rise times of bit line (V_{DS}) and word line (V_{GS}), which have to be driven to the necessary voltage levels.

The program operation time for a NOR array can be estimated based on the following parameter:

- Number of PGM pulses (estimation rules are introduced in the performance chapter);
- PGM voltage plateau time—minimum time for both high voltages at the cell terminals;
- Worst case charge and discharge times of bit and word lines;

Table 2.4 shows typical voltage and timing conditions for channel hot electron flash programming.

The program performance can be increased by reducing the number of pulses required to shift each cell V_{th} above the program verify level or by increasing the number of cells, which could be programmed in parallel. The program parallelism of a NOR array is limited by:

- The number of sense amplifier, which could access the same word line in parallel;
- The maximum limited product current due to the CHE injection operation principle;

Channel Hot Electron injection requires significant I_{DS} current values to be supplied out of pumped high voltages. The NOR programming requires a low resistive bit line path including the local Y-select circuits which results into large select devices. This array optimization limits the parallelism especially along the shrink roadmap, because bit lines are becoming smaller and the resistance increases.

High voltages are applied along the word line and along the local bit line. Every program operation within an array will disturb all non-selected cells along these bit and word lines with high voltage conditions shown in Fig. 2.33. The design and technology target is to balance cell parameter and applied voltage levels in such a way that already programmed cell states cannot be destroyed within the specified application limits valid for program and read operation using the same bit and word lines.

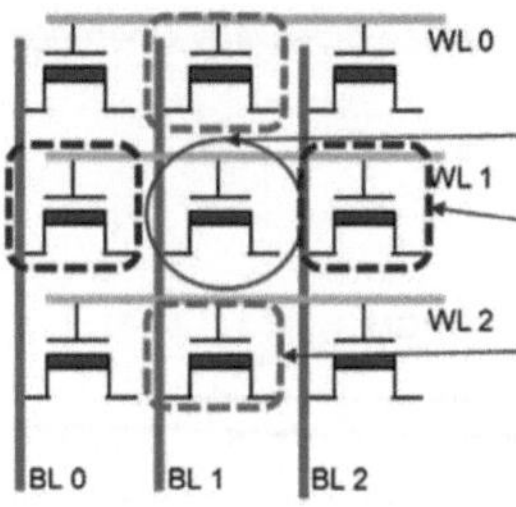

PGM conditions	WL	BL
Program cell operation at WL 1 & BL 1	9.0V selected	4.0V Selected
Gate Disturbance along WL 1	**9.0V**	0V non-selected
Drain Disturbance along BL 1	0V non-selected	**4.0V**
No Impact	0V non-selected	0V non-selected

Fig. 2.33 NOR program disturbance along WL and BL for the non-selected cells

2.4.2.3 NOR Erase Operation and Array Effects

Two different terms—Erase sector or Erase block—are known describing the array sub-area, which shifts the V_{th} of all cells belonging to this area within one erase operation of a flash memory. The wording memory array block is used to describe this array segmentation in this work.

The erase operation within a NOR array architecture is executed within a memory array block for all cells in parallel → **Segmented Sector Operation Principle**.

Different implementations can be selected to erase a NOR memory array block. A positive high voltage could be applied only to the common source of all cells within the block, or a negative gate voltage could be applied to the word lines and a positive to the source or a positive to the bulk. The FN tunnelling bulk erase is preferred due to smaller cell size and better reliability behaviour.

FN tunnelling requires a long negative high voltage pulse in the range of mill seconds. The programmed cells have a higher threshold voltage by 3–4 V compared to not programmed cells and therefore the electrical field for the programmed cells is high enough to enforce the tunnel process to shift the cells done into the target region. The shift of the cells during the erase process reduces the effective electrical field and the tunnel process slows done. The FN erase process is for a given timing a theoretically self-limiting process. The effectiveness of this self-limitation depends strongly on all the technology deviation. The width of the erase distribution is a **quality parameter** for the production technology and has a significant impact on the flash algorithm and the achievable program and erase performance.

The lower edge of the threshold voltage erase distribution (V_{th_LL}) is a sensitive parameter for NOR flash memories. Figure 2.34 shows a NOR flash array V_{th} distribution including cells with an erratic bit behaviour.

Erased cells can be erased over and over and they are shifted outside of the V_{th_LL}. A limited number of cells can show an erratic behaviour. Their V_{th} values are much lower compared to the applied erase voltage conditions.

Cells on the left side of the V_{th_LL} border impact the sensing accuracy of a NOR array. Their higher leakage current (higher compared to the design target) is added to the sense current of other cells to be read. The countermeasures to maintain the erase distribution within the specified limits will be discussed in the algorithm chapter.

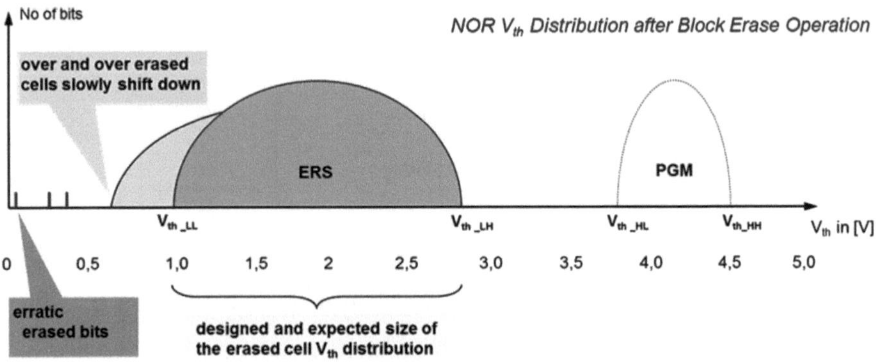

Fig. 2.34 V_{th} Distribution after Erase for a NOR type memory

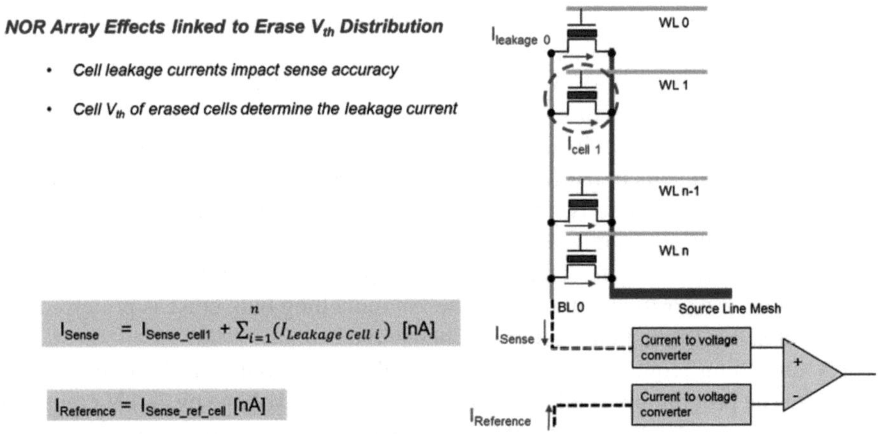

Fig. 2.35 NOR array erase weakness—impact on sensing accuracy

The strong benefit of a NOR cell array—the fast direct access to each cell—is the reason for a serious drawback of the NOR array assessing the erase operation. Along a bit line all cells are connected to this bit line, the leakage current of each cell will be added to the sense current of the target cell.

Figure 2.35 illustrates the principle issue of the leakage current impacting the sense current within a NOR array.

The NOR erase operation requires at the end an erase verify to ensure that all cells are within the target erase distribution width. The V_{th_LL}edge of the erased distribution requires an active control.

The long erase time of a NOR flash (>500 ms) is a performance weakness of the NOR array. The erase time is significantly longer in relation to any read or program operation. During the erase time the memory is completely blocked and therefore an erase suspend mode was introduced to increase the response time of the NOR flash memory.

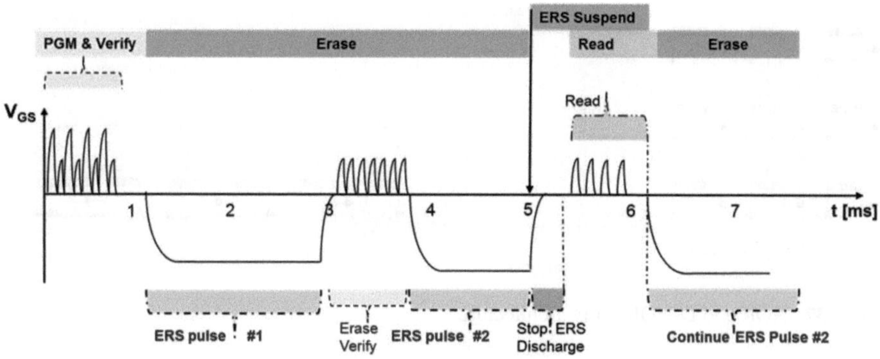

Fig. 2.36 Erase timing for a NOR type memory with erase suspend

Figure 2.36 shows a program, read and erase operation time sequence. The erase operation for one NOR erase block is suspend—stopped in the middle of an erase pulse—within the typical erase operation time. A read operation is executed and the erase is afterwards continued.

The control of the erase distribution width impacts directly all reliability parameter of a flash memory. The width and the position of the erase distribution are key parameters of a flash memory.

2.4.2.4 NOR Array Architecture: Physical Layout Consideration

The NOR memory array architecture targets a dense connection of all cells with less process steps.

The classical NOR array accesses each cell using all cell terminals. Every cell requires a bit line contact which increases the effective cell size by the size of the contact. The idea to share one bit line contact for two cells reduces the overhead for contacts within a NOR array architecture.

The **Di**vided bit line **NOR** (**DINOR**) concept was introduced which combines the local bit line (poly silicon) creating the contact with the concept of a global bit line (metal) shared by two local bit lines. The classical NOR concept (on the left side) is compared in Fig. 2.37 with the DINOR array.

The development of the Self Aligned Contact (SAC) increases the density of the DINOR array again. Figure 2.38 shows the DINOR Array architecture including the global or main bit line and the select transistor to switch to the divided local bit lines. The cross section is showing the contact.

The combination out of local short bit lines (e.g. polylines) and global long metal bit lines reduces the overall bit line capacity and resistance, which is required and ensures the specified fast read access.

Figure 2.39 illustrates the cell and array matching within a DINOR array. The density of an array is defined by the minimal pitch—defined by lithography—and

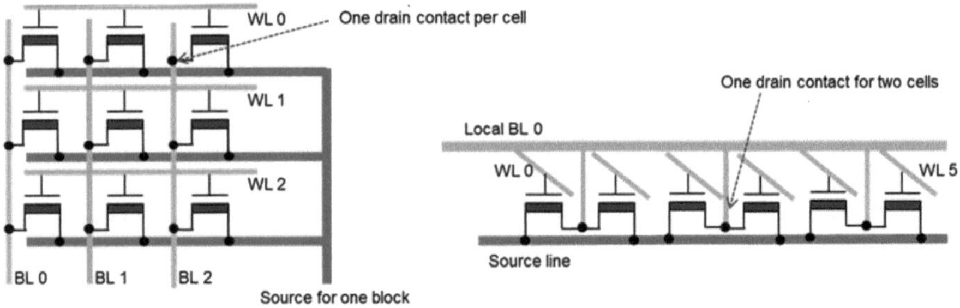

Fig. 2.37 NOR and DINOR array architecture

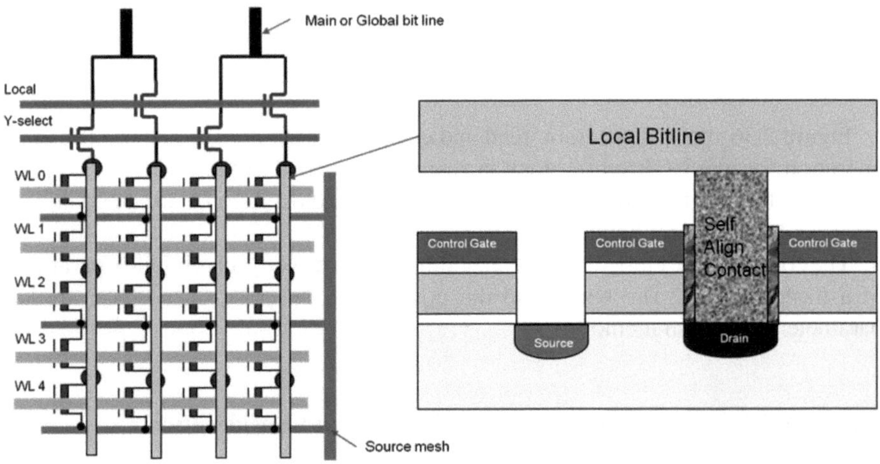

Fig. 2.38 DINOR flash array combined—ETOX® cell and self aligned contact

the number of contacts required to access the cell, the bit line and the source line and the well. A very dense logical array schematic shown in Figure 2.29 is translated into a physical layout, which is optimized for highest density.

The layout optimization influences the applicable cell operations. The development of self-aligned bit line and source line contacts enforces a channel bulk erase operation. Less source line contacts and higher array efficiency is translated into more resistance deviation within the array. These differences in array resistance values can be compensated by design, the impact will always be visible in the long time reliability behaviour of the cells.

Different NOR array architectures—NOR with dedicated bit line and source line contacts per cell, DINOR sharing one drain cell contact between two cells and VG-NOR using buried bit lines combined with the virtual ground array concept are shown in Fig. 2.40.

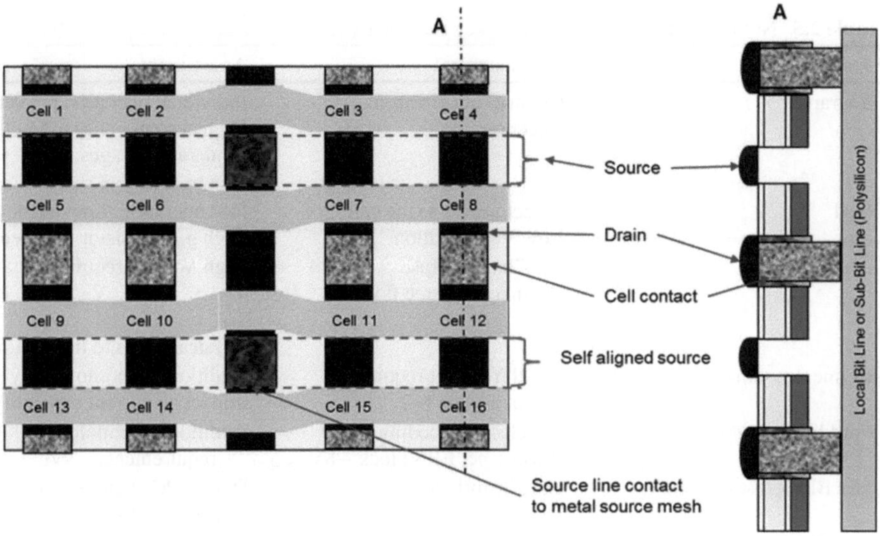

Fig. 2.39 DINOR array—layout density and cross section

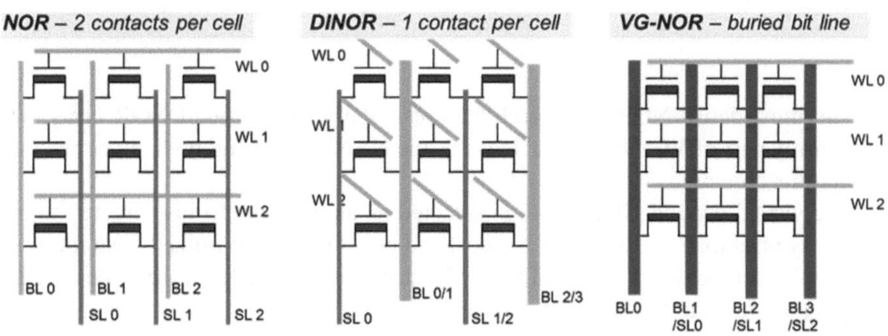

Fig. 2.40 NOR, DINOR and VG-NOR array architectures showing the number of contacts per cell

The virtual ground NOR array architecture is the most efficient implementation and is assessed in detail in the next chapter focusing on the cell and array matching topic.

2.4.2.5 Summary of NOR Cell and Array

The Assessment of the NOR array and the floating gate cell is summarized in Table 2.5, which includes all parameters used for the memory concept assessment.

Table 2.5 NOR cell and array summary—key statements

Operation	Key statement	Assessment
Program	Channel hot electron Injection	Individual operation per cell High current ~100 μA/cell; Moderate voltages: $V_{GS} = 9\,V$, $V_{DS} = 4\,V$
Read	Direct access to the cell Low V_{th} operation	Fast operation, Low latency High g_m transistor required
Erase	FN Channel erase Drain and Source floating	High Voltage required $V_G = -8\,V$, $V_{Substrate} = 7\,V$ Physical stress to the Bottox
Bit line decoding	Local Y-Select required 1–8 or 1–4 or 1–2	Additional inhomogeneity Impact on current capability
Word line decoding	X-Select—decoding of 9 V Passing per ERS block −8V	Additional technology requirements
Intra Block disturbance	Gate disturbance Drain disturbance	During PGM and Read During PGM and Read
Inter Block disturbance	Dependent on block architecture	During Erase
Parasitic effects	Neighbour and sense effects	Dependent on sense concept

The most dense NOR array architecture is the virtual ground NOR array. The inner array part of VG-NOR is as dense as the NAND array and both are contactless within this core area.

The virtual ground NOR array is selected as an example to investigate the cell and array principles introduced for a cost and performance optimized flash memory. Two topics are getting a special focus. The parasitic array effects are impacting the sense operation and the cell and array matching.

2.4.3 Virtual Ground NOR Array

The virtual ground NOR array has no common ground. Every buried bit line could be shared as source or drain line. Figure 2.41 shows the principal array structure. The array is very dense—no contacts required for the cells—bit line-selects are required on both sides to connect the global bit line i to drain and the global bit line i+1 to the source of the cell to be read or to be programmed.

The virtual ground NOR array architecture allows single bit, row and column operation and is the perfect array concept for a non-volatile cell with localized charge storage on each side of the cell transistor, because source and drain can be switched by the selection of different potentials at the corresponding global bit lines.

The different operation modes for a virtual ground NOR flash architectures are introduced and compared to the classical NOR or DINOR flash array architecture.

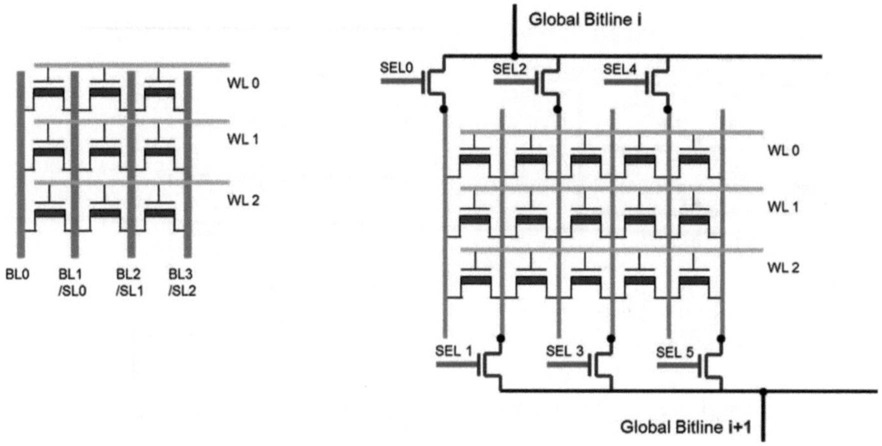

Fig. 2.41 Virtual ground NOR array architecture

2.4.3.1 VG-NOR Read Operation and Sensing Principles

The NOR read operation principles are fully applicable for the virtual ground NOR array. The common ground NOR array architecture requires a drain-side sensing concept, because all cells have a common source. The virtual ground array is based on a floating ground. Each buried bit line can be switched either to source or to drain of the corresponding cell to be read. This freedom gives the opportunity to design a drain-side or a source-side sensing for a VG-NOR flash memory.

Figure 2.42 illustrates the additional parasitic effects driven by the VG-NOR array architecture. The sensing within a NOR array is impacted by the parasitic leakage currents along the target bit line to be sensed known from the NOR discussion see Fig. 2.35. The bit line leakage effect can be controlled by the width of the erased V_{th} distribution which is ensured by the corresponding erase algorithm.

The sensing within a VG-NOR array is additionally impacted by the leakage current along the word line, due to the fact that the other neighbour bit lines are floating at the beginning of the sense operation (virtual ground array concept).

The sense current through the cell and the parasitic currents on both sides of the cell are shown in Fig. 2.43. The drain potential will create a significant additional current going into the floating (e.g. left) part of the accessed buried bit line during read. The same effect can be also seen during the CHE program at the beginning of the program pulse and can create an additional program disturbance.

The source side of the target cell to be read is in the first order immune to this effect. In a more detailed analysis the history of the last operation can leave a certain potential on the floating bit lines on the (e.g. right) side of the cell, which will cause an accumulated additional current during the sense operation. For long local buried bit lines the sense current creates a certain voltage drop on the source path, which will create an additional parasitic current which is added to the sense current.

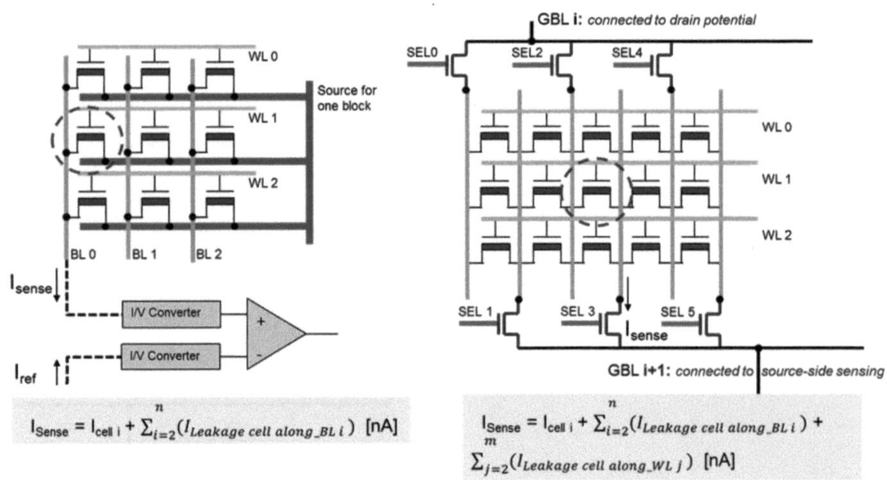

Fig. 2.42 Parasitic effects compared for NOR and virtual ground-NOR sensing architecture

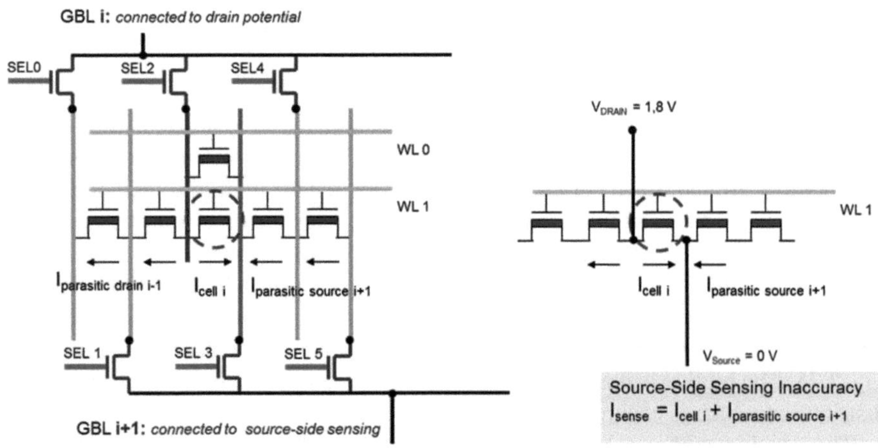

Fig. 2.43 Source-side sensing for VG-NOR Compensation of parasitic effects

All these effects have to be compensated by the sense amplifier circuit and the applied timing sequence to charge and discharge all corresponding lines.

The second concept for VG-NOR is drain side sensing which is successfully applied to NOR flash designs. The challenge of the VG-NOR array compared to the DINOR array is immediately visible; as the drain contact is shared by two cells (and more) using the same word line. The floating bit lines could be assumed as close to ground, which would result under worst case conditions into a sense current nearly twice as much as the cell current of the target cell. Applying the same voltage potential to more than one neighbour bit line on the drain side of the cell is the design solution to suppress the parasitic leakage on the drain node [42].

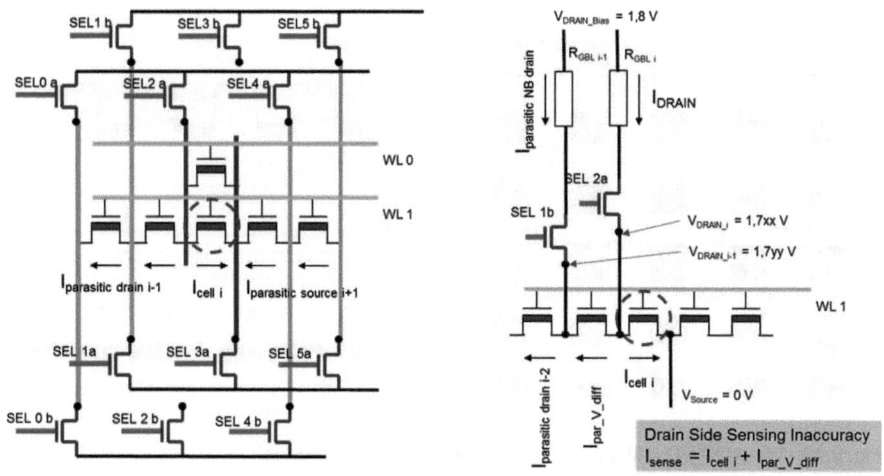

Fig. 2.44 Drain-side sensing for VG-NOR with compensation of drain side leakage currents

The parasitic current path and the design solution for drain side sensing are shown in Fig. 2.44. The additional select transistors required to supply the neighbour bit lines of the drain sensing node with the correct drain potential are marked with a small b. The implemented product sensing concept is more complicated; the voltage drop enforced from the sense current along the bit line has to be compensated too. The voltage potential supplied to the next bit lines has to be slightly different and well controlled.

The accuracy of the drain side sensing could be ensured by design innovation and will be judged as comparable to the classical simple drain side sensing applied to the NOR array architecture. The bit line select design and layout overhead are added. The expected benefit of the VG-NOR array in terms of smaller cell size without contacts is reduced by an increased bit line select and sense circuit adder.

- No contacts within the array, but twice or three times more select transistors.

The first rule learned from the NOR and VG-NOR cell and array size comparison is:

> *The complete array overhead has to be part of the assessment comparing different memory concepts and is an intrinsic part of the key parameter used for the overall performance modelling described in chapter 6 in this work.*

This work adds in Sect. 5.2.2 an **Array Efficiency** parameter to benchmark the complete array density always with a dedicated key performance parameter.

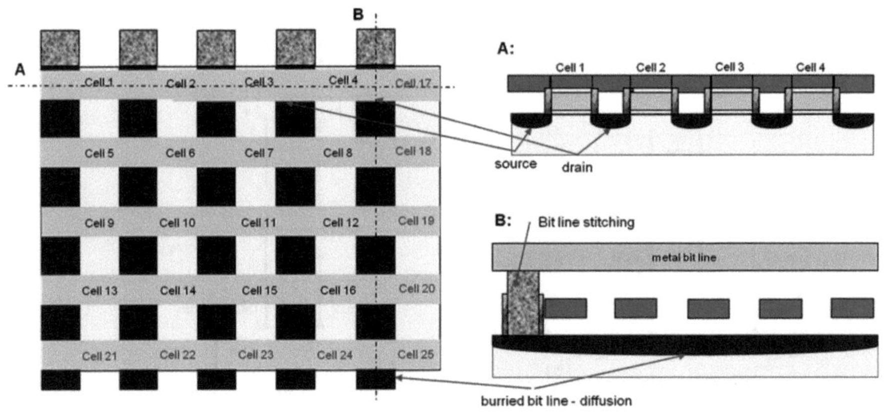

Fig. 2.45 VG-NOR FG cell array—layout density and cross section

2.4.3.2 VG-NOR Operation Principles and Array and Cell Matching

Program and erase operation can be applied exactly as described for the NOR array. The VG-NOR array can be combined with floating gate and charge trapping cells. A couple of program and erase principles are successful applied within the VG-NOR array architecture CHE injection and band-to-band HHI for programming and FN tunnelling or HHI for erase [43–46].

The VG-NOR array architecture is a dense and competitive flash memory concept. The DINOR array—shown in Fig. 2.39—and the VG-NOR array—shown in Fig. 2.45—illustrate the difference of both concepts. A regular structure out of buried bit lines and word lines without any contact and without any irregularity is shown in Fig. 2.45.

The elimination of the specific source ground lines and contacts increases the density of the array by a factor of 1.56 (25 cells for VG-NOR compared to 16 cells for DINOR).

The inner memory core density is improved by 56 % based on the above assumption. The advanced sensing and the additional effort for the select gates comes on top and the overall benefit for a floating gate cell based VG NOR array has to be calculated design specific in detail. A first estimation indicates that the benefit on product level for the VG-NOR array architecture concept is less than the numbers assumed above.

The complicated select gate structure requires a considerable design overhead. The timing for fast switching of all select devices has to be balanced to compensate all parasitic effects. A die size optimized VG-NOR layout therefore requires irregular select structures which influence the homogeneity of the memory array as well as the capability to shrink.

As an example the CHE programming indicates the additional effects created by the virtual ground array architecture. The program operation requires a high voltage at the drain side of the cell. The parasitic current into the floating part along the word line is significant and depends on:

- Program timing sequence, which cell was programmed and verified the cycle before;
- Program data pattern along the word line already programmed;
- Location of the target cell, how many cells are in the floating pipe in total;

Along the shrink roadmap the buried bit lines become smaller and the resistance increases from technology node to node. The influence of the effects described above increases. The increased resistance of the buried bit line creates a voltage drop due to parasitic array currents which becomes more visible as an irregularity within the array and results into a larger erase distribution width. Every irregular structure enforces additional noises due to lithography effects every which are impacting the threshold voltage window stronger than expected.

The second rule learned from NOR and VG-NOR cell and array comparison is:

The complete array overhead has to be analysed in terms of parasitic effects and disturbances to assess very precisely the capability for the storage of more than one bit per cell.

2.4.3.3 VG-NOR Array and localized Charge Trap 2 bit/Cell Matching

The combination of floating gate cell with virtual ground NOR array improves the bit density by a factor of one point five. The additional dies size and design overhead for local Y-selects and enhanced sensing is consuming the increased VG-NOR array density. The floating gate encapsulation requires the same technology complexity.

A charge trapping cell—nitride layer replaces the floating gate as storage element—simplifies the technology and adds the capability to store two physical bits per transistor. This cell architecture increases the bit density for a VG-NOR array by a factor of 3.1 compared to a FG NOR based memory (50 bits for 2 bit/cell SONOS VG-NOR compared to 16 cells for 1bit/cell FG DINOR) shown in Fig. 2.46.

Flash products based on this concept were commercially distributed under different names for example NROM™, MirrorBit®, TwinFlash®. [1]

The Virtual Ground NOR array combined with two bit per cell SONOS technology increases the cell efficiency significantly, simplifies the technology for the inner core array and combines a compact and very dense array with a fast read access and a short program cycle time.

The localized trapped charge per cell requires an optimized erase operation. The commercially used principle is the band-to-band hot hole injection erase. A slice architecture shown in Fig. 2.47 can be introduced to limit the erase operation to cells within one slice and reduce the erase block size and the erase time.

[1] NROM™ invented by Saifun Technlogies, MirrorBit® used by AMD and Spansion, TwinFlash® used by Infineon Technologies Flash GmbH.

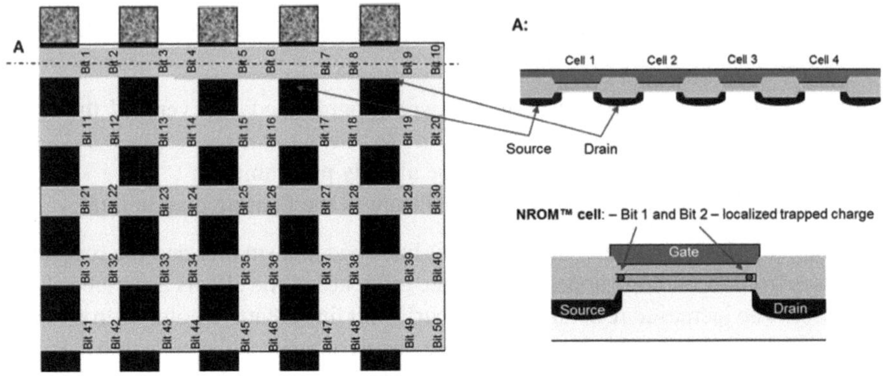

Fig. 2.46 VG-NOR NROM cell array—layout density and cross section

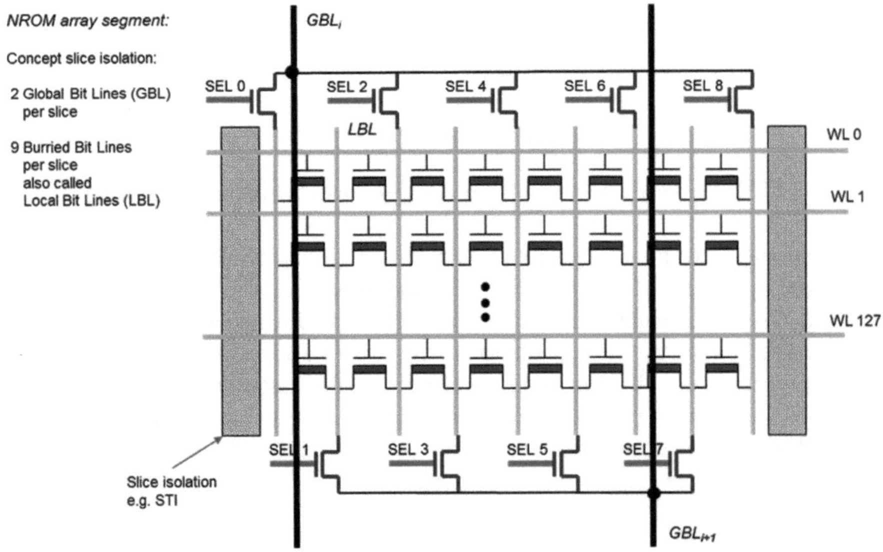

Fig. 2.47 NROM array—segmentation based on the slice principle [47]

In addition the slice architecture is one option to manage the parallel access to many cells belonging to one word line during the read and verify operation, to limit the parasitic currents especially if the source side sensing architecture is applied.

The match between cell and array architecture is different for charge trapping and floating gate cells shown in this example for a Virtual Ground NOR array. The charge trapping cell—based on localized bit storage per transistor—utilizes all benefits of a virtual ground (bit line) array architecture.

The switching between source and drain is required for the two bit storage and increases the memory density by two. The inner array is a competitive dense flash memory solution.

The assessment of the complete array including all select gates and the required area for the slice will have a different conclusion. Especially the stitching of the buried bit line has to be well balanced between low resistance for sense and program operation and less stitching to increase array efficiency of the inner core array.

A weakness of a virtual-ground NOR array is the disturbance sensitiveness. The inter- and intra-block disturbances during **all** operations have to be compensated by appropriate design measures.

2.4.3.4 Summary of VG NOR Cell and Array

An assessment of VG NOR and a two-bit per cell NROM cell is summarized in Table 2.6, which includes all parameters used for the memory concept assessment.

The VG-NOR array combines an excellent low latency read capability with a high density inner array core capability, especially if the array is combined with a two bit per cell charge trapping cell.

This cell and array combination creates excellent application specific solutions in which both code execution and large data storage are functional requirements.

The next chapter introduces the NAND array architecture which has the same inner core array density like the VG-NOR array. In contrast to VG-NOR the NAND array cannot be combined with a two bit per cell (physically separated) charge trapping cell, because cells are not directly accessible for localized programming.

Table 2.6 VG NOR cell and array summary—key statements

Operation	Key statement	Assessment
Program	Channel hot electron Injection	Individual operation per cell Relatively high current Moderate voltages: $V_{GS} = 9\,V$, $V_{DS} = 4\,V$
Read	Direct access to the cell Low V_{th} operation	Fast operation, Low latency High gm transistor, required
Erase	Hot Hole Injection Source floating	High Voltage required $V_G = -6\,V$, $V_{DB} = 6\,V$
Bit line decoding	Local Y-Select required 1–8 or 1–4 or 1–2	Additional inhomogeneity's Impact on current capability
Word line decoding	X-Select—decoding of 9 V Passing per ERS block −6 V	Additional technology requirements
Intra Block disturbance	Gate disturbance Drain disturbance	During PGM and Read During PGM, Erase and Read
Inter Block disturbance	Depend on block architecture	During Erase
Parasitic effects	Neighbour and sense effects	Dependent on sense concept Source versus Drain Side sensing

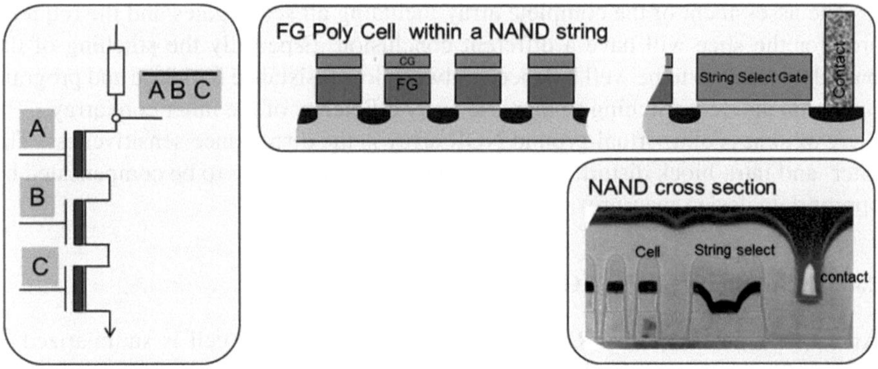

Fig. 2.48 NAND string array principle and cross section [49]

2.4.4 NAND Array: Indirect Cell Access

NAND flash was introduced as memory array architecture by Toshiba in 1989 [48]. The following definition describes the logical array behavior:

NAND is defined as logical operator that consists of a **logical** *AND* followed by a **logical** *NOT* and returns a false value **only if both** operands are **true.**

Figure 2.48 shows a logical NAND schematic, a floating gate cell within a NAND string ending with the select gate and the bit line contact. A cross section shows the real dimensions of the cells connected in series (end of the NAND string with bit line contact).

A slice of a NAND erase block includes bit line contacts, string select transistors, 32 NAND cells, ground select transistors and the source line. Figure 2.49 compares the NAND string layout view—left side—with the logical structure of a NAND erase block with 32 cells—right side. The logical circuit includes the bit lines connected through the bit line contact to two string select transistors.

2.4.4.1 NAND Read Operation and Sensing Principles

The NAND read operation principle is sensing a cell within a chain of 32 or 64 in series connected cells the so called NAND string. This indirect cell access impacts the sense operating point, the sense performance and the sense accuracy of NAND flash.

Sensing of the target cell within a NAND string has to be translated into sensing the current of the string modulated by the V_{th} of the cell to be read. Figure 2.50 shows the sensing principle for the cell. The gate word line is stabilized at 0V, an erased cell will allow a current flow through the string and a programmed cell will stop this current flow.

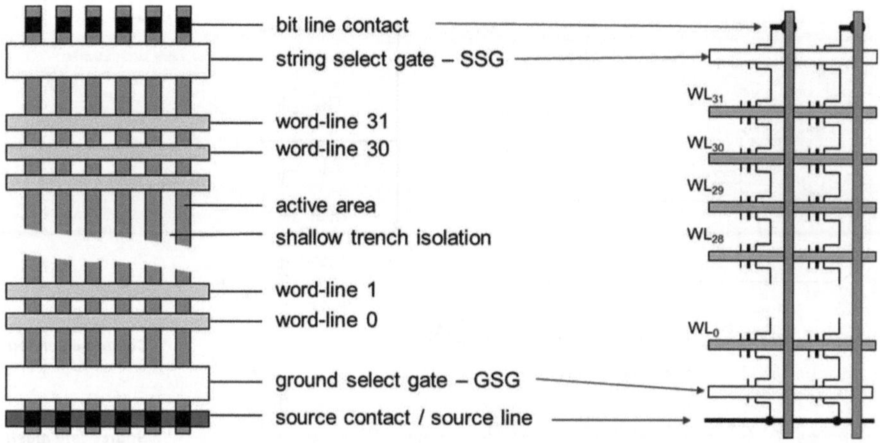

Fig. 2.49 NAND layout and logical overview—NAND string with 32 cells

All other cells belonging to the NAND string have to be conductive during the sense operation to ensure the current integration onto the BL capacity connected to the target cell. All gates of the other cells in the string are supplied with a read PASS word line voltage (V_{RD_PASS}) shown in Fig. 2.51. This overdrive voltage has to be high enough to open all programmed cells within the string. On the other side this voltage (V_{RD_PASS}) has to be low enough not to disturb the cell content during repetitive read operations.

The schematic on the right side of Fig. 2.50 shows the sense current path—the cell current is equal to the string current—and the main parasitic effects impacting the sensing within a NAND string:

- Impact of the resistance of other cells within NAND string (overdrive operating conditions need to open all cells irrespective of charge stored);
- Leakage current of the drain select gates of 2047 erase blocks in parallel;
- The voltage drop on the common source line collecting approximately 16.000–32.000 cell currents.

The accuracy of the NAND read is sensitive to the threshold voltage of other cells within the string. A small program V_{th}distribution is required without outlier on top. Otherwise the pass overdrive voltage is not able to open the cell for passing the sense current of other cells within this string.

The NAND array voltage conditions during read on word line 30 are shown in Fig. 2.51. The overdrive voltage $V_{RD_PASS}$is applied to all word lines except the word line to be read. The sensing circuits—called page buffers—are connected at the end of the NAND array to the selected bit lines.

The NAND page buffer consists of sense amplifier circuits and data latches to support the read and program algorithm. The NAND sense principles are introduced and discussed in Sect. 2.5.

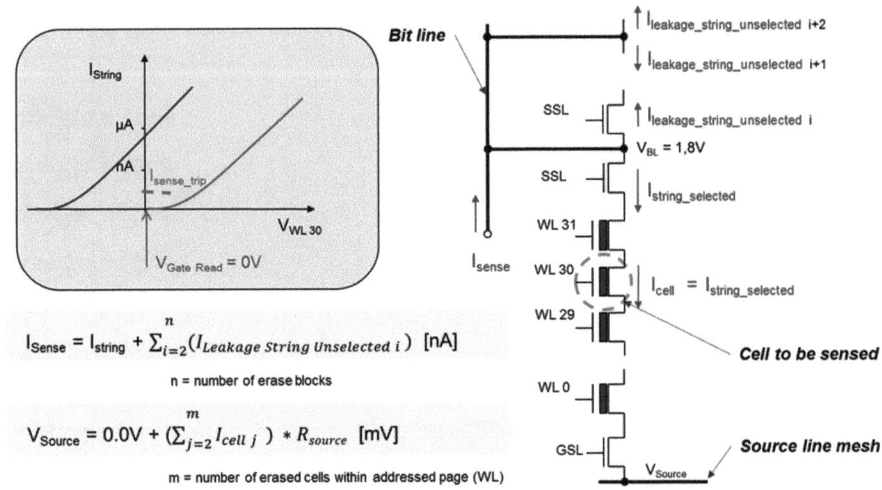

Fig. 2.50 NAND sensing scheme – string and cell current path

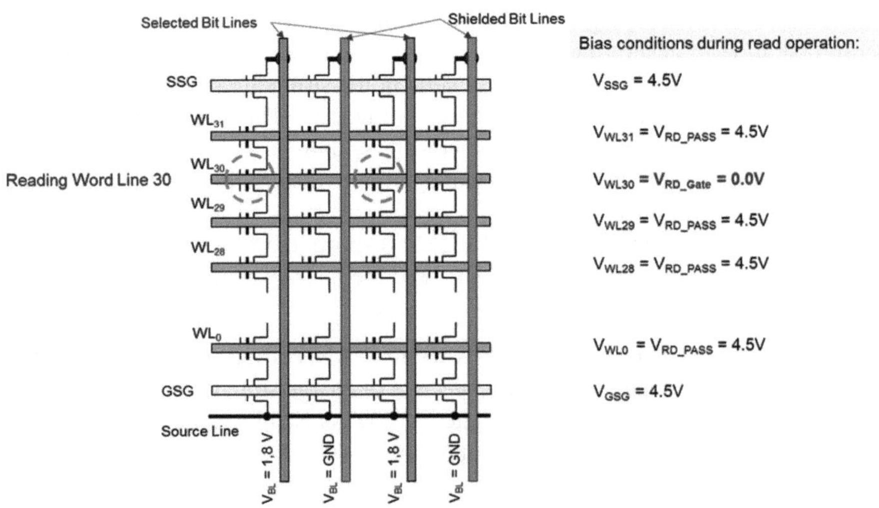

Fig. 2.51 NAND Bias conditions during read operation for shielded BL architecture

2.4.4.2 NAND Program Operation: Program Inhibit Concept

The program operation within a NAND array is executed at all cells along the word line. The NAND array supports the **M**ultiple **C**ell **O**peration **P**rinciple, which is important for the program performance calculation of NAND based memories.

The shift of the V_{th} of the cell in a NAND array is based on the Fowler Nordheim tunnel process. FN tunnelling has two benefits the excellent linearity between the program voltage and the cell V_{th} and the very low program current required to

Table 2.7 NAND array program voltage conditions

NAND array	NAND erase block	Target Cell	Voltage [V]	Current [μA]	Time [μs]
Target word line	Vpp program pulse sequence	Gate	20–24V	< nA	18
Inhibit word lines	Self-boost voltage sequence		9.0V	< nA	20
Target bit line	Target NAND string	Drain	0V		22
Inhibit bit line	Inhibit NAND string		Vcc		22
Source line mesh	Source line	Source	0V		

charge the cell. The program operation is based on a number of program pulses. The control of the program pulse is achieved by an accurate word line voltage control. Programming a NAND array is done always by a

- Gate stepping → gate voltage is increased step by step, drain and source are tied to ground.

The drain voltage of the string controlled by the **S**tring **S**elect **G**ate (**SSG**) is used as parameter to inhibit all cells along the word line, which are not allowed to be programmed. Old NAND designs have driven a fixed voltage (5.0 V) into all bit lines, which are intended not to be programmed. With NAND devices working at 3.3 V supply voltage this static inhibit scheme would consume too much power, because all inhibited bit lines would have to be driven by a pumped voltage.

The self-boosted NAND inhibit technique was a disruptive innovation solved the above described issue and enabled the roadmap for a low energy NAND program operation. The self-boosted inhibit creates the inhibit voltage condition by capacitive coupling of the voltages applied to all word lines. The WL decoder circuitry required for the read pass voltage overdrive is already designed in. The capacitive coupling of the pass word line voltage (V_{PROG_PASS}) into the strings to be inhibited are ensured by a correct timing sequence of the string select gates related to the ramp of the pass voltage on all non-selected word lines [50, 51].

Accurate control of all word line voltages and all string bias conditions within the selected NAND block is a pre-condition to execute a program operation within a NAND array, which is obviously more complex than a program operation within a NOR array. The timing sequence of string select gate related to the ramp of the pass word line voltages are critical parameters to establish a stable self-boosted inhibit. A self-boosted inhibit is limited in time, which restricts the program pulse shape modulation during advanced multi-level program sequences.

Table 2.7 shows the typical voltage conditions during a NAND program operation.

NAND array voltage conditions during program are shown in Fig. 2.52. Cells belonging to word line 30 and belonging to a NAND string driven with 0 V from the drain select side get the full program voltage across the cell and the FN tunnelling operation conditions are fulfilled.

The program operation within a NAND array creates automatically two disturbance effects shown in Fig. 2.53:

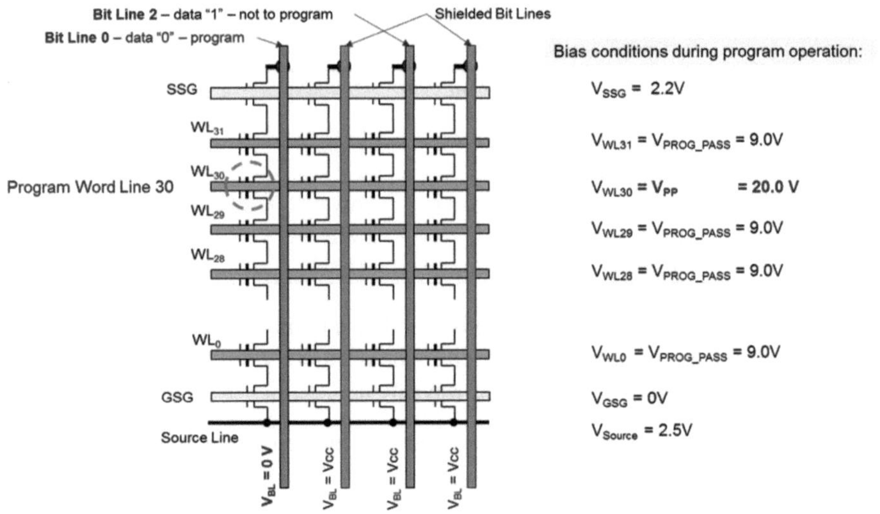

Fig. 2.52 NAND array voltage conditions during program

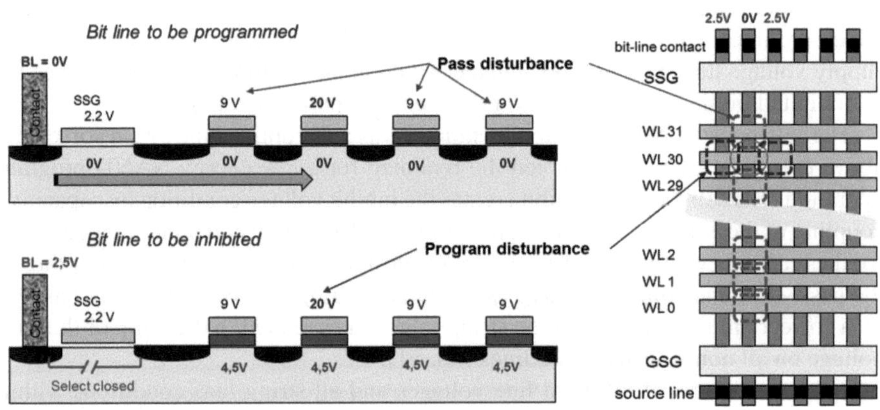

Fig. 2.53 NAND PGM inhibit and disturbance—cross section

- Program pass voltage (V_{PROG_PASS}) disturbance on all neighbour word lines
  - If V_{PROG_PASS} to high and
- Program disturbance on the neighbour bit lines belonging to the target word line
  - If channel potential to low.

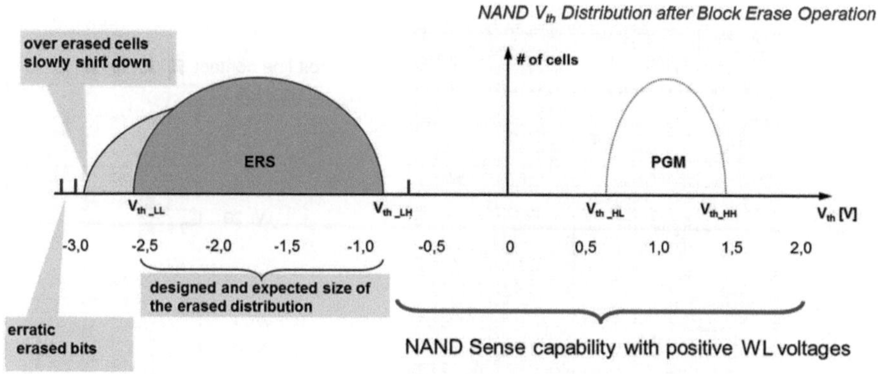

Fig. 2.54 Cell V_{th} Distribution after block erase for a NAND type memory

2.4.4.3 NAND Erase Operation: V_{th} Window Margin

The NAND array architecture defines per design the erase memory block size in WL direction limited by the word line decoder and in BL direction defined by the NAND string length and limited by the string select line (SSL) and the ground select line (GSL). The erase operation is executed on all cells using the FN tunnelling principle

- Segmented Sector Operation Principle.

The default erase implementation applies a positive voltage to the well and ground the all word lines belonging to the memory block to be erased. The erase procedure is simple. A long negative high voltage pulse (1–2 ms) will discharge all programmed cells and shift them back below the zero voltage V_{th} state.

The erase operation seems uncritical, because over erased bits have no influence on the accuracy of the NAND sense operation. This assumption is true for the sensing, but not for the V_{th} operation window margin definition. The first issue of the NAND erase control is the limited capability of a positive voltage word line sensing. The V_{th} visibility of the sense operation is limited to a small negative V_{th} range as shown in Fig. 2.54.

The V_{th_LL} edge of the erased distribution is a key parameter for NAND flash too. Especially the lower edge influences indirectly the effective coupling of these over erased cells onto the direct neighbour cells in the case this cell is programmed. In contrast to NOR the V_{th_LL} edge cannot be verified by the default word line voltage and sense amplifier circuits.

The erase distribution width and the position become an important factor for MLC NAND memories. The detailed discussion about program and erase algorithm and margin and reliability considerations is done in the corresponding Sects. 2.6 and 4.4.

The NAND array architecture was developed to keep the design as simple as possible. The applied erase operation does not enforce the design to provide a negative word line voltage. Instead the word lines of the target erase block are grounded and

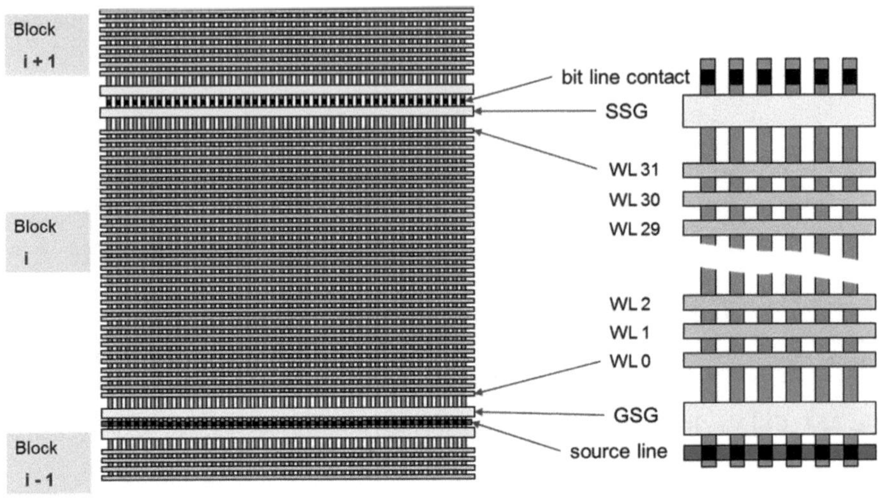

Fig. 2.55 NAND array layout—erase block layout based on NAND string

all other word lines have to float. The well is connected to a positive high voltage, and only for the grounded erase block the high voltage values are applied to all cells belonging to this erase block. As a consequence no separation by wells is necessary for the NAND block erase operation. This concept ensures excellent array efficiency.

The virtually grounded erase concept generates an issue at the border of the floating array parts, which are shifted to high voltage levels. All floating sectors are indirectly connected to the well supplied with the erase voltage conditions. Therefore the NAND Y-mux layout area has to be designed in such a way to fulfil isolation and blocking criteria up to 20–25 V.

2.4.4.4 NAND: Array and Cell Matching

The matching between the floating gate cell and the NAND array is a perfect array and cell combination, which is one of the reasons for the successful development of NAND memories.

The dense memory core array is the NAND string region itself, encapsulated by the select gates—on top by the string select gate to the bit line contact and on the other side by the ground select gate to the common source line mesh net. Due to the fact that each NAND string is encapsulated by the select gates this ensures disturb free operation of other NAND blocks. Another benefit for cell and array matching are the voltage driven NAND memory operations. Therefore the select gates could be designed in pitch with the cells and the bit line spacing.

The Fig. 2.55 illustrates this optimal cell, select gate and array matching for NAND.

80 % of the layout of a NAND memory chip looks like the left side of Fig. 2.55. The erase blocks are only separated by ground select lines followed by the source line and the string select lines followed by bit line contacts. The source line and the well have to be contacted within a certain distance and two metal meshes are distributed over the complete NAND array with regular stitching contacts.

The NAND string area—the inner part of the erase block—is free of any contact. This block is placed block by block on the memory chip in a mirrored way to utilize the bit line contact and the source line shared for two erase blocks. Long metal bit lines in cell pitch are required by this concept and are possible due to the slow sense concept through the NAND string.

The technology process to produce such long metal lines and especially the bit line contacts to each string in pitch (minimal feature size) is a challenge. In addition this has to be made by double and triple patterning because there is still no direct lithography technology ready for volume production.

The voltage driven FN tunneling process is achieving an excellent selective programming behavior by a voltage driven self-boosted inhibit. This combination makes the NAND flash array architecture unique.

The high voltage difference between different word lines during programming is a technology challenge in terms of voltage isolation capabilities especially within the word line decoder structure and between the first word line and the select gates. An established design measure is the usage of a dummy word line at both edges of the string, which was applied at the transition from 32 cells to 64 cells per string [52].

2.4.4.5 Summary of NAND Cell and Array

The Assessment of the NAND array and the floating gate cell is summarized in Table 2.8, which includes all parameters used for the memory concept assessment.

FG cell and NAND array architecture are an excellent combination for non-volatile memories targeting the data storage applications. The high parallelism achievable during read and program allows a continuously increasing data bandwidth with a very low and constant current consumption.

The missing capability to use a cell with multi bit (stored locally) per cell capability is well compensated by the multi-level per cell approach described in deep detail for MLC NAND flash in the corresponding chapter.

The memory core array architectures of NAND and of VG_NOR have the same effective cell size $4F^2$, in which F is the minimal feature size of the technology node. Both concepts are compared in chapter 6 applying the developed performance indicator methodology.

Table 2.8 NAND cell and array architecture summary—key statements

Operation	Key statement	Assessment
Program	FN tunnelling	Low current operation Massive parallel cell operation Highest program throughput
Read	Indirect access Number of sense amplifier = 1/2 number of bit lines	Slow operation, high latency, Massive parallel data cache available Highest read throughput
Erase	FN tunnelling	Low current operation Slow process (2–5 ms)
Bit line decoding	No local y-decoder Global Y-decoder at the array border (HV-devices)	Global Y-decoder depends on design concept (odd/even or All Bit Line) Die area consuming feature (HV)
Word line decoding	X-decoder up to Vpp_max + V_{th} of HV trans (up to 28 V)	Layout and fan-out define the number of cells per string
Intra Block disturbance	Gate disturbance Pass disturbance	Compensate by self-boosted inhibit During program and read
Inter Block disturbance	None	During erase
Parasitic effects	Leakage of select gates Self-boosted inhibit timing	
Cell to Cell interference	High FG to FG coupling	Compensated by program algorithm

2.4.5 Summary: Flash Array Architecture

Three different array architectures were investigated and a short summary is derived. The ideal non-volatile array architecture has to combine the following parameters:

- A dense memory core array (quantitative better than 4 F^2)
  - An array cell combination without contacts in the core block area
- A cell/string/array combination which allows a very robust sensing scheme
  - Deviations forced by lithography and systematic, deterministic interdependencies can be compensated by read and verify operation
- A decoding scheme which results into a dense word and bit line decoder structure to address memory cells for read and write
- An intrinsic physically segmented array architecture to reduce the impact of disturbance
- Low voltage control is preferred for all bit line and word line operations
  - Low voltage is defined as max 2x Vcc for non-volatile memory arrays;

- If high voltage operations are required, only under the following conditions:
  - High voltage paths only in one direction—e.g. word line decoder
  - Low voltage operation in bit line direction

The increase of memory density is ensured by the reduction of feature size along the shrink roadmap. The reduction of feature size enforces an increase of parasitic capacitances and sheet resistances of the long lines required for bit lines and word lines. Therefore the ideal non-volatile memory cell/array combination has to handle this by:

- Main operations read and write are controlled by voltages
  - Physical principles to write (PGM) and overwrite (ERS & PGM) are voltage driven.
- The cell/array ratio in terms of distances between cells and line spacing has to maintain the principles of shrinking with a reasonable margin factor.
  - The voltage difference between two word lines in the NAND array has to be always lower than the breakdown voltage of the material used to isolate them.
  - The maximum voltage applied to the cells has to be reduced along the shrink roadmap or an inhibit scheme has to be in place to fulfil the above requirements.

The 193 nm immersion lithography is combined with techniques like self-aligned double patterning (SADP) and self-aligned quadruple patterning (SAQP) to fulfil the lithography roadmap. As long as SADP and SAQP are applied the compensation of asymmetric deviations of bit line to bit line and word line to word line spacing is a strong requirement for the cell/array combination.

- A cell/array combination which supports a high parallelism during all write operations is an effective way to compensate the above described systematic and deterministic interdependencies.

The increase of parasitic capacitances and sheet resistances can be compensated by material innovation, but the array design structure has to over compensate the increased rise time values (BL and WL) by an increase of parallelism. This conclusion indicates the strong requirement for a simple sense structure, to ensure the ability to compensate the loss of performance due to increasing resistance and capacitance values by an increase of parallelism of parallel read and write operation. The ratio of the number of sense amplifiers linked to the total number of bit lines is becoming the important value compared to the classical judgement of sense duration time multiplied by the number of sense amplifiers used.

- Memory cell/array structures with a high number of sense elements—ideally for every bit line—are the target array to fulfil the performance requirements

The limitations of 2D memory array structures are highlighted since more than 10 years on every international research and development conference. In contrast to these scenarios NAND flash is currently produced with 19 nm feature size based on 193 nm lithography supported with SADP and SAQP.

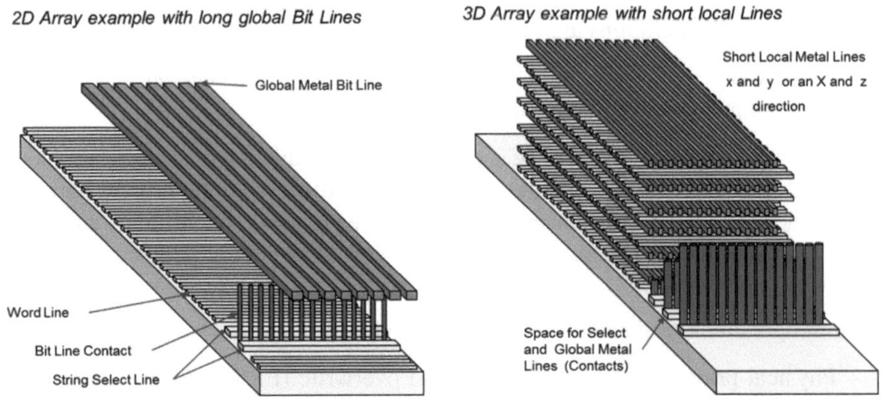

Fig. 2.56 3-D Memory array—disruptive innovation—3D array example in the style of a pack of lines [53]

The introduced performance indicator methodology shows in Chap. 7 a continuously growing gap between the performance capability of the flash memory array architecture and the application based performance indicator trend line along the 2D shrink roadmap. The third dimension offers the required flexibility for new innovations of known and well established memory array architectures. Figure 2.56 visualizes this additional flexibility of future commercially successful memory architectures.

The following chapters are developed for 2-D memory architectures and the assessment and optimization of these concepts. The principles developed in chapter 6 can be extended and modified to guide the 3-D memory array structure development process.

The next chapter will introduce the important circuit blocks to design a flash memory. This basic knowledge is required for an adequate memory array model development considering all circuit restrictions along the shrink roadmap.

2.5 Memory Building Blocks

The memory functionality is realized by different logic blocks shown as an example in Fig. 2.57 for a NAND flash memory. Flash memories require high voltages and the decoder architecture has to include additional functionality compared to standard volatile random access memories. This chapter focuses on blocks responsible for high voltage generation and decoding. The sensing and data buffer architecture is described to get an overview about the die size requirements and layout considerations for the building blocks surrounding the memory array.

A memory product is characterized by the ratio between the memory arrays versus the complete die size—this parameter is defined as cell efficiency. The memory array

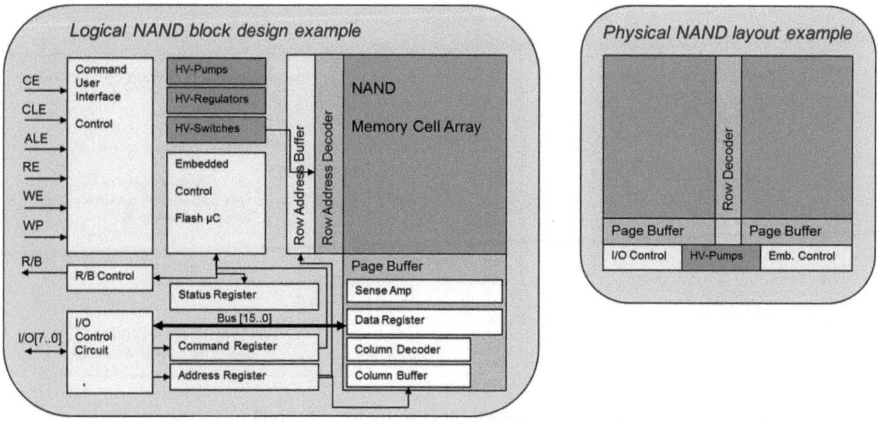

Fig. 2.57 Logical and physical memory block structure—NAND memory

has to utilize more than 50 % of the die size otherwise we consider it as a logic device with large embedded memory.

The minimal feature size is reduced for every technology node along the CMOS shrink roadmap. The minimal pitch for the memory array reduces in both directions and the number of cells is doubled. The surrounding logic blocks—like decoder structures—have to be doubled too but cannot utilize the reduced feature size due to high voltage requirements of flash memories.

The flash memory development roadmap has to overcome this dilemma node by node based on different strategies:

- Circuit innovations have to be developed to reduce the surrounding logic block sizes;
- Cell innovations have to be introduced to reduce the voltage requirements for program and erase operations;
- Material innovations have to be made ready for high volume production to ensure a working FN tunnelling process of the applied cell architecture.

2.5.1 Row Decoder: Global and Local X-Decoder

The word line decoder—typically called Row Decoder or X-Decoder—has to be developed with lowest number of high voltage devices and circuit techniques working properly without a triple well technology to reduce technology cost.

The word line decoder circuit and layout has to combine the following functionalities:

- Segmentation of the memory array in word line length to realize the required array performance values for time critical word line operations.

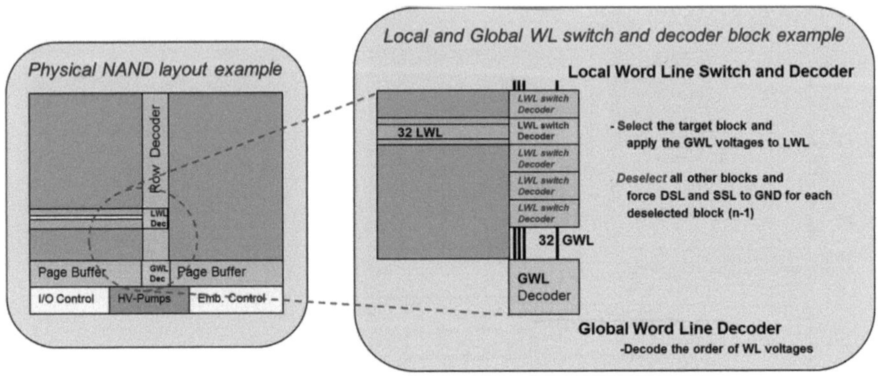

Fig. 2.58 Local and global word line decoder hierarchy for a NAND memory

  - For NOR memories the word line decoder is applied for an electrical and logical grouping of flash cells into array blocks.
  - The NAND memory segmentation works with long word lines. Single or double sided row decoder circuits are driven by layout considerations to achieve die size optimized fan-outs fitting to the NAND string length.
- Level shift functionality has to be implemented to shift the decoding control signals from 1,8 or 3,3 V to 9 V for NOR or to 28 V for NAND memories.
  - Different circuit approaches are published like ratioed circuits, cross coupled level shifter or feedback circuits [54, 55].
- Local and global row decoding circuits are applied to ensure the required word line performance values and achieve target row decoder block size requirements.
  - Die size optimized design innovations are the target for the row decoder circuit.
- Hot switching functionality has to be embedded to generate the required pulse shape during the step pulse programming algorithm.

A word line decoder top view for NAND is shown in Fig. 2.58 including the high voltage path consisting of **L**ocal **W**ord **L**ine (**LWL**) decoder, **G**lobal **W**ord **L**ine (**GWL**) metal lines and High Voltage support lines and **GWL** decoder.

The basic concept is reducing the number of circuit blocks required to select, switch and decode high voltages. The local word line decoder unique to each block selects the target block and deselects all others—same functionality is utilized for erase. The global word line decoder is placed once per die and is common to all blocks.

A cost optimized control of high voltage word line operations requires on one side simplification to reduce die size and on the other side flexibility to pass a certain number of different WL voltage combinations to the selected NAND erase block. Figure 2.59 shows the number and the order of required word line voltages (V_{PROG_PASS})

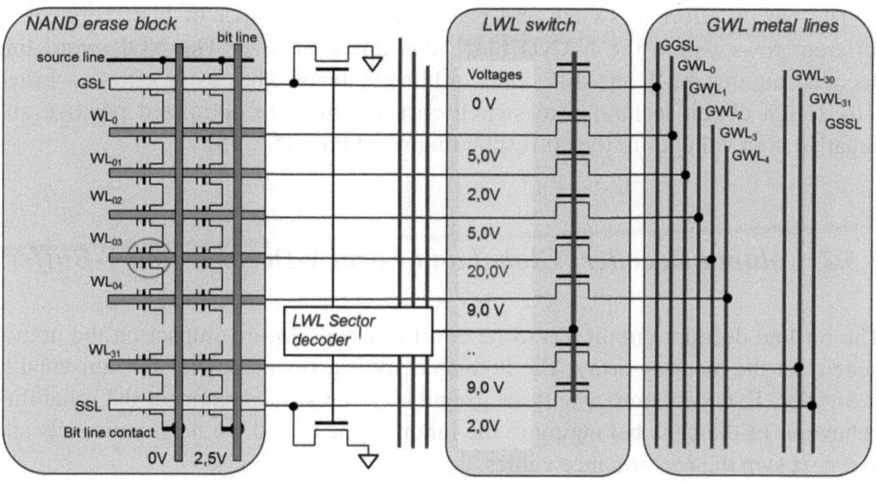

Fig. 2.59 Local word line switch and decoder and variety of V_{PROG_PASS} word line voltages

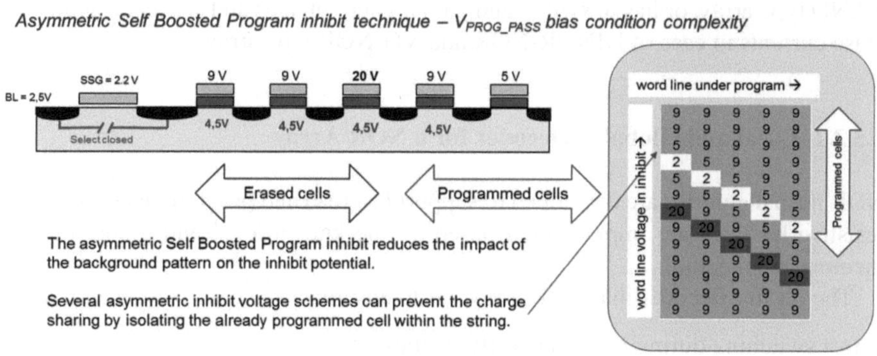

Fig. 2.60 Asymmetric self boosted program inhibit technique to suppress the background pattern dependency

to generate the self-boosted program inhibit for sub-50 nm NAND product designs. The design implementation collects different bias voltages into groups, which are supplied by voltage pumps and regulators already designed for read operations.

This example is used in this book to highlight the difference between the principles of self-boosted inhibit – introduced in Sect. 2.4.4.2—and the optimized product design implementation shown in Fig. 2.59. The applied technique is called asymmetric self-boosted program inhibit which improves the inhibit potential and stability by suppressing the background pattern dependency [56].

The order of different passing word lines is modified, depending on the location of the programmed word line as shown in Fig. 2.60. The GWL decoder has to decode 5–8 different voltages in an exact predefined order. Next to the select gates the sequence has to be modified and requires characterization effort.

The design effort to switch and decode a certain number of high voltages to different rows within the NAND block consumes die size. The NOR word line decoder requires much less different word line voltages. The design effort is shifted into design circuit technique to switch cost and die size optimized positive and negative high voltages to the corresponding word lines [54, 55].

2.5.2 Column Decoder: Global and Local Y-Decoder and Y-Buffer

The bit line decoder circuit has to be developed with lowest impact on the homogeneity of the memory array. The homogeneity is a key parameter for non-volatile memories. Every inhomogeneity or irregularity can strongly impact the reliability behaviour of the cells belonging to the inhomogeneity and the neighbour cells and in a next step the performance values.

Depending on the array type and the cell operation principle the bit line decoder has to switch and pass the required voltages for read and verify operations in case of NAND type array, or has to switch and pass higher voltages and sense, program and erase currents in case of DINOR-NOR and VG-NOR type array.

2.5.2.1 Local and Global Y-Decoder for a NOR Array

NOR flash memories are developed to support fast read and program operation. Low resistive bit lines are implemented based on the known local and global bit line memory architecture.

The bit line decoder in a NOR array has to ensure

- fast switching during read and verify operation;
- capability to pass high voltages during HHI erase for SONOS memories;
- capability to pass high voltages (4–6 V) during CHE program and
- low resistive path to pass the program current (CHE program) without significantly voltage drop along the complete bit line path.

A cascaded bit line decoding architectures is the design solution enforced by the timing requirements for read and verify as well as program operations. Short local bit lines are connected via the local Y-select to the global bit lines, which are connected directly or via a global Y-select to the sense amplifier circuitry.

The high program current values—up to 100 μA per cell—are limiting the shrink potential of the local Y-selects in a NOR flash memory architecture. A low resistive column decoding scheme is a must, which defines the size of the transistors used in the local Y-select circuit and layout. The local Y-select design and layout has to guarantee the isolation criterion higher than 6.0 V between all select devices.

The local Y select logic is part of the sensing path and every select path has to have exactly the same resistive and capacitive load. Area optimized Y-select implementations could not compensate all in homogeneities.

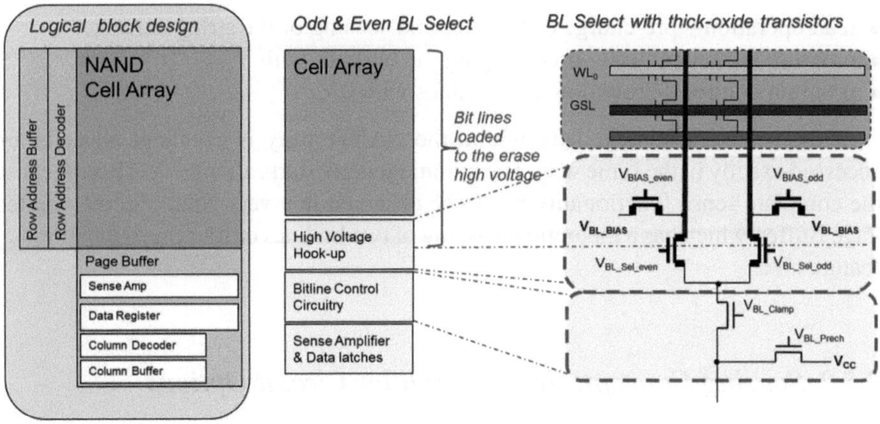

Fig. 2.61 Y-Select for NAND—odd and even BL select with thick oxide devices—HV Hook-up

2.5.2.2 Global Y-Decoder for a NAND Array

NAND memories are page oriented flash memories designed for data storage. In principle every cell along the word line can be accessed for read and program operation in parallel. This is a strong benefit for higher cell efficiency and much higher array homogeneity of NAND based flash memories.

The shielded bit line sensing method was introduced 1994 [57] and consequently a one out of two bit line decoding scheme was applied. The odd and even bit line select circuit is built out of transistors with thick gate oxide to block high voltages. During the NAND erase operation the bit lines are connected via the bit line contacts to the bulk and follow the bulk potential to the erase voltages (20–24 V). The high voltage isolation between the array and the low voltage sense amplifier and bit line control circuits is ensured by the NAND Y-Mux shown in Fig. 2.61. The die size overhead required for four thick oxide devices and the corresponding isolation distances is big and impacts the scalability of NAND. A new approach to shift the high voltage blocking functionality into the bit line control circuit [58] is published with a 33 % size reduction of this block and illustrates the principle of innovations required to shrink all decoder blocks following the NAND cell and array shrink roadmap.

The NAND Y-Mux has to ensure the following functions:

- Decoding of bit lines (for shielded bit line architecture in odd and even);
- Biasing the un-selected bit line to a predefined fixed voltage potential (e.g. Vss) to shield;
- Blocking of the high voltage during erase operations;
- Layout zone with additional lithography requirements to change from dense array pattern into a less dense decoder and sense amplifier lithography pattern;

The bit line path in the NAND architecture is functionally used only for low voltage functions:

- read operation—pre-charge of bit lines and sensing of the string current;
- program inhibit control—pre-charging the bit lines with V_{CC};
- program control—grounding the bit lines with 0 V;

If the achieved homogeneity within the NAND array is excellent all cells are accessed exactly in the same way. The inhomogeneity starts after the Y-Mux, because the complete sense functionality has to be designed in a very small circuit—called Page Buffer, which has a theoretical spacing of two bit lines or four times the minimal feature size.

2.5.3 Sensing Concept: Sense Amplifier Circuit Options

The sense concept and the corresponding sense amplifier circuit as well as the layout parasitics of the sense amplifier define the achievable accuracy over the complete environmental specification range and the achievable performance and reliability values of the memory product. An optimized sensing concept can improve reliability parameters by one order of magnitude for the same cell and array architecture.

For multi-bit or multi-level cell memories the sensing performance and accuracy are becoming the key parameter, and the following sub-chapters introduce different sensing concepts in more detail.

2.5.3.1 Principles of Sensing Concepts: Sense Amplifier for MLC NOR

The principle of sensing can be described as a parametric measurement unit (4 quadrant V/I source). The measurement of the cell current can be applied as a direct current measurement for a fixed gate voltage or as indirect measurement—a current to voltage converter with a second stage comparing the sense voltage.

The current to voltage converter can be based on the memory array using the bit line capacity as a large capacitor for the sense current integration. The cell current discharges the bit line capacity to a voltage level which is compared with the sense amplifier threshold level. This concept can be applied only for slow read operation in a time scale of μs and will be discussed in section NAND sense.

NOR memories require a fast read operation and are using a differential sense amplifier concept. Figure 2.62 shows the principle elements and I/V curves for a sense operation based on reference cells and a differential sense amplifier.

Four parameters can be derived for a sense operation, which have a significant influence on the sense amplifier design, the accuracy and the speed of the sense operation.

- The **bit line and drain bias** control and the **bit line pre-charge time** necessary to achieve a certain voltage value independent of the position of the cell within the array segmentation.

- The **time for the sense operation** itself, the integration time which is required to convert the cell current to the corresponding voltage level to be compared.
- The **reference current** required at the sensing point to decide if the target array cell is above or below a certain V_{th} level derived from the applied constant gate voltage.

Different optimized sense concepts can be applied to the NOR and especially the VG NOR and they are discussed in detail in the following literature [47, 55, 59].

- The read access is specified within 20–90 ns and therefore the NOR sensing scheme is based on a differential sensing approach. Target and reference cell are read in parallel. The voltage difference between these two operations enables a stable sense decision.
- For Multi-level cell flash different sense concepts are published and applied on MLC NOR or Multi-Bit-Cell (MBC) VG-NOR flash products:
  - Fixed gate voltage sensing—compare the cell current in parallel versus different reference currents;
  - Sensing in the time domain;
  - Fixed current sensing—compare the cell current at the end of each gate voltage level.

Figure 2.63 shows the four V_{th} distributions of a NOR Multi-Level-Cell flash memory. The sense margin for each read level is well defined between the sense operating point—read verify (RV)—and the program verify level—the lower edge of the corresponding programmed distribution—$V_{th\_L1\_HL}$. The sense margin between the upper edge of the programmed distribution—$V_{th\_L1\_HH}$—and the sense operating

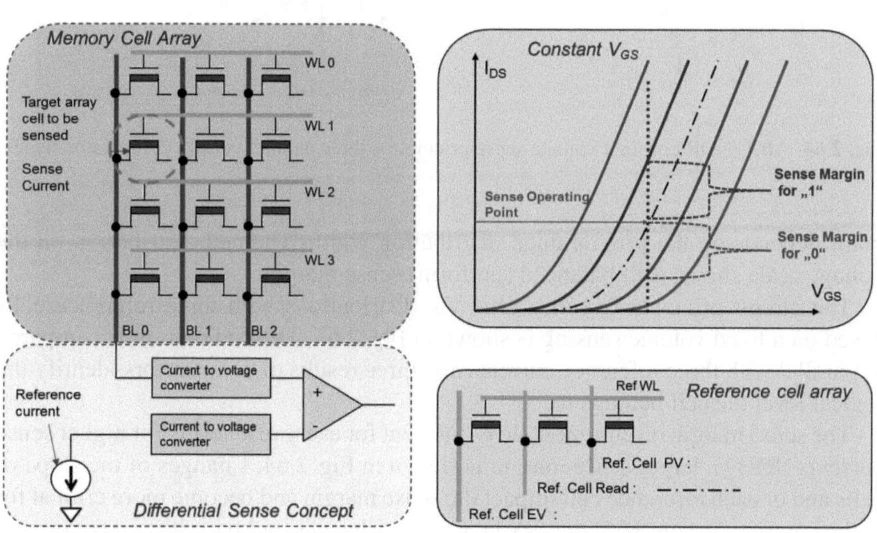

Fig. 2.62 Differential sense concept overview—sense operating point and reference cell array

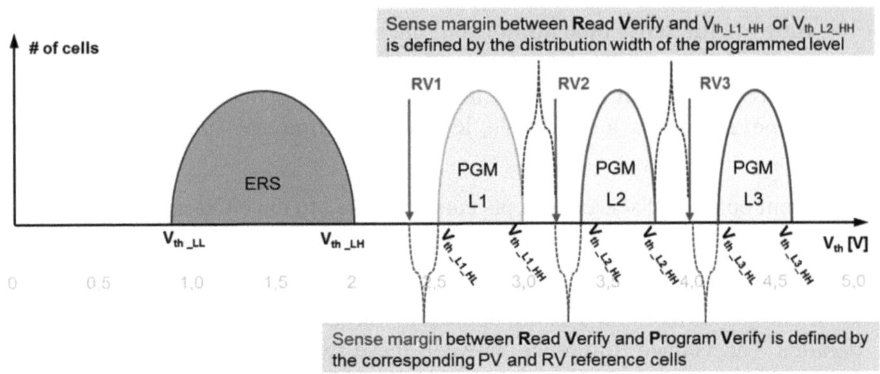

Fig. 2.63 NOR Multi-level-cell V_{th} distribution

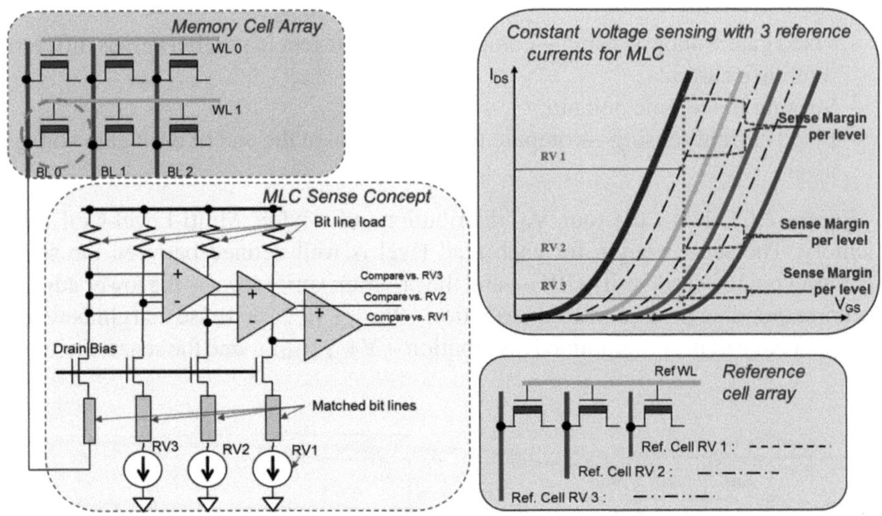

Fig. 2.64 MLC NOR constant voltage sense principle—three parallel voltage comparator stages

point depends on the programmed distribution width. The cell distribution on the voltage scale shows well balanced (uniform) sense margin.

The circuit principle to sense four V_{th} distributions with three reference cells based on a fixed voltage sensing is shown in Fig. 2.64. The cell current is compared in parallel with three reference currents and three results of comparators identify the logical level the cell belongs to.

The sense margin on current scale is different for every read level. For higher sense currents (RV1) a larger sense margin is shown in Fig. 2.64. Changes of the slope of cells and of each reference cell impact the sense margin and become more critical for higher sense currents (RV2 and RV1). Cell g_m changes and parasitic array currents limit the constant voltage sensing concept. The impact and the counter measures are discussed in Sect. 5.3. The change of the cell g_m is shown in Fig. 5.12.

Reference cell concepts were developed with separate single reference cells per sense amplifier, with reference cell arrays and mirrored common reference sense currents to all sense amplifiers, and with reference cells placed into the main area and cycled together with array cells.

Multi-Level cell differential sense amplifiers require matched bit lines to have same R and C values for target and reference cell. Additional die size is needed for sense circuits and reference cells.

The constant current sensing already introduced in Fig. 2.27 would allow a much higher accuracy, but requires an approximately three times longer sense time. This concept will be discussed in detail together with the NAND sense amplifier concept.

2.5.3.2 NAND Page Buffer Circuit- Sense Amplifier for MLC NAND

The cell current is limited by the NAND string resistance and the sense concept is developed to measure the cell current through the NAND string. The sense amplifier circuit has to be as small as possible due to the fact that every second bit line requires a complete page buffer circuit—the name for the NAND sense amplifier and data latch circuit.

Therefore the NAND sensing concept has to be a single ended sensing scheme. The current to voltage conversion is achieved by discharging the bit line capacitance. The limited cell current capability is combined with a long integration time to ensure the required accuracy.

The NAND sensing principle is shown in Fig. 2.65:

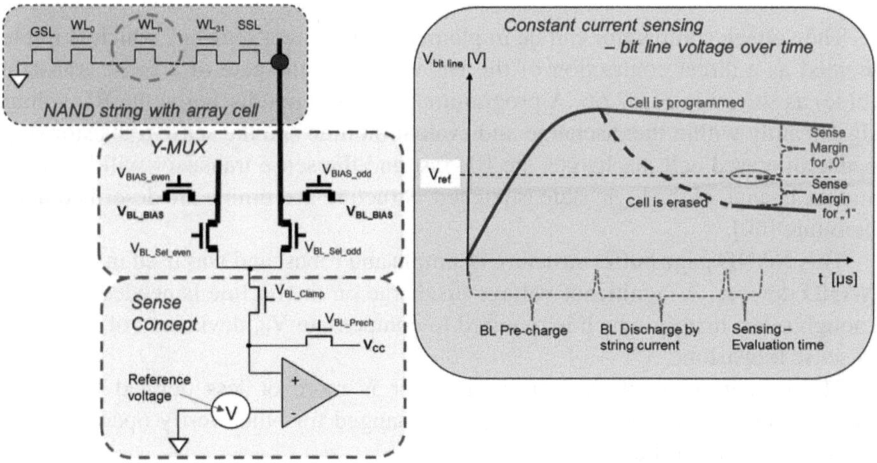

Fig. 2.65 NAND sensing principle and BL potential over time for shielded bit line array architecture

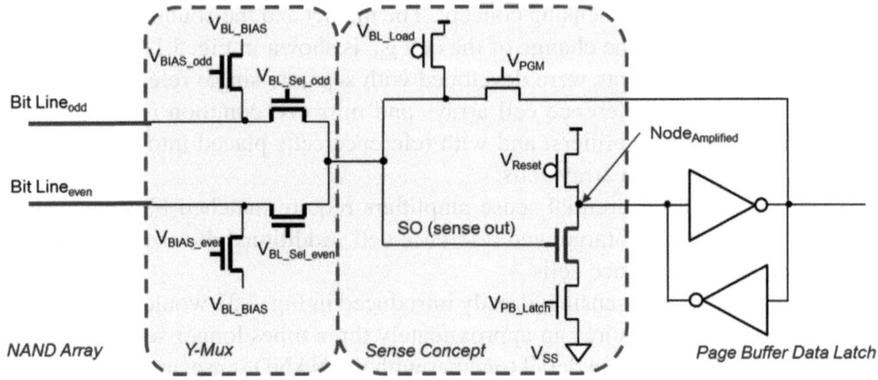

Fig. 2.66 NAND page buffer structure with direct gate sensing scheme

The NAND string requires an adaption of the sensing requirements:

- No direct access to cell terminals within a NAND array, therefore sensing accesses the complete string. Bit line biasing has the same meaning as drain biasing for direct cell sensing.
- The cell current is limited by the string resistance and the conductivity of all passing cells during the read operation. A higher string current capability requires a higher read pass voltage and enforces an increased read disturbance of the complete NAND sector.
- The source line and the source mesh have to be low resistive to limit the voltage drop on the source. An erased NAND sector with a few programmed pages is the worst case condition for the voltage drop above the source mesh impacting the sense accuracy.

The voltage comparator can be implemented as "direct sensing" which is implemented as a direct connection of the BL voltage to the gate of a sense transistor (blue) as shown in Fig. 2.66. A programmed cell will not discharge the BL voltage significantly within the discharge and evaluation time and the sense transistor stays open, an erased cell discharges the BL fast and the sense transistor will be closed and the change of the logic state is latched. Structure and timing are described in the literature [60].

This NAND page buffer structure is simple and robust and was used in most SLC NAND devices. A significant voltage discharge on the bit line is needed to ensure enough sense margin which is required to compensate V_{th} deviations of thousands of sense transistors.

The trigger point of the sense transistor is more or less defined by design (device technology) and cannot be easily changed for other verify operation conditions required for the erase algorithm.

For MLC-NAND the sense operation has to be executed multiple times. An acceleration of the above described sense concept can be achieved reducing the

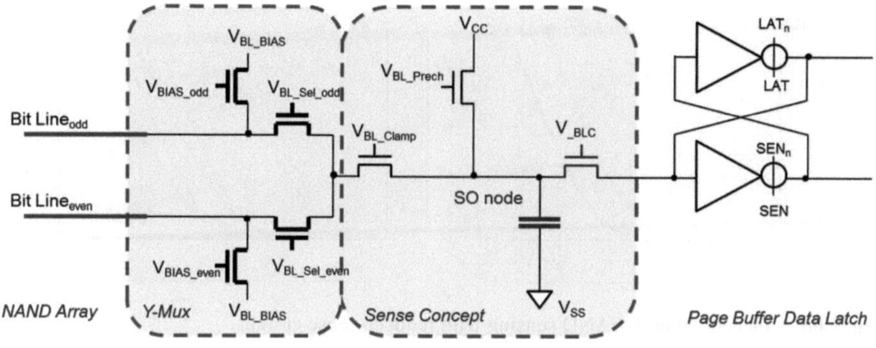

Fig. 2.67 NAND page buffer structure with SO node capacitor [61]

voltage value to be discharged during the sensing phase. A sense technique to detect small voltage changes is achieved introducing an additional small capacity—the SO (Sense Out) node capacity. A page buffer using this sense principle is shown in Fig. 2.67.

This sensing scheme is based on a sequential discharge of two capacitances bit line and the SO node [61]. The trigger point of this sense circuit can be changed by varying the below described voltage and timing conditions. The bit line is pre-charged with a voltage defined by $\mathbf{V}_{\mathrm{BL\_Clamp}} = \mathbf{0.7\,V} + \mathrm{V}_{\mathrm{th}}$. After bit line pre-charge is successfully done $\mathrm{V}_{\mathrm{BL\_Clamp}}$ is grounded. The sense current can slowly discharge the bit line potential. In parallel the SO node is pre-charged to a certain potential $\mathrm{V}_{\mathrm{SO}} = 2.5\,\mathrm{V}$. During the sense operation $\mathbf{V}_{\mathrm{BL\_Clamp}} = \mathbf{0.6\,V} + \mathrm{V}_{\mathrm{th}}$ is applied for a short period of time.

- If the BL potential was reduced by an erased cell more than 100 mV the small SO node will be discharged to the bit line potential.
- If the BL potential can be reduced only by the leakage of the other drain select transistors because the cell is programmed, the small SO node capacity will not be discharged.

A NAND page buffer includes the complete logic functionality necessary for algorithmic operations during read, program and program after erase. Every transistor is used for more than one purpose to keep the number of transistors for a complete MLC NAND page buffer circuit design below 30. A MLC BL control circuit and the timing sequence are described in [62]. The reuse of transistors and latches during the MSB programming is the important design and layout efficiency criterion.

The inhomogeneity in NAND flash memory designs starts in the page buffer layout area. The page buffer layout has to be as small as possible and at least four of them are staggered above each other. The sense amplifier layout and the shielding of all the important lines within the staggered layout is a unique design challenge increasing from technology node to node.

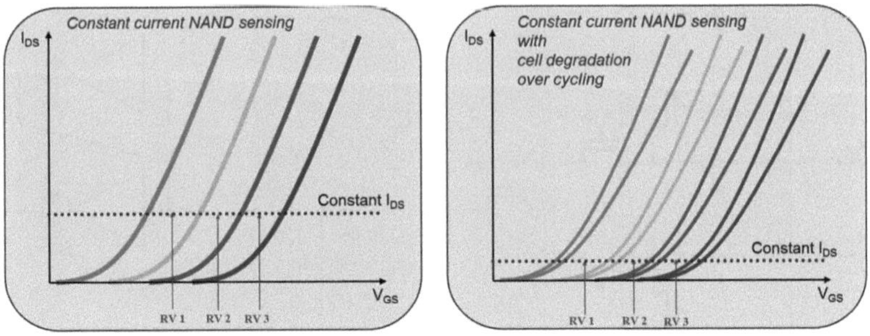

Fig. 2.68 Constant voltage NAND sensing with reduced sense current

2.5.3.3 NAND Page Buffer: Sensing Innovation for 3 bit and 4 bit Per Cell

For 3 and 4 bit multi-level cell NAND designs the sense operation has to be accelerated and the sense accuracy has to be increased in parallel significantly. The reliability chapter will discuss the available V_{th} sense margin for MLC NAND flash designs including array and reliability effects in detail. One major conclusion is already derived now: The accuracy of sensing a cell within a NAND string can be increased by reducing the cell current required for the sense operating point—shown in Fig. 2.68.

The design challenge is now to achieve with a significant smaller sense current a faster sensing, which requires a sensing without the time consuming discharge of the large bit line capacitance.

The NOR sensing drain biasing technique has to be adapted for the single ended NAND sensing. A page buffer structure shown in Fig. 2.67 with the SO capacity node can be modified, so that a continuous current flows from the SO node into the BL representing the cell current. Now the sense circuit can measure the discharge current of the small SO node capacity which is equal to the cell current within the NAND string. This faster current sensing approach was published [63] and applied together with the **A**ll **B**it **L**ine NAND architecture [64] to accelerate the program verify operation.

The elimination of the bit line discharge (and its correlated crosstalk) during the sense operation improves the robustness of the NAND sensing concept in the same way, if a shielded bit line or all bit line NAND architecture is applied to a product design.

The combination out of constant current sensing—the most robust scheme for multi-level sensing—a reduced sense operating point and the above described fast current sensing approach are a perfect setup for high accuracy and fast repetitive sense operations.

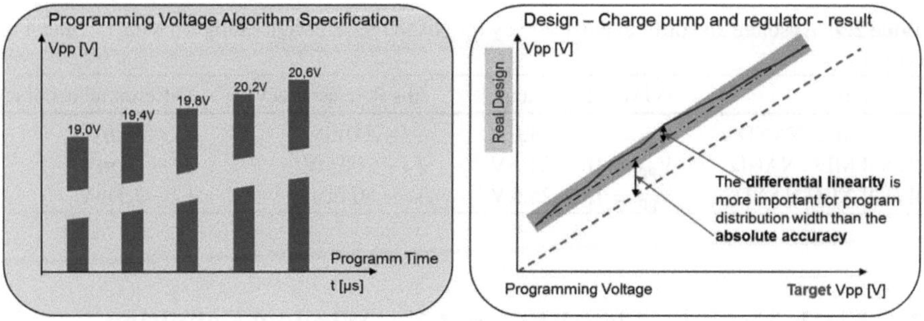

Fig. 2.69 Program Voltage (Vpp) absolute and differential accuracy

2.5.4 High Voltage Generation and Accuracy

Flash memories require high voltages for program and erase operations on chip. The internal high voltages are generated by charge pump circuits under a wide supply voltage range, typically from 2.7 to 3.6 V for NAND and from 1.8 to 3.3V for NOR and embedded flash memories.

The design challenge for the charge pump circuit is to guarantee the charge pump efficiency for lower values of the supply voltage and limit the current consumption for higher values of the supply voltage. Charge pump architectures based on the Dickson charge pump [65] are known and efficiency optimizations of high voltage generation are published [66, 67]. Charge pump design guidelines [68] are summarized in the literature. We are focusing on the accuracy definition and requirements derived from multi-level cell margin considerations—which are discussed in detail in the corresponding chapter.

The accuracy requirements to the programming voltage are extremely high, with specified accuracy values of 10–100 mV for high voltages generated by charge pumps in the range of 18–28 V.

Figure 2.69 shows a typical program voltage ladder and illustrates on the right side the accuracy definition for the FN tunneling program algorithm. The absolute accuracy of the voltage is important, but can be easily compensated by a lower start value of the first V_{pp} voltage algorithm. The differential accuracy between program voltage pulses—the differential linearity—is the important accuracy value. The differential inaccuracy has a direct impact on the resulting cell V_{th} distributions width as discussed in Sect. 2.7.4. Table 2.9 highlights the increased accuracy requirements along the multi-level cell development path to increase the number of bits per cell up to 4 bit per cell—called **XLC** (e**X**tended **L**evel **C**ell) NAND in this work.

The differential linearity has to be guaranteed for all program pulses for different load conditions—the number of cells to be programmed is reduced from pulse to pulse—and for all reasonable environmental conditions (supply voltage and temperature). This high accuracy conditions can be ensured by excellent design techniques combined with application rules which have to ensure a continuous execution of the operations.

Table 2.9 Absolute and differential accuracy for NAND flash design examples with 1, 2 and 4 bit per cell

Flash type	PGM voltage range	Absolute accuracy	Differential accuracy
1b/cell SLC NAND	$V_{pp} = 18 - 24.0\,V$	+/−200 mV	+/−50 mV
2b/cell MLC NAND	$V_{pp} = 18 - 24.5\,V$	+/− 100 mV	+/− 10 mV
4b/cell XLC NAND	$V_{pp} = 18 - 25.5\,V$	+/− 50 mV	+/− 2,5 mV

2.6 Flash Memory Algorithm and V_{th} Window Definition

The different components required to design a flash memory are introduced in the last section. The interaction between cell architecture, memory array, sensing concept and manufacturing technology during all the product operation over lifetime is the key knowledge of the non-volatile memory development. In contrast to volatile memory architectures the design of the program, read and erase algorithm—in most flash designs a well optimized embedded software—will have a strong influence on performance and lifetime of the non-volatile memory.

The analysis of effects impacting each edge of the programmed and erased threshold voltage distributions starts the margin definition. The margin definition is the core knowledge for the design of an optimized flash algorithm and for a reliable lifetime of a non-volatile memory product.

This section describes the link between physical cell behaviour, process and statistical deviations and threshold voltage distributions. The flash algorithm concept is introduced and procedures are shown to overcome the discussed influences.

2.6.1 Flash Threshold Voltage Window: Margin Setup and Accuracy

The basis of the flash algorithm development is a threshold voltage (V_{th}) window margin calculation. The quantitative impact of different algorithm parameters during program, read and erase onto the read sense margin is analysed. The accuracy of the applied voltages to program and erase and the accuracy of the sensing are incorporated in the second distribution in Fig. 2.70.

The flash cell architecture and technology define a maximal usable V_{th} window. This window space can be divided into the three regions:

- the erased window space—cell distribution below V_{th_LH}
- the programmed window space—cell distribution above V_{th_HL} and
- the window space between V_{th_LH} and V_{th_HL}, to ensure a distance called read window including the complete reliability margin.

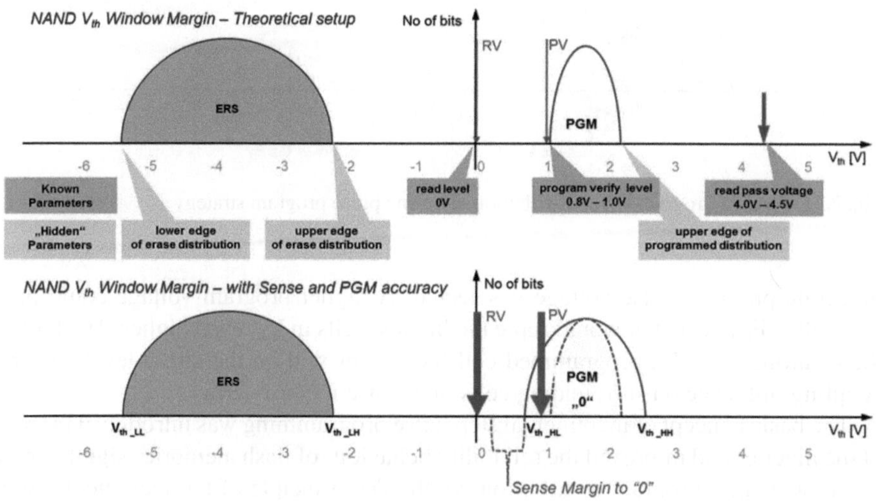

Fig. 2.70 NAND V_{th} Window Margin—known and unknown parameter—with accuracy consideration

The program algorithm creates the program distribution width and position and the erase algorithm is responsible to maintain the erase distribution width within the assumed limits. The width of both distributions is as important as the read window, and these dependencies will be introduced step by step together with the corresponding counter measure.

The next chapters will introduce the principles for NAND program and erase algorithms.

2.6.2 Principles of Flash Program Algorithm

The target of the program algorithm is to shift all cells to be programmed above the V_{th_HL} level, which will result into a program distribution width due to the statistical nature of the cells. In contrast to all volatile memories a single write operation—translated into a single program pulse—would shift the V_{th} levels of the flash cell only statistically into the program window space.

The simplest program operation would be theoretically a one pulse program strategy, which should result into two main benefits:

→ Minimized high voltage operation
→ Minimized stress to the bottom oxide of the cells.

Figure 2.71 shows the threshold voltage distribution of the erased and the programmed distribution after a one pulse strategy was executed. A very wide program distribution is visible and not all cells have passed the program verify level if a

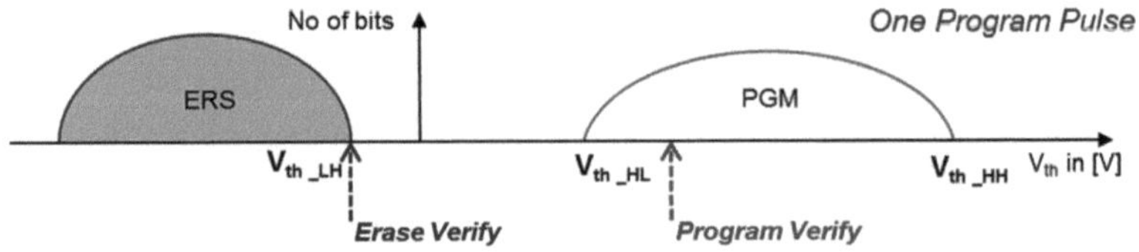

Fig. 2.71 NAND Program—V_{th} distribution after one pulse program strategy

moderate program pulse voltage is selected. A higher program voltage could program all cells above, but would force the highest cells in V_{th} even higher. The higher the resulting V_{th} of a programmed cell the higher will be the stress level and the coupling influence on surrounding cells in a dense memory array.

The basic concept of incremental step pulse programming was introduced [4] mid of the nineties and improved the reliability behaviour of flash memories significantly due to a smaller program distribution width. The principle of the incremental step pulse programming is a program pulse followed by a verify operation, which is a read operation with read reference level at V_{th_HL}. In case this target level—called program verify (PV)—is not achieved the cell would need and get another program pulse followed by the next verify operation.

The logical consequence is a program algorithm applying a couple of program pulses—each pulse increments the program voltage (Vpp) by a delta Vpp of 400 mV shown in Fig. 2.69—which creates automatically a 400 mV wide program distribution width. In reality the resulting distribution width is wider due to the program voltage inaccuracy and the sense inaccuracies during each verify.

Figures 2.72 and 2.73 illustrate the principle how a wide initial program distribution after the first pulse is compressed into a small target distribution.

The last program pulse and verify operation have to ensure that the target program distribution width is achieved—shown in Fig. 2.73.

The program principle is described with a logical flow. An example is given in Fig. 2.74 on the left side. Based on the above described program principle the following conclusions can be made:

- A smaller delta Vpp results into a smaller program distribution width;
  - A reduced Vpp step increment can be translated into an increased number of program pulses. The longer program time results into more read window margin.
- The smallest theoretically achievable program distribution width is equal to the delta Vpp increment only if the program algorithm—for example FN tunnelling applied to FG cells—operates in the linear cell characteristics—as shown in Fig. 2.74 on the right side.

For a dense cell array with interferences between neighbour cells the conclusion is that the above discussed program distribution width can only be achieved if all cells could be programmed in parallel. This is theoretically only doable for a test pattern—all word lines get the same Vpp and all bit lines are grounded. Program algorithms

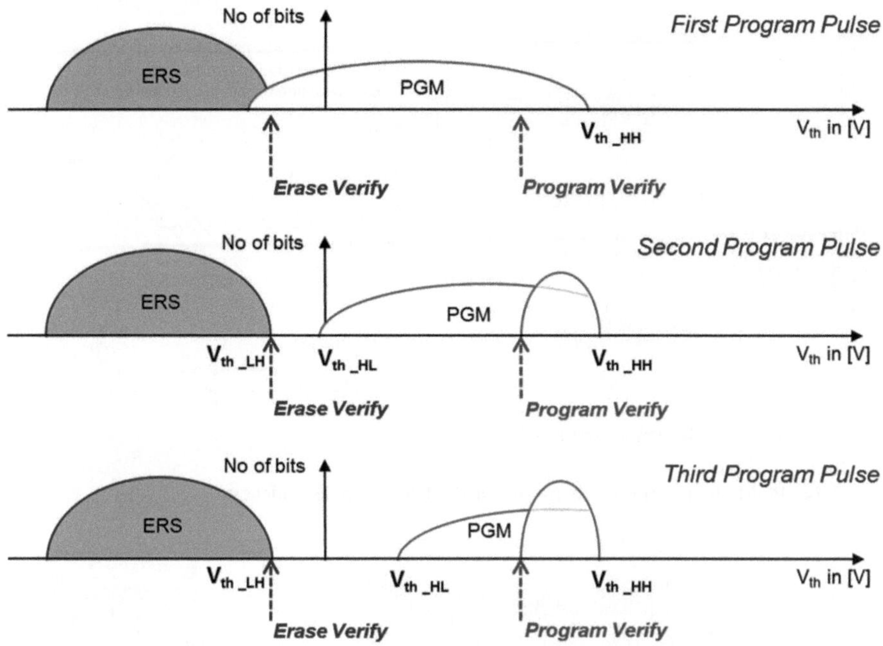

Fig. 2.72 Incremental step pulse programming Algorithm—V_{th} distribution after each program pulse

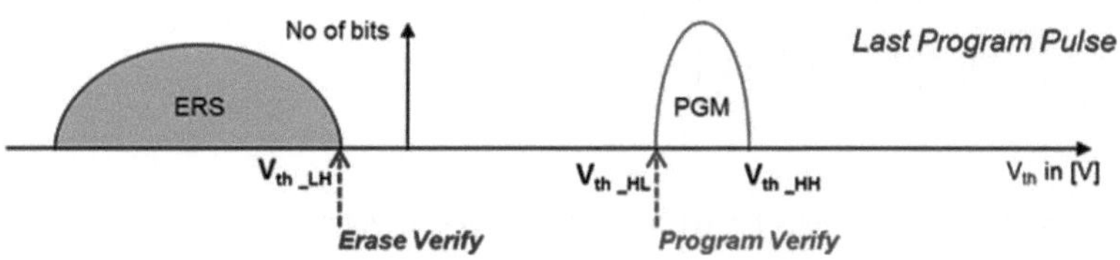

Fig. 2.73 NAND program algorithm—V_{th} distribution after complete algorithm execution

will be introduced for NAND flash memories in Sect. 2.6.4. These advanced flash algorithms achieve almost the above described theoretical target.

2.6.3 NAND Cell Interference: FG to FG Coupling

Statistical fluctuations and cell to cell interferences are increasing along the CMOS shrink roadmap. Floating gate to floating gate coupling—FG to FG—in a NAND array is selected to investigate the link between cell, array and program algorithm.

FG to FG coupling was highlighted as challenge for the shrink roadmap on conferences [69, 70]. Emerging non-volatile memory technologies including charge

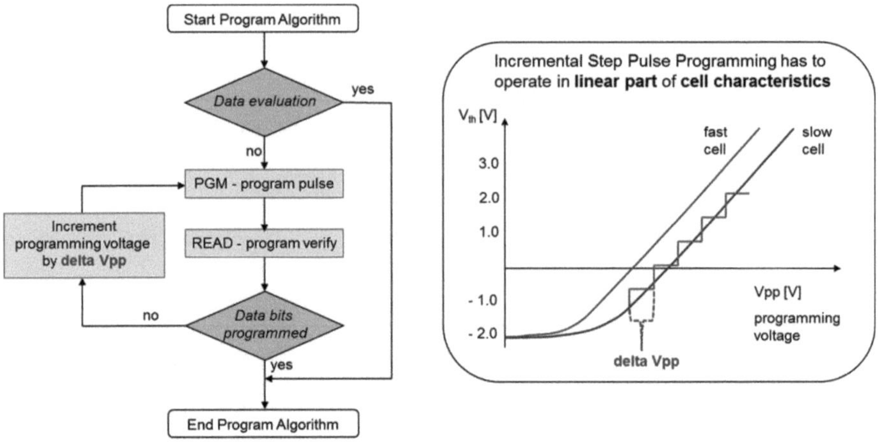

Fig. 2.74 PGM algorithm flow chart—linearity of V_{th} Vpp dependency

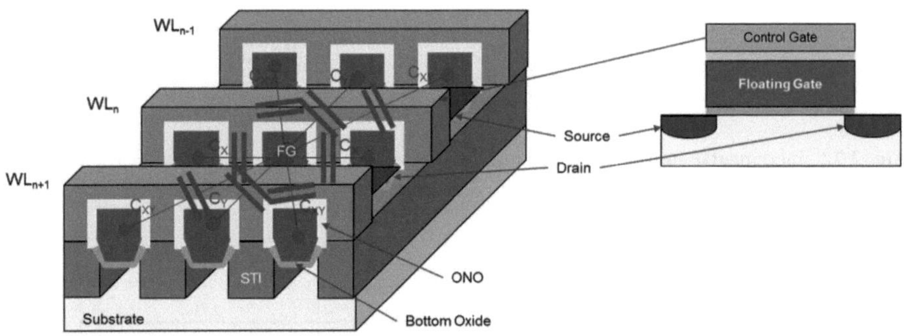

Fig. 2.75 NAND cross section of 3 WL's—cell to cell FG coupling

trapping are presented already for 2D NAND memories to reduce the strongly increased FG to FG coupling for sub 40 nm nodes.

We will introduce the interferences and will address step by step the mathematics to solve the interference challenge if certain conditions are fulfilled. The cell to cell interference is a key challenge in all non-volatile memory arrays. The capacitive coupling is the major challenge for the NAND array, in array types like VG-NOR leakage pathes have to be analysed and compensated too.

The NAND array is one of the most dense array structures in the semiconductor industry. Figure 2.75 illustrates the dependencies between all nine investigated NAND array cells. The target cell in the middle is surrounded by eight neighbor cells. The major interference on the target cell is the direct capacitive coupling in BL direction C_y—face to face coupling between two cells in two adjacent word lines—and in WL direction C_x. The diagonal cell coupling C_{xy} is a factor of 10 lower than C_x and C_y.

The FG coupling impact on the target cell is summarized in the following formula:

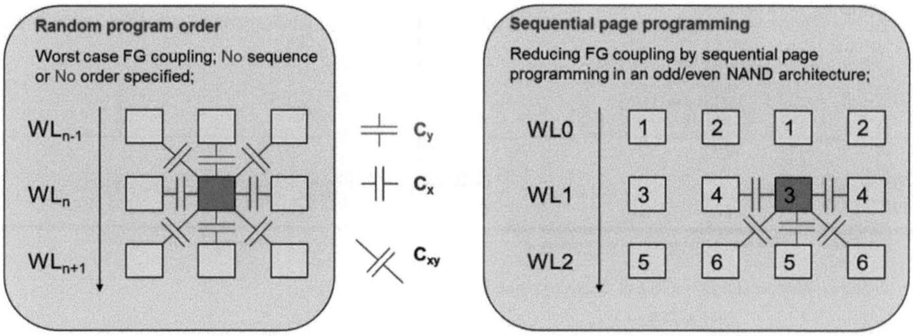

Fig. 2.76 NAND Cell FG to FG Coupling—*Top view* and Sequential Page Program order

$$C_{WorstCase\_Random} = 2 * C_X + 2 * C_Y + 4 * C_{xy}$$

This FG to FG coupling shifts the V_{th} of the target cell by volts based on the worst case which assumes a V_{th} shift of the surrounding 8 cells by 6–8 V each. An accurate V_{th} position within the cell V_{th} operation window is hard to achieve during a program of an addressed cell.

To illustrate this important effect we are starting the analysis of the typical flash operation mode. The NAND memory is typically filled page by page with data. Random byte write operations are only specified for NOR flash memories. The pages 1 and 2 marked per cell on the right side in Fig. 2.76 are already programmed before target page 3—marked with yellow—is programmed and cannot impact the V_{th} of cells in page 3. The corresponding calculation for NAND flash programmed by a sequential page order reduces the impact of coupling already significantly [14]:

$$C_{WorstCase_{default}PagePGM} = 2 * C_X + C_Y + 2 * C_{xy}$$

Figure 2.76 illustrates the reduction of the FG to FG coupling simply by the normal page program order, which is the default for data storage in any case. Two different situations are analyzed for the odd/even page NAND memory [71] architecture. The coupling calculation gives two different values for odd and even pages in terms of worst case impact of FG to FG interference:

$$C_{WorstCase_{default}PagePGM\_odd} = C_X + 2 * C_Y + 2 * C_{xy}$$

$$C_{WorstCase_{default}PagePGM\_even} = C_X + 2 * C_{xy}$$

The even page (the second in the program sequence) has again a more reduced impact of coupling. Approximately 75 % of the impact of FG-FG coupling disappears for even pages due to a fixed program order. The **program sequence** could be specified as a **product specification feature** and has a major impact on the strength of cell to cell interferences.

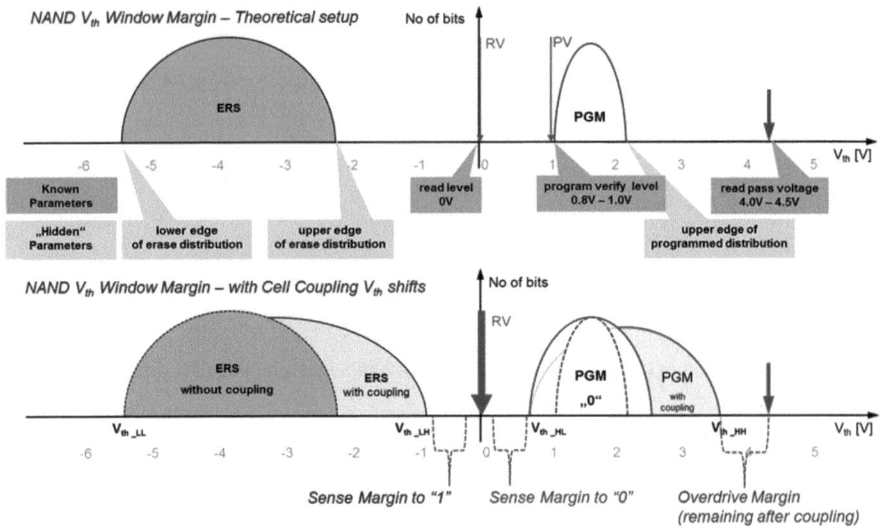

Fig. 2.77 NAND V_{th} Window Margin—with and without floating gate coupling V_{th} shift

The NAND cell to cell interferences analysis has to consider the following summarized parameters which impact the strength of the coupling on the target cell V_{th}, and has to compensate the corresponding V_{th} shifts in the V_{th} window margin analysis shown in Fig. 2.77 with a counter measure.

- FG to FG coupling factors derived from technology parameter per node C_x, C_y and C_{xy};
- Maximum allowed V_{th}-shift of cells during last or final programming sequence;
  - The cell at the lower edge of the erased distribution is programmed to the assumed upper edge of the programmed distribution.

$$V_{th\_WorstCase_{default}PagePGM\_Shift} = V_{th\_HH} - V_{th\_LL}$$

- Specified sequence of programming: after a cell is finally programmed, which surrounding cells are allowed to be programmed randomly;
  - Predefined program sequence like sequential page programming;

An increase of cell and technology driven FG coupling coefficients along the shrink roadmap is acceptable if the effective coupling could be reduced by the two other parameters introduced above.

The reduction of the maximum allowed V_{th}-shift is not that simple, because a large total window is required for reliability margin and especially for multi-level cell products. The program algorithm can reduce even the maximum shift by orders of magnitude, as discussed in the next Sect. 2.6.4.

The other important parameter is the position and width of the erase distribution, which strongly impact flash algorithm parameters. This is slightly hidden and will be analyzed in the erase chapter.

The FG to FG coupling can be assessed as a strong drawback of the floating gate cell NAND array architecture or part of the solution to extend the shrink roadmap even longer. A strong coupling of the floating gate cells is becoming the strength of this memory because the coupling between the eight surroundings cells is responsible for more than 50 % of the V_{th}position of the target cell. The reduced number of electrons which could be stored within the FG along the CMOS shrink roadmap could be addressed again by the algorithm techniques. A loss of electrons within nine cells will have a predictable imprint. A stronger coupling between these cells improves a correct adjusted error detection and error correction technique in such a way that the correct data can be always recovered.

Independent of some potential side effects the stronger coupling between cells has to be included in the mathematics of the algorithm development, which will be covered in the next chapter.

2.6.4 Program Algorithm Part II: Incorporating Cell to Cell Interferences

A couple of rules have to be added to the product specification to reduce the effect of cell to cell interferences. The logical pages have to be programmed only once—preferred is a complete usage to balance the number of programmed and erased cells automatically by the user pattern. On chip randomizer can reduce the probability of worst case data pattern statistically [72].

The second parameter is the program algorithm combined with the corresponding page buffer features in the flash memory design. The order of odd and even pages in a NAND memory reduces the FG coupling as shown in Fig. 2.76 and eliminates three coupling capacitances ($1 * C_y + 2 * C_{xy}$).

The All Bit Line NAND array and page buffer architecture programs all cells belonging to a physical word line together and reduces the cell to cell interference dramatically. Figure 2.78 shows the result: Now five couplings are eliminated ($2 * C_x + 1 * C_y + 2 * C_{xy}$).

The yellow marked cell on the right side in Figure 2.78 will only see one F2F coupling and two diagonal couplings (appr. 10 times lower). The coupling impact is reduced down to 25 % by programming sequence and complete word line access. The NAND array is a preferred architecture to implement a complete word line access which improves in parallel the read and the program data throughput.

The next important parameter which we have to attack is the maximum allowed V_{th}-shift during the last or the final programming sequence.

There are two possible ways to reduce the V_{th}-shift of the cells during programming. The first one is to reduce the erase distribution width for example from 4.0 V width to 2.0 V, which reduces the maximum V_{th}-shift by 2.0 V. The remaining program shift will still be in the range of 6.0 V.

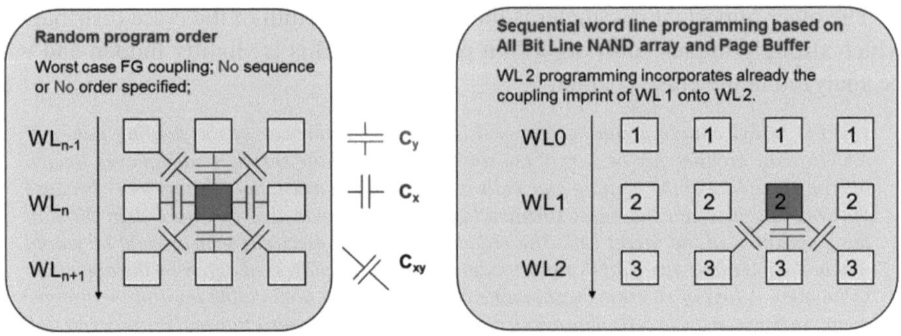

Fig. 2.78 NAND cell coupling—*top view* and sequential word line program order

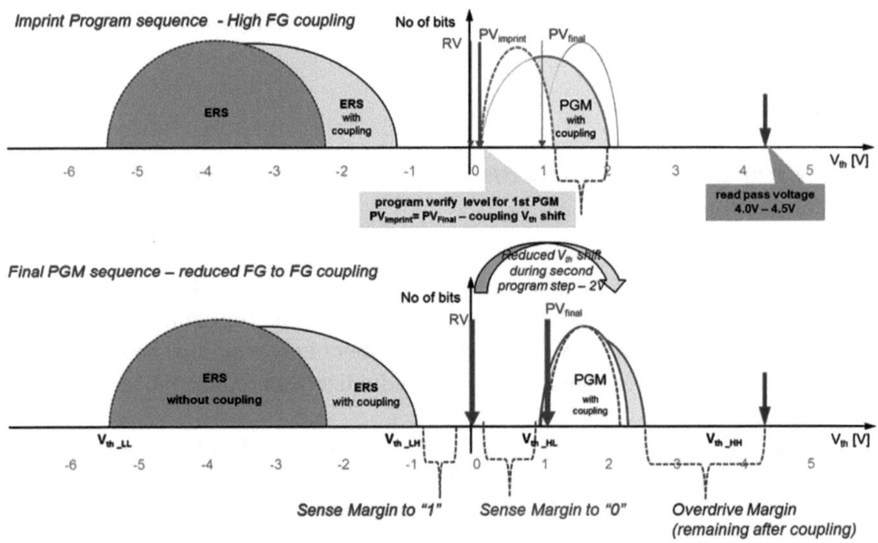

Fig. 2.79 NAND V_{th} Window Margin—two phase imprint programming with reduced FG coupling

A consequent cancelation of the FG to FG coupling is achieved by the imprint programming method. The logical data for at least two physical word lines are stored in an advanced page buffer cache and the first word line is programmed to the $PV_{imprint}$ Level, than the second word line is programmed to the $PV_{imprint}$ Level and finally the first word line is programmed to the PV_{final} Level. This sequence is repeated up to the end of the block. Figure 2.79 illustrates the principle of a two sequence programming for an SLC NAND V_{th} window margin analysis. The remaining read overdrive margin is significantly increased compared to Fig. 2.77.

The remaining maximum V_{th}-shift for the final programming on the neighbor word line is reduced to less than 2.0 V within an All Bit Line NAND array. The FG coupling impact is reduced from four direct neighbor cells shifting with 8.0 V (illustrated in Fig. 2.77) to one direct neighbor cell shifting with less than 2.0 V shown in Fig. 2.79.

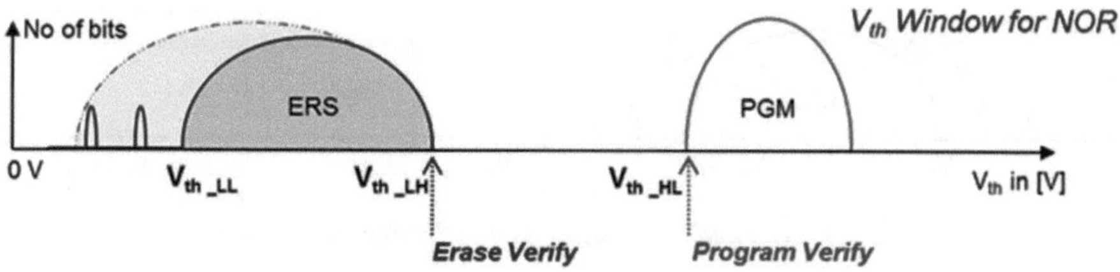

Fig. 2.80 Over erased cells (bits) within a V_{th} distribution of a NOR flash

2.6.5 Principles of Flash Erase Algorithm

Flash memories specify a block erase shifting all cells of a complete erase block—for example 4,096,000 cells for a NAND erase block—by one or more than one erase pulse into the erased V_{th} state. This simple pulse erase operation is like a flash, therefore these devices are called "Flash" memories. The erase distribution is the result of the erase algorithm and defines the starting point for any program operation. Memory array structures (NAND, DINOR, VG-NOR) have different parasitic effects (neighbour current, cell interference) and impact the final result of the erase.

The **position** and the **width** of the erase distribution have a significant influence onto the parameters:

- Program speed
- Program failures
- Read accuracy
- Retention margin over long term storage

Therefore erase algorithms will be introduced for both NAND and NOR flash memories and the consequences of wider erase distributions and countermeasures are discussed within next chapters.

2.6.5.1 Erase Algorithm for NOR Flash Memories

In a NOR flash array all cells in bit line direction are connected to one bit line (NOR) or to two bit lines (VG-NOR). The sensing operation is designed in such a way, that only the selected cell contributes to the sense current. In a large NOR array the V_{th} of the erased cells is statistically distributed and all these parallel connected cells will create a leakage current. The leakage currents influence the sense accuracy for the target cell. Therefore the forbidden cell state in a NOR array is the over erased cell state—which creates an increased leakage current even if the word line is not selected.

Figure 2.80 shows two over erased cells below the target erased distribution width. As lower the V_{th} of the over erased cells as higher the leakage of these cells.

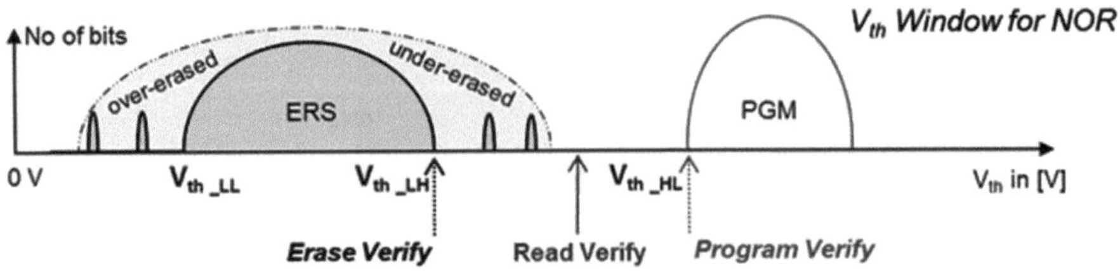

Fig. 2.81 Over and under erased cells (bits) within a V_{th} distribution of a NOR flash

The complete problem is becoming visible if the following case it assumed. The erase verify operation must detect all cells below V_{th_LH}. During an erase operation in a NOR flash some cells are becoming strongly over erased, as shown in Fig. 2.80. The V_{th} of these cells is much below V_{th_LL}. The leakage of these cells would result into the effect that other cells above V_{th_LH} are sensed to be erased enough due to the leakage current of the over erased cells added up to the sense currents of the cells belonging to the target word line. Figure 2.81 shows the physical result of this phenomenon, the erase distribution becomes wider on both sides.

The erase verify operation by itself cannot detect this behaviour, which would result in a lot of programming failures or bit failures during operation. The erase algorithm has to take care of these cells that are becoming only visible after the over erased cells are programmed back into the main erase distribution or above. One algorithmic solution is the intermediate Program After Erase executed immediately after each erase pulse shown in Fig. 2.82.

The erase algorithm applied within NOR flash memories is therefore a combination of program pulses to all cells, erase pulses to all cells and program after erase pulses to all or selected word lines. The target of this combined operation approach is to move the over erased cells back into the predefined limits of the erased distribution.

- Program **B**efore **E**rase—**PBE**
  - Program all cells with a medium high program voltage to a V_{th} close to V_{th_HL}, so that already programmed cells are not shifted and erased cells are shifted above erase verify. This operation improves the reliability significantly. Without PBE cells which are not programmed by the logical data over a certain time get always erase pulses which would force them to shift deeper and become strongly over erased.

- Program **A**fter **E**rase—**PAE**
  - After the final erase pulse the erase operation could be stopped. Every over erased cell and every cell shifted too deep by it's erratic behaviour would force sense failures and can impact the data along the bit line. Therefore a PAE is executed per block or per word line with or without verify.

Two algorithm options are shown in Fig. 2.82:

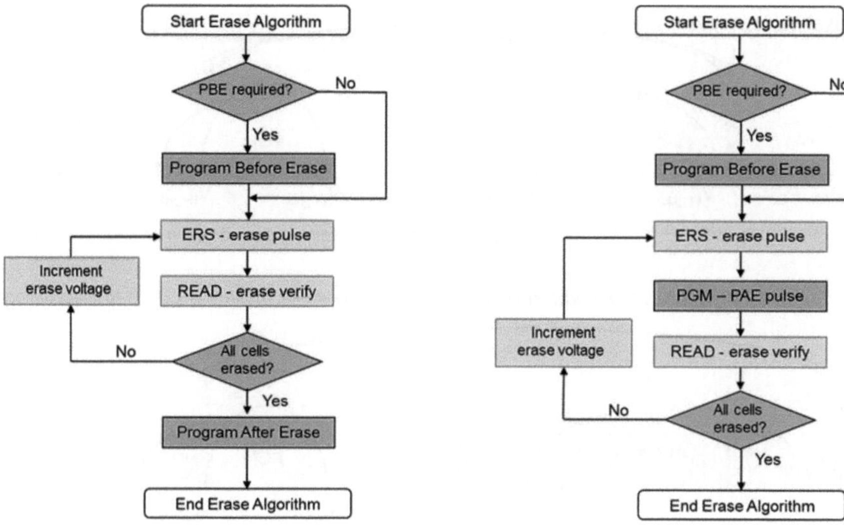

Fig. 2.82 NOR Erase algorithm—with final PAE (*left*) and with intermediate PAE (*right*)

The erase operation is a combination of program before erase, erase and program after erase operations, which are executed dependent on the specific implemented algorithm. The execution time is dependent from the logical data programmed, the temperature and the stress level (endurance) of the erase block. Therefore only typical numbers are specified for this value in the product specification and can vary over time.

The NOR flash erase operation for small block sizes is time consuming compared to fast erase times of large NAND blocks. The long erase times are one drawback of the NOR array architecture especially for memory sub-systems for data storage application.

2.6.5.2 Erase Algorithm for NAND Flash Memories

The NAND array offers by design a straightforward erase strategy. All cells connected within a NAND string belong to a NAND erase block separated by the select gates. The array architecture ensures that over erased cells cannot create leakage paths. The erase algorithm is an incremental step pulse ladder of erase pulses followed by a verify to shift all cells below the erase verify level.

This simple erase strategy ensures short values of 1–2 ms for the erase time in the product specification. Typical values are given here, because the number of pulses could vary dependent from data history and environmental conditions.

The width of the erased distribution impacts the maximum allowed V_{th}-shift during the program operation directly. A larger erase distribution width results into effectively more initial program shift of cells, which are located near to V_{th_LL}.

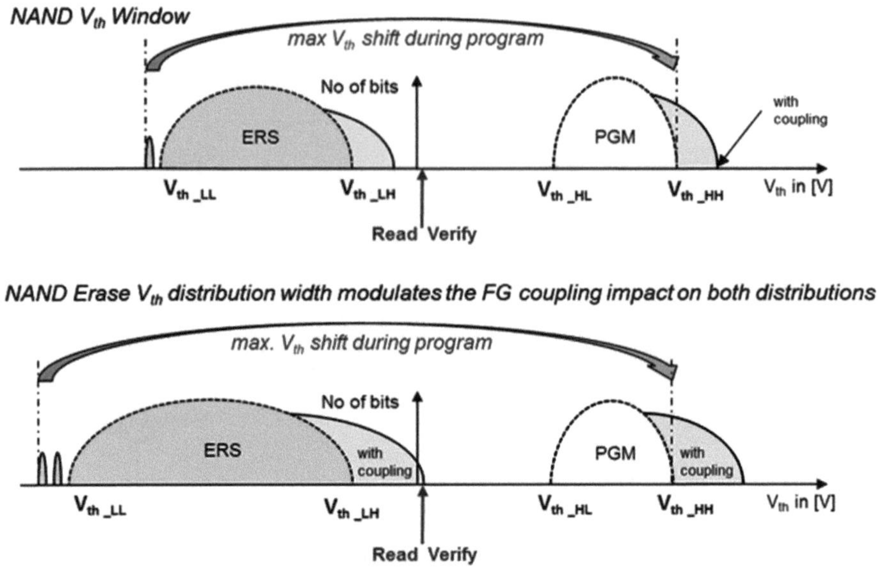

Fig. 2.83 NAND V_{th} window—Erase distribution width impacts max V_{th} shift and FG coupling

A widening of the initial erase distribution from 2 V to approximately 4 V adds 2 V more initial program shift. The FG coupling impacts neighbor cells and this becomes an issue for the erased distribution.

The coupling shift of the erased cells cannot be compensated with any kind of algorithm, because they have to stay below the erase verify level. This is becoming a major issue along the shrink roadmap, because the erase distribution width has to become smaller exactly by the amount of increased coupling shift from technology node to node.

Figure 2.83 illustrates the impact of erase distribution widening to the NAND V_{th} window margin.

The following conclusions have to be made for performance and margin optimization linked to the NAND erase operation and the erase distribution width:

- The smaller the erase distribution can be maintained over time the less negative the V_{th_LH} has to become. The design challenge is here to implement a sense concept which can verify the V_{th} of the erased cells between −2.0 and −1.0 V with 0.0 V as gate voltage.
- The initial distribution width caused by intrinsic variations of process technology—we call it the **Intrinsic Cell Distribution**—has to be as small as possible. This intrinsic cell distribution width defines the achievable erase performance and impacts nearly all reliability parameter, which will be discussed in the reliability chapter.
- An erase algorithm especially with PAE can reduce the target erase distribution width and improves the FG coupling impact onto the program distribution width,

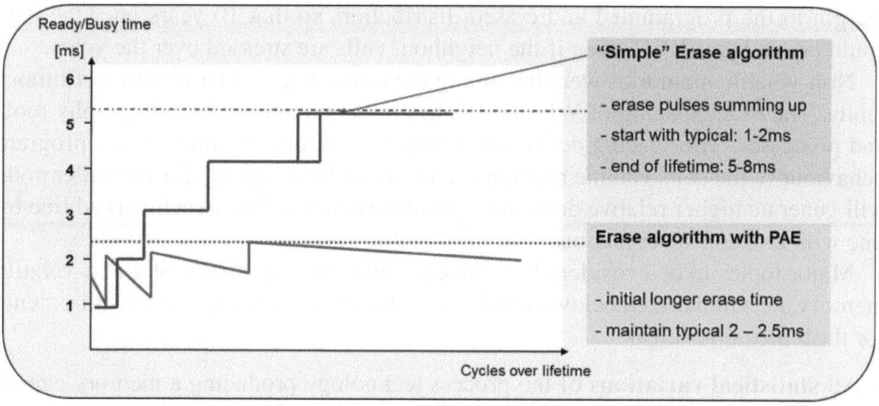

Fig. 2.84 NAND Erase algorithm results—with and without PAE

but a PAE can never reduce the FG coupling impact onto the erase distribution width itself. This is due to the fact that the PAE itself results in cross coupling onto the erased distribution.

A NAND product parameter specification assessment without a deep knowledge of the impact of erase distribution width and implemented countermeasures could enforce the selection of a product, which does not fulfil the specific system application requirements.

An application example compares two NAND memories from different vendors—one specifies a typical erase time of 1.5 ms the other 2.5 ms. A flash product characterization of the erase time development over the specified endurance range is shown in Fig. 2.84 and the result could impact the choice of the better fitting NAND memory product.

The erase operation is an excellent example for the link between product specification, algorithm behaviour and sense capability. The hidden dependency between erase algorithm and program performance has to be understood to make conclusions, as the better specification in terms of initial erase performance is not improving the system performance over lifetime.

2.6.6 Algorithm Summary: The Statistical Nature of Flash Memories

All integrated circuits have a statistical nature. The difference between CMOS logic, volatile memories and non-volatile memories is the worst case combination out of density, accuracy, requirements in life time and active degradation during operation. A write operation to a non-volatile memory at 55°C has to ensure the accuracy and margin to the programmed and erased distribution, so that 10 years later the data could be read at $-10°C$, even if the neighbour cells are stressed over the years.

Non-volatile memories were becoming the technology leader in terms of lithography. The exact spacing of WL and BL has to be guaranteed by lithography tools and processes. The spacing deviation of WL and BL directly impacts the program behaviour of the cells via line resistance and capacitive coupling. Each smaller node will generate higher relative deviation, but the product will be benchmarked line by line with existing specifications.

Major topics to be considered during each algorithm operation of a non-volatile memory are summarized below to make the difference and the specific requirements for flash memories clear:

- All **statistical variations** of the process technology producing a memory can be measured by the **Intrinsic Cell Distribution** width. The initial V_{th} of each cell identifies the starting point for all operations and for reliability considerations.
  - This initial behaviour can stay stable or could become erratic for some cell concepts.
  - This initial behaviour—in this work called "***initial technology imprint***"—could be modified for some specific cell concepts over life time, which would result into a feature to extend lifetime based on a special algorithm.
- The **physical stress** due to program and erase operation—especially for tunnelling physics—will degrade the bottom oxide. Additional traps will be created and charge will be trapped in the Bottom oxide and will de-trap over life time. A threshold voltage drift in both directions can be foreseen for different flash cell architectures and is a known phenomenon.
- The **history of operation** per cell in terms of repetitive write or read operation—linked to imprint and trap density—influence the achievable effective V_{th} shift during a program or erase operation. Not all cells are always in the linear program region—V_{th} follows linearly the program voltage Vpp step increase—and therefore some non-linear program regions have to be considered during the start of the algorithm.
- The V_{th} window margin analysis of the **neighbourhood**—data stored in neighbour cells will influence the result of the achievable effective V_{th}—has to incorporate the statistical nature of interferences between cells, word lines and bit lines.
- The read and verify operation requires an **absolute accuracy** in terms of voltage accuracy.
  - The accuracy for program and erase operations is specified as relative or differential voltage accuracy. The absolute values are not that important, because the repetitiveness and the stepping are considered. The inaccuracy between read and verify is fully accountable as an absolute value.
- The **change of the behaviour** (I/V-curve) of the non-volatile element itself over cycling.

– Due to the fact that the slope of each cell (transistor) will be changed over cycling—a fast and low current sensing is required—to shift the operating point in the region with lowest changes over life time.

The issues listed above force the flash memory design into a theoretical target, which can be described like "all cells should see the same stress level for each operation over lifetime". This condition requires a different dedicated V_{th} window for each cell and for each moment in lifetime.

The consequence is that every memory product will see different stress levels on cells and on blocks and an overall system optimization approach has to incorporate this real behaviour. The differences in achievable endurance values between cells and blocks can be one or two orders of magnitude and these differences increase over lifetime, in case the right counter measures are not in place on system and algorithm level. Key knowledge of system design engineers is how to implement efficient identification and diagnosis procedures for all flash blocks under control and criteria to monitor their behaviour continuously.

Based on the statistical assessment of non-volatile cells the following statements can be made:

- Design for reliable operations of non-volatile memories have to be based on statistical considerations.
- Program and erase timing parameters are an indication for typical values based on random data under typical condition—like program, erase and read operation at same temperature level within a typical time frame—days or weeks.

The principles of flash window margin analysis are focusing the decision process on the important parameters of flash product specifications. The complexity of program and erase algorithm can have side effects and a basic memory characterization is strongly recommended after the decision was made or to support the selection process between different product specifications or different memory architectures.

For verification purpose a number of test runs are defined to show the evidence, that the selected memory fulfils the application specific requirements. A plan for a flash memory characterization can look like:

- *Memory margin and retention characterization*
  - *Read and Write Margin analysis*
  - *Dependency from x-y-location within the memory array (inhomogeneity)*
  - *Dependency from the commands executed before (history)*
  - *Dependency from content of the memory cells before (history)*
  - *Dependency from content of all neighbour memory cells (interference)*
  - *Data retention of data stored in the memory array with and without disturbance*

Factors influencing width and stability of the programmed and erased distribution are described in the literature [14, 61] and will be discussed in more detail in the reliability chapter.

Especially for charge trapping cells a movement of the V_{th} distribution over time is well known and the read level is adapted during life time. In such a case the margin calculation becomes a dynamic procedure over lifetime. The read algorithm innovations will be discussed in the corresponding section.

2.7 Multiple Bits per Cell Area: Flash Memory Concepts

This work targets economic principles of memory and memory system optimization. The main development target is the reduction of cost per bit. The CMOS shrink roadmap targets the continuous reduction of feature sizes and enables smaller cell sizes and therefore larger memory densities. Every new technology node requires next generation of innovative lithography tools.

Flash memories offer the capability to store more than one bit per cell. The following chapters introduce a subset of concepts to increase the bit density for non-volatile memories.

The first approach is the concept of multiple bit storage per cell:

- The **Multi-Level Cell Memory concept** increases the number of programmed distributions placed over the total V_{th} window space of the flash memory. One bit could be achieved by two distributions, 1.5 bit by three distributions, 2 bits by four distributions, 3 bits by 8 distributions and 4 bits by 16 distributions.
- The **Multi-Bit Cell Memory concept** stores multiple bits at different physically separated positions in a cell. An assessment of one concept is made with the focus on important application specific parameter changes enforced by a multi-bit concept per cell.

The second approach is the concept of three dimensional—3D—memory cell and array architecture:

- In principle Single-Level Cell as well Multi-Level Cell operations can be applied to increase the bit density in real 3D non-volatile memory concepts.
- Cross point arrays offer an excellent 3D capability.

Major challenges are described for different concepts which were in high volume production to derive important performance parameters for system optimization decisions.

2.7.1 Principles of Multi-Level Cell

The doubling of bits per cell based on a multi-level cell requires additional verify and read levels and a logical to physical address mapping to address the additional bits. Figure 2.85 illustrates both topics.

The concept of mapping two bits to logical pages reduces the number of reads to decode the bits.

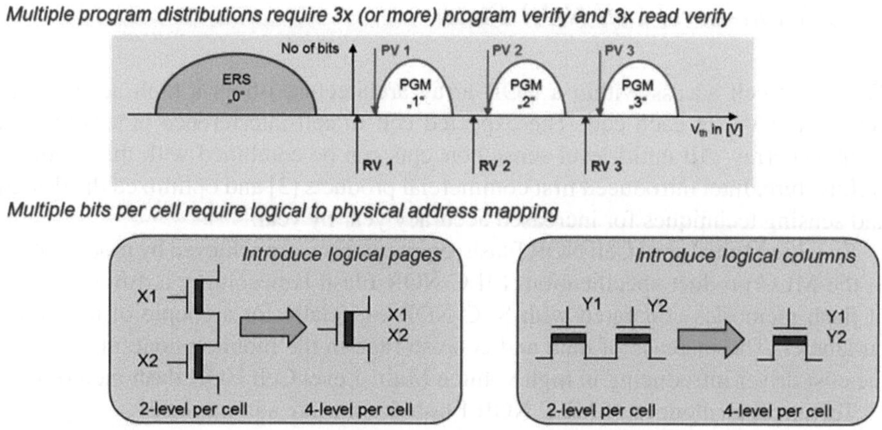

Fig. 2.85 Multi-Level Cell [MLC] principles—logical to physical address mapping for logical pages or columns

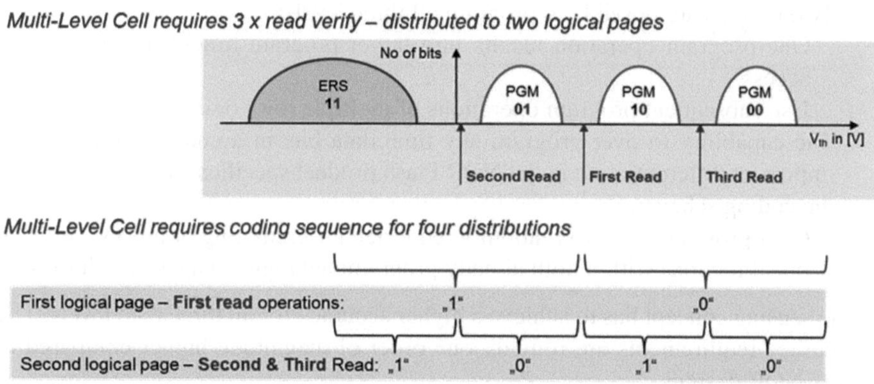

Fig. 2.86 Multi Level Cell [MLC] principles—coding scheme for two logical pages

The multi-page MLC read is in average on 1.5 times slower than the SLC read and two times faster than the multi-column MLC read access. Figure 2.86 shows the principle of decoding.

The Multi-Level Cell principle requires a very precise programming sequence. The already discussed differential accuracy of the program voltage is becoming one of THE important parameters. The same is now valid for the sense accuracy and the definition of the sense operating point. The sensing circuit has to be able to memorize the decoding results for already finished sense operations belonging to one or more logical page addresses.

The threshold window margin analysis has to be done now for three sense windows and every noise effect in cell, design and technology has to be understood in deep detail. Effects influencing the stability and the widening of the V_{th} distribution have to be eliminated by the corresponding design and technology innovation.

2.7.2 Multi-Level Cell NOR Flash

The direct cell access within a NOR array architecture offers a high accuracy to control the V_{th} of each cell. The expected cell to cell interference is less than in a NAND array. All multi-level sense concepts can be combined with this memory architecture. Intel introduced first commercial products [3] and optimized the design and sensing techniques for increased accuracy year by year.

Specific Multi-Level Cell NOR Flash requirements were achieved by modification in the MLC product specification. MLC NOR Flash is becoming a different class of flash memories compared with SLC NOR especially for a couple of reliability parameter. The increase of data and code storage in the mobile phone market was the cost driver introducing in high volume Multi-Level Cell NOR flash memories.

Technical challenges of MLC NOR Flash designs are summarized:

- Mapping between logical and physical addressing has to be defined for MLC flash products.
  - Two options are possible to program all three levels:
    One program operation results into larger program times and longer read access.
    Use subsequent program operations of multiple rows or columns.
  - The capability to over program any time data bits in a code memory is one important differentiation in the NOR Flash product specification and influences the coding scheme.
    Over program means a randomly requested programming of bits (V_{th} shift in one direction) within both already programmed logical pages or columns.
- The sensing concept has to achieve a higher accuracy for all three read levels. The accuracy requirements are roughly one order of magnitude higher compared to SLC NOR designs.
  - The NOR Flash erase algorithm with PBE and PAE was the second important innovation to guarantee a small erased distribution, and consequently reducing all kinds of leakage currents in the array as much as possible.
- The constant voltage sensing concept introduced in the sense circuit chapter for MLC sensing cannot always fulfil the accuracy requirements along the shrink roadmap. Therefore stepped gate voltage or a ramped gate voltage sensing concepts shown in Fig. 2.87 were introduced to achieve the accuracy requirements
  - Multi-level cell NOR flash sensing operations require longer read cycle times based on above concepts and the asynchronous read access increases by a factor of two–three.

The remaining challenge for MLC NOR flash are the statistical deviations of the reference cells, of all program verify and read verify levels and the statistical Gaussian distribution of the programmed cells. The reliability effects have to be added to the small read window (e.g. erratic cell behaviour, drain disturbs [73]) The window size

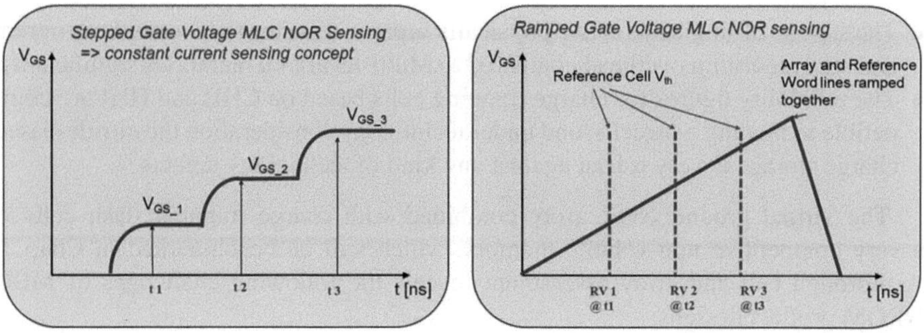

Fig. 2.87 NOR MLC sensing—stepped gate voltage and ramped gate voltage sensing principle [55]

is limited due to the increasing read disturb for higher gate voltages required to read the highest window. The probability that cells will be read on the wrong side of the sense level increases with the density of the memory designs (1 Gbit or 2 Gbit). Error correction techniques can be integrated, but are not that competitive compared to MLC NAND flash.

- The error tolerant NAND flash product specification is becoming a hard argument in the competition with MLC NOR Flash for data and code non-volatile memory sub-systems.

The direct competition with the dense NAND array and the upcoming MLC NAND was accelerating the NOR array and efficiency optimization reducing the logic required for additional operations. Novel program and read operations are discussed in the literature to enable virtual ground NOR—also called NOR Virtual Ground (NVG™) [74]—as the more competitive array architecture for high density flash operations due to the contact-less core array [75].

2.7.3 *Multi-Bit Nitride ROM Virtual Ground NOR Flash: 2–4-Bit/Cell*

2.7.3.1 2 Bit/Cell Multi-Bit Flash Memories

The localized bit storage nitride ROM flash concept combines the direct access to a cell with a dense virtual ground NOR array and adds a storage layer, which stores two bits without the time consuming programming of three levels.

- The virtual ground NOR memory array concept reduces the number of contacts required in the NOR array significantly as discussed in the section of memory array.

- The storage of single bits at two physically separated positions reduces the program and read algorithm overhead compared to Multi-level Cell memories significantly.
- The reliability figures for charge trapping cells based on CHE and HHI are compatible to floating gate cells, and under technology consideration the nitride-based charge storage is very robust against any kind of technology defects.

The virtual ground NOR array combined with charge trapping flash cells is a very competitive non-volatile memory, which will be benchmarked in Chap. 7. A thorough cell and array assessment reveals the following challenges of MBC NROM product designs:

- The CHE programming has a less linear dependency between V_{pp} and V_{th} compared to FG cells. Every new technology requires a time consuming characterization process to determine the optimized combination out of technology parameters and program algorithm settings. Programming the first bit influences the program behaviour of the second one.

- The erase algorithm time for larger block sizes is becoming a bottleneck impacting the write throughput of the memory.
- The different reliability behaviour of charge trapping enables a recovery technique to modulate the cell behaviour with an adaptive algorithm and ensure an application specific optimization in a very wide range.
  - In contrast to other cell concepts a charge trapping layer could be modulated based on algorithmic techniques. The window widening has to be controlled and the movement has to be followed by adaptive sensing techniques. The movement is called "Flying windows" describing the idea of moving windows in between moving distributions.

- In case the erase blocks are not physically separated within a large array structure, operation within the selected block impacts other sectors due to disturbances. The robustness can be achieved by voltage inhibit schemes, which are critical to control along the shrink roadmap.
- The physical separation of the two bits has to be controlled along the shrink roadmap.

Figure 2.88 illustrates the behaviour of programmed and erased distributions for a nitride based multi-bit cell. Both distributions move up on the V_{th} scale after a pre-cycling and a retention time. The principle of distribution movement is a hidden unexpected part of charge trapping flash—different for electrons and holes [76]- and has to be compensated over lifetime.

Different solutions are thinkable to address the weaknesses of charge trapping multi-bit cells:

- A solution can try to eliminate the distribution widening by advanced erase algorithm as long as possible. This algorithmic requires adaptive read algorithm techniques to detect the position of the read window [47].

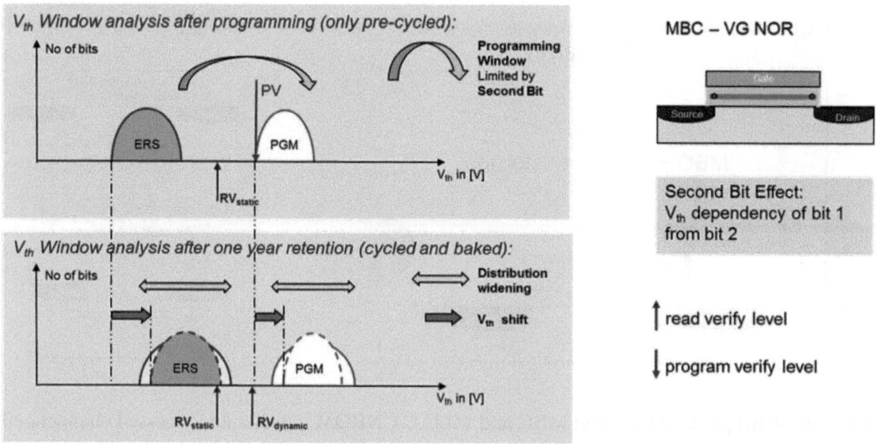

Fig. 2.88 MBC nitride ROM V_{th} Operation Window analysis—distribution shift and widening

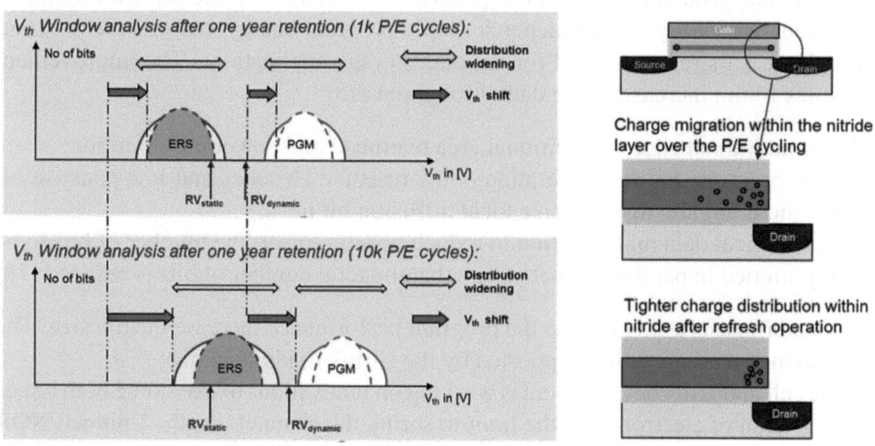

Fig. 2.89 V_{th} window behavior for flying windows combined with refresh

- A kind of refresh can be introduced for the charge trapping layer to reset the cycling behaviour to an initial state comparable to less cycled cell behaviour. Up to 1 million program and erase cycles are demonstrated with this extended endurance mode [77].

Figure 2.89 shows the principle behaviour over time for different pre-cycling conditions and illustrates the complexity behind Multi-Bit Cell non-volatile memory concepts. The read operation has to detect the window, has to quantify the quality of the window and influence the next step of the adaptive read algorithm strategy. The data integrity has to be always maintained within the specified timeframe typically over 5–20 years.

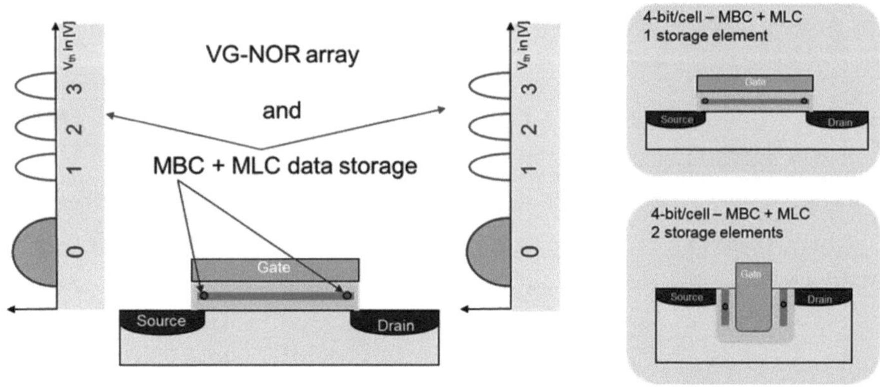

Fig. 2.90 4-Bit per Cell based on MBC and MLC CT NROM—planar and recessed channel cell

The virtual ground NOR array can program and verify only one cell within a physically separated array slice. Independent of a program energy analysis the program throughput is therefore limited for the data storage applications. The improvement measures for an increase of the data throughput are:

- More additional slices—additional area overhead and power consumption;
- Faster program and read operation—low resistive switches and low resistive bit lines including the low resistive local diffusion bit lines.
- Smart logical data manipulation to reduce power—invert the number of bits to be programmed in parallel in such a way that the total number of bits is reduced [78]

Countermeasures to increase the program performance increase the die size. The performance increase is not supported by the shrink roadmap alone.

The cell and array development is working on innovations to overcome both issues the migration of electrons and the limit to shrink the channel length. 2-bit/cell NOR type SONOS flash memory cells with spacer-type storage node on recessed channel structures are combined with FN tunnelling erase. This combination would improve most of the discussed issues of charge trapping based flash cells with localized charge storage [46].

2.7.3.2 4 Bit/Cell Multi-Bit Flash Memories

Targeting the increase of bit density per die size the nitride ROM cell architecture enables the unique combination out of Multi-Bit Cell and Multi-level Cell within one device. Figure 2.90 shows the principle of the 4-bit per cell concept (MBC + MLC) based on charge trapping flash cells.

This combination enables a threshold voltage operation window for program and read with more V_{th} margin compared to 4-bit per cell flash designs based on 15

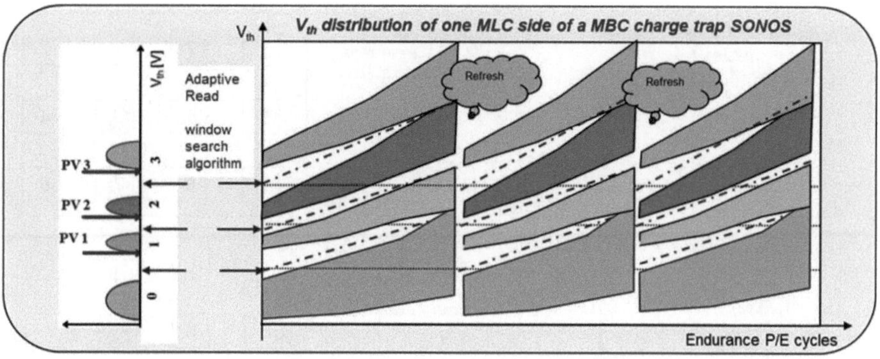

Fig. 2.91 Flying windows for 4-Bit MBC and MLC NROM with refresh

program level. Innovative cell concepts with 2 nitride storage elements are required to extend the endurance window.

The distribution widening and the distribution shift have to be addressed properly by a 4-bit per cell flash design based on the virtual ground NOR array. The reliability issue of moving distribution are highlighted in Fig. 2.91. The distribution widening and shift is the limiting effects regarding the usable maximum cycle count of this MBC and MLC nitride ROM array architecture.

The increased bit density based on localized charge trapped in nitride ROM cells and multi-level programming results into a cost effective flash memory. This 4 bit/cell concept has good reliability values for limited endurance (less than 50) and fast read access cycle. On system level the decision can be made if these benefits can compensate slow program throughput and high program current.

- 4 bit/cell multi-bit cell NVM memory products offer fast read access times for a reliable one time data storage solutions, which achieves an excellent cost per bit position.

2.7.4 Multi-Level Cell NAND Flash: 2–4 Bit/Cell

Floating Gate based NAND flash offers a competitive and dense array structure, a physical separation between erase blocks and a cell with a linear dependency between FN tunnelling program voltage and the V_{th} of the cell. The NAND product specification was developed in such a way that page and block operations have a 100 % fit with the physical array structure and are incorporating bit errors during read covered by ECC and algorithm errors during program and erase operations covered by the complete mapping concept of a so called "bad" block.

The long program ($\gg$100 μs) and read ($\gg$10 μs) operation (cycle) times are preparing the application for even longer program (>600 μs) and read (40 μs)

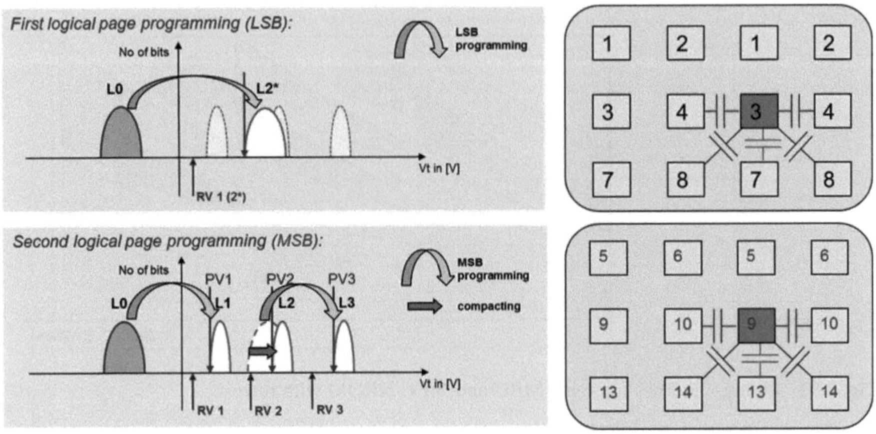

Fig. 2.92 NAND V_{th} Window analysis—MLC LSB and MSB program algorithm

operation times. MLC NAND introduces the logical page addressing which reduces the average operation times. The long operation times are fully compensated by the available parallelism and highest program and read data bandwidth are offered by today's MLC NAND flash products. The performance indicator chapter is focusing on a program bandwidth assessment for SLC and MLC (2 and 4 bit/cell) NAND memories in detail.

The Multi-Level Cell NAND development is addressing array margin losses (FG to FG coupling, back pattern noise, Source line noise) and is trying to eliminate them. A MLC program algorithm illustrates how MLC NAND programming incorporates automatically all required countermeasures to reduce the FG to FG cell coupling—introduced and discussed in detail in the algorithm chapter.

> The first level—called L2* in Fig. 2.92—is programmed during the LSB programming on the first four logical pages corresponding to the first two physical word lines. The maximum V_{th} shift for the following MLC programming of the second—called L1—and third—called L3—level is significantly reduced. The second level—L2—is compacted during the second programming operation, so that the coupling V_{th} shift is incorporated in the V_{th} window margin calculation.

We try to simplify the MLC innovation to identify the key parameters to ensure a reliable product:

- An accessible V_{th} Window space larger than 3.5 V.
- An application note defining a strong page program order and a page buffer concept allowing programming and compacting of three levels in parallel solves mostly all coupling issues of Floating Gate MLC NAND within a certain technology range −70 to 45 nm.
- A Floating Gate cell with high quality tunnel oxide ensures highest V_{th} stability over time.

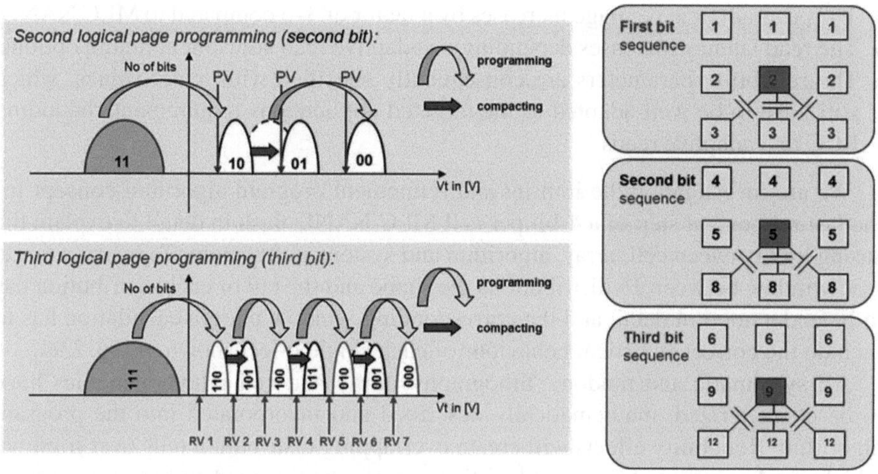

Fig. 2.93 NAND V_{th} window analysis for 3-bit/cell and programming sequence

The challenge to place accurately more program distributions crystallizes the following learning:

- Cell to Cell interferences are solved by mathematics within a regular memory array.
- Programming one more bit—or two levels more—requires additional algorithmic techniques which automatically ensure accuracy and compensate in parallel array and noise effects, but only in case all noise effects are fully understood and the mathematics is correctly applied.

There are more details on design, cell and technology behind, but the basic concept can be applied again—3 bits corresponding to 8 levels—and—4 bits corresponding to 16 levels. The most important effect of designing a MLC flash memory capable programming 15 levels—published first time in 70 nm [79]—is the learning effect during product characterization and qualification.

We apply the introduced algorithm principles; combine 8 levels with the All Bit Line NAND array architecture and a three phase programming sequence to reduce the max V_{th} shift of the last impacting programming one step more—shown in Fig. 2.93.

The V_{th} window margin analysis is becoming the key knowledge for 2 and 3 bit/cell MLC NAND flash designs. Floating gate MLC flash requires a kind of iterative adjusting algorithm approach which automatically programs every word line three to five times multiplied by the number of high voltage program pulses. The implementation of the self-boosted program inhibit scheme is becoming the success factor for more than 2 bit/cell designs.

The development and the usage of 3- and 4-bit per cell MLC NAND designs are well described in the literature [80–82]. The e**X**tended **L**evel per **C**ell (XLC) NAND devices need longer program algorithm execution times, which is compensated by larger page sizes.

- The program time per page increases by a factor of 3–5 compared to MLC NAND.
- The read latency increases depending on adaptive read adjustment tuning options.
- The reliability parameters are consequently specified with lower values, which still have to be well adapted to the targeted applications requirements including ECC and adaptive reads.

We are investigating the imprint and refinement program algorithm concept for the last refinement step of a 4-bit per cell MLC NAND flash in detail to explain the strong link between cell, array, algorithm and system architecture. To insure a target read window between 16 distributions the shape and the tail of each distribution has to be understood in detail and the corresponding window margin calculation has to include the correct statistical behaviour of each single effect shown in Fig. 2.94.

All systematic and random (lithography) array and cell inhomogeneities have to be characterized, mathematically described and incorporated into the program algorithm. Reliability effects will create overlapping distribution tails over lifetime, which can be corrected by a combination out of adaptive read and error correction techniques dynamically optimized for the dominant overlapping effect.

Key success factors of floating gate MLC NAND are the shrink roadmap, the excellent and predictable operation performance, the straightforward algorithm approach for a floating gate cell combined with pure voltage operation modes for read, program and erase.

- The long program operation time focuses the MLC NAND design on highest parallelism.
- The read and a program throughput increase outperform other flash technologies.
- Failure tolerant specification and adaptive error correction concepts are mandatory.

The 4-bit per cell MLC NAND can be judged as one of the most cost and energy competitive non-volatile memory concepts as of today.

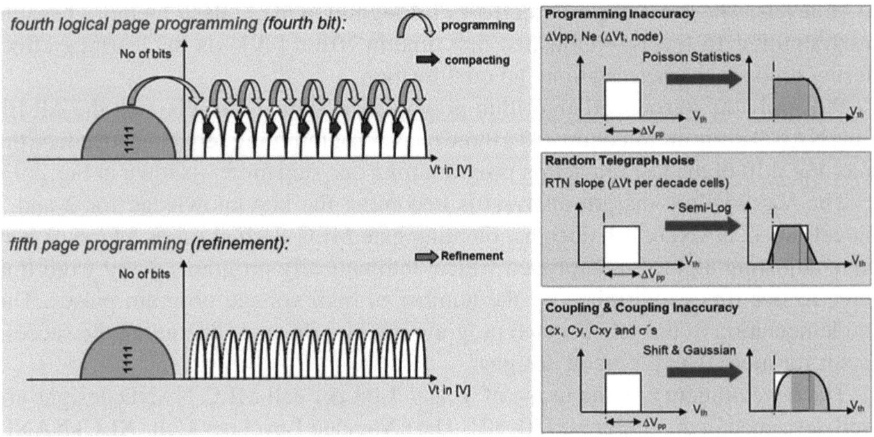

Fig. 2.94 NAND V_{th} window for 4-bit/cell and analysis of distribution shapes

2.7.5 Multi-Level and Multi-Bit Summary

Multi-Bit and Multi-level Cell flash memory concepts and their operation principles are introduced.

The multi-bit flash memory is characterized by a simple architecture, but the charge trapping storage element introduces reliability issues impacting the threshold voltage operation window and CHE programming limits the program data throughput increase.

The multi-level per cell NAND flash memory is the dominant non-volatile memory architecture. The e**X**tended **L**evel per **C**ell program algorithm incorporates more than 90 % of all array interferences. Advanced All Bit Line NAND array concepts compensate technology variation enforced by lithography effects.

For all MLC NAND memories with more than 2 bit per cell the naming **XLC** NAND is used in this work, in case no specific separation between 3 and 4 bit per cell techniques is required.

2.8 Summary of NVM Fundamentals

The memory cell and the physics of the storage element, the memory array and the select and bit and word line construction define together the performance of the memory array architecture.

The introduced cell and memory array combinations are belonging on a higher abstraction level to two main classes of memory array architectures:

- A memory array in which all bit lines or every second bit line is connected to a sensing circuit.
  - Multiple cell operation principle
  - Voltage driven operation
  - Contact less array core

    Memory performance characterized by the page size of the array.
- A memory array in which the access to the cell is optimized and a limited number of cells are connected to the sensing circuit.
  - Single cell operation principle
  - Parasitic leakage currents within the array core

    Memory performance characterized by the number of sensing circuits.

The performance parameters improvement of different memory array architectures along the shrink roadmap is the focus of the second part of this work. The interactions between obvious and hidden memory array weaknesses and the application requirements are defining the success of a memory based system.

Cost per bit is the main decision parameter for high volume semiconductor memories. Non-volatile memories have the capability to store more bits per cell and reduce the cost per bit further down. Multi-Level Cell (MLC) and Multi-Bit Cell (MBC) memory concepts are introduced for floating gate and charge trap cell based flash memories. The required additional design (sense circuits) and algorithm effort introduce a first impression of the complexity of multi-bit per cell memories.

Memory performance and cost parameter as well as flash memory reliability and durability parameter are introduced and applied to the two main memory array architecture classes combined with MLC and MBC techniques.

Performance and durability parameter for volatile and non-volatile memories are introduced in the next chapter.

References

1. K. Kahng, S. Sze, A floating gate and its application to memory devices. IEEE Trans. Electron Dev. Bd. **46**, 629 (1967)
2. F. Masuoka, M. Asano, H. Iwahashi, T. Komuro, S. Tanaka, A new Flash E2PROM cell using triple polysilicon technology, in *IEEE IEDM Techn Digest*, pp. 464–467, Washington, 1984.
3. M. Bauer, R. Alexis, G. Atwood, B. Baltar, A. Fazio, K. Frary, M. Hensel, M. Ishac, J. Javanifard, M. Landgraf, D. Leak, K. Loe, D. Mills, P. Ruby, R. Rozman, S. Sweha, S. Talreja, K. Wojciechowski, A multilevel-cell 32Mb flash memory, in *ISSCC Digest of Technical Papers*, pp. 132–133, San Francisco, 1995.
4. K.-D. Suh, B.-H. Suh, Y.-H. Um, J.-K. Kim, Y.-J. Choi, Y.-N. Koh, S.-S. Lee, S.-C. Kwon, B.-S. Choi, J.-S. Yum, J.-H. Choi, J.-R. Kim, H.-K. Lim, A 3.3V 32Mb NAND flash memory with incremental step pulse programming scheme, in *ISSCC Digest of Technical Papers*, pp. 128–129, San Francisco, 1995.
5. P. Xuan, M. She, B. Harteneck, A. Liddle, J. Bokor, T.-J. King, FinFET SONOS flash memory for embedded applications, in *IEDM Technical Digest*, pp. 609–612, Washington, 2003.
6. M. Stadele, R. Luyken, M. Roosz, M. Specht, W. Rosner, L. Dreeskornfeld, J. Hartwich, F. Hofmann, J. Kretz, E. Landgraf, L. Risch, A comprehensive study of corner effects in tri-gate transistors, in Solid-State Device Research conference, Procedings ESSDERC, pp. 165–168, 2004.
7. G.E. Moore, Cramming more components onto integrated circuits. Electron. Bd. **38**(8), 114–117 (1965)
8. S. Mukherjee, T. Chang, R. Pang, M. Knecht, D. Hu, A single transistor EEPROM cell and its implementation in a 512 K CMOS EEPROM, in *IEDM Technical Digest*, pp. 616–619, Washington, 1985.
9. J. Lee, V. Dham, Design considerations for scaling FLOTOX E2PROM cell, in *IEDM Technical Digest*, pp. 589–592, Washington, 1983.
10. J.E. Brewer, G. Manzur, *Nonvolatile Memory Technologies with Emphasis on Flash: a Comprehensive Guide to Understanding and Using NVM Devices, Auflage*, 1st edn. (Wiley, Hoboken, 2008)
11. R.J. Baker, in *Memory circuits-pheripheral circuits*. CMOS Circuit Design, Layout, and Simulation (IEEE Press, Wiley Interscience, Piscataway, 2005), pp. 448–456.
12. B. Prince, *Semiconductor Memories: A Handbook of Design, Manufacture, and Application*, 2nd edn. (Wiley, New York, 1991)
13. E. Snow, Fowler-Nordheim tunneling in SiO2 films. Solid State Commun. **5**, 813–815 (1967)
14. C. Friederich, Multi level programming scheme with reduced cross coupling, in *Control of harmful effects in program operation of NAND flash* (Aachen, Shaker, 2011), pp. 62–66
15. R. Bez, D. Cantarelli und S. S. The channel hot electron programming of a floating gate MOSFET: an analytical study, in *12th Nonvolatile Semiconductor Memory Workshop*, Monterey, California, 1992.

16. M.-S. Liang, Memory cell having hot-hole injection erase mode, USA Patent 4472491, 1985.
17. T. Chan, J. Chen, P. Ko, C. Hu, The impact of gate-induced drain leakage current on MOSFET scaling, in *IEDM Technical Digist*, pp. 718–721, Washington, 1987.
18. P. Cappelletti, C. Golla, P. Olivio, E. Zanoni, Physical aspects of cell operation and reliability, in *Flash Memories* (Kluwer-Academic Publisher, Dordrecht, 1999), p. 171
19. S. Sze, *Physics of Semiconductor Devices* (Wiley, New York, 1981)
20. H. Wegener, A. Lincoln, H. Pao, M. O'Connell, R. Oleksiak, H. Lawrence, The variable threshold transistor, a new electrically-alterable, non-destructive read-only storage device, in *IEDM, Technical Digest*, vol. 13, pp. 70–70, Washington, 1967.
21. P. Chen, Threshold-alterable Si-gate MOS devices, transactions on electron devices. IEEE **24**(5), 584–586 (1977)
22. F. MIENO, SANOS Memory Cell Structure, United States Patent US 2010/0001353 A1, 07 Jan 2010.
23. Y. Park, J. Choi, C. Kang, C. Lee, Y. Shin, B. Choi, J. Kim, S. Jeon, J. Sel, J. Park, K. Choi, T. Yoo, J. Sim, K. Kim, Highly manufacturable 32Gb multi-level NAND flash memory with 0.0098 μm2 Cell size using TANOS(Si-Oxide- Al2O3-TaN) cell technology, in *IEDM, Technical Digest*, pp. 1–4, Washington, 2006.
24. C. Kang, J. Choi, J. Sim, C. Lee, Y. Shin, J. Park, J. Sel, S. Jeon, Y. Park, K. Kim, Effects of lateral charge spreading on the reliability of TANOS (TaN/AlO/SiN/Oxide/Si) NAND flash memory, in *IEEE Reliability physics symposium, proceedings. 45th*, pp. 167–170, Phoenix, AZ, 2007.
25. C. Friederich, M. Specht, T. Lutz, F. Hofinann, L. Dreeskornfeld, W. Weber, J. Kretz, T. Melde, W. Rosner, E. Landgraf, J. Hartwich, M. Stadele, L. Risch, D. Richter, Multi-level p+ tri-gate SONOS NAND string arrays, in *IEDM Technical Digest*, pp. 1–4, Washington, 2006.
26. M. Specht, U. Dorda, L. Dreeskornfeld, J. Kretz, F. Hofinann, M. Stadele, R. Luyken, W. Rosner, H. Reisinger, E. Landgraf, T. Schulz, J. Hartwich, R. Kommling, L. Risch, 20 nm tri-gate SONOS memory cells with multi-level operation, in *IEDM Technical Digest*, pp. 1083–1085, Washington, 2004.
27. Y.-H. Shih, H.-T. Lue, K.-Y. Hsieh, R. Liu, C.-Y. Lu, A novel 2-bit/cell nitride storage flash memory with greater than 1M P/E-cycle endurance, in *IEDM Technical Digest*, pp. 881–884, Washington, 13–15 Dec 2004.
28. E. Stein, V. Kamienski, M. Isler, T. Mikolajick, C. Ludwig, N. Schulze, N. Nagel, S. Riedel, J. Wilier, K.-H. Kusters, An Overview on Twin Flash Technology, in *Non-Volatile Memory Technology Symposium*, Dallas, 2005.
29. A. Shappir, E. Lusky, G. Cohen, B. Eitan, NROM Window Sensing for 2 and 4-bits per cell Products, in *NVSMWS* (Monterey, CA, 2006)
30. M. Janai, Threshold voltage fluctuations in localized charge-trapping nonvolatile memory devices. IEEE Trans. Electron Dev. **59**, 596–601 (2012)
31. J.-G. Yun, I.H. Park, S. Cho, J.H. Lee, D.-H. Kim, G.S. Lee, Y. Kim, J.D. Lee, B.-G. Par, A 2-bit recessed channel nonvolatile memory device with a lifted charge-trapping node. IEEE Trans. Nanotechnol. **8**, 111–115 (2009)
32. L. Hai, M. Takahashi, S. Sakai, Downsizing of ferroelectric-gate-field-effect-transistors for ferroelectric-NAND flash memory, in *3rd IEEE International Memory Workshop (IMW)*, pp. 1–4, Monterey, CA, June 2011.
33. J. Müller, E. Yurchuk, T. Schlosser, J. Paul, R. Hoffmann, S. Muller, D. Martin, S. Slesazeck, P. Polakowski, J. Sundqvist, M. Czernohorsky, K. Seidel, P. Kucher, R. Boschke, M. Trentzsch, K. Gebauer, U. Schroder, T. Mikolajick, Ferroelectricity in HfO2 enables nonvolatile data storage in 28 nm HKMG," in *Symposium on VLSI Technology (VLSIT)*, 2012, Honolulu, 2012.
34. M. Durlam, P.J. Naji, A. Omair, M. DeHerrera, J. Calder, J.M. Slaughter, B.N. Engel, A 1-Mbit MRAM based on 1T1MTJ bit cell integrated with copper interconnects. IEEE J. Solid-State Circuits **38**, 769–773 (2003)
35. Y. Huai, Y. Zhou, I. Tudosa, R. Malmhall, R. Ranjan, J. Zhang, Progress and outlook for STT-MRAM, in *International Conference on Computer-Aided Design (ICCAD), IEEE/ACM*, pp. 235, San Jose, 2011.

36. S. Lai, Current status of the phase change memory and its future, in *IEDM Technical Digest*, pp. 10.1.1-10.1.4, Washington, 2003.
37. K. Byeungchul, S. Yoonjong, A. Sujin, K. Younseon, J. Hoon, A. Dongho, N. Seokwoo, J. Gitae, C. Chilhee, Current status and future prospect of Phase Change Memory, in *IEEE 9th International Conference on ASIC (ASICON)*, 2011, pp. 279–282, Xiamen, 2011.
38. F. Bedeschi, R. Fackenthal, C. Resta, E. Donze, M. Jagasivamani, E. Buda, F. Pellizzer, D. Chow, A. Cabrini, G. Calvi, R. Faravelli, A. Fantini, G. Torelli, D. Mills, R. Gastaldi, G. Casagrande, A Bipolar-selected Phase Change Memory featuring Multi-Level Cell Storage. IEEE J. Solid-State Circuits Bd. 1(1, Jan 2001), 217–227 (2009).
39. M. Specht, R. Kommling, F. Hofmann, V. Klandzievski, L. Dreeskornfeld, W. Weber, J. Kretz, E. Landgraf, T. Schulz, J. Hartwich, W. Rosner, M. Stadele, R. Luyken, H. Reisinger, A. Graham, E. Hartmann, L. Risch, Novel dual bit tri-gate charge trapping memory devices. IEEE Electron Device Lett. **25**, 810–812 (2004)
40. A. Fazio, Non-volatile memory technology: present and future trends, in *ISSCC, Tutorial F1*, San Francisco, 2007.
41. J. Javanifard, T. Tanadi, H. Giduturi, K. Loe, R. Melcher, S. Khabiri, N. Hendrickson, A. Proescholdt, D. Ward, M. Taylor, A 45 nm Self-Aligned-Contact Process 1 Gb NOR Flash with 5 MB/s Program Speed, in *ISSCC Digest of Technical Papers*, pp. 424–426, San Francisco, 2008.
42. B. Le, M. Achter, C. G. Chng, X. Guo, L. Cleveland, P.-L. Chen, M. Van Buskirk, R. Dutton, Virtual-ground sensing techniqus for a 49ns/200MHz access time 1.8V 256Mb 2-bit-per-cell flash memory. IEEE J. Solid-State Circuits 2014–2023 (2004).
43. R. Koval, V. Bhachawat, C. Chang, M. Hajra, D. Kencke, Y. Kim, C. Kuo, T. Parent, M. Wei, B. Woo, A. Fazio, Flash ETOX(TM) virtual ground architecture: a future scaling directions, in *VLSI Technology, Digest of Technical Papers*, pp. 204–205, Kyoto, 2005.
44. C. Yeh, W. Tsai, T. Lu, Y. Liao, N. Zous, H. Chen, T. Wang, W. Ting, J. Ku, C.-Y. Lu, Reliability and device scaling challenges of trapping charge flash Mmemories, in *Proceedings of 11th IPFA*, pp. 247–250, Taiwan, 2004.
45. N. Ito, Y. Yamauchi, N. Ueda, K. Yamamoto, Y. Sugita, T. Mineyama, A. Ishihama, K. Moritani, A novel program and read architecture for contact-less virtual ground NOR flash memory for high density Application, in *Symposium on VLSI Circuits Digist of Technical Paper*, pp. 116–117, Honolulu, 2006.
46. K.-R. Han, H.-A.-R. Jung, J.-H.L. Lee, Band-to-Band Hot-hole Erase Characterisitcs of 2-Bit/cell NOR type SONOS flash memory cell with spacer-type storage node on recessed channel structure. Japan. J. Appl. Phys. **33**, 798–800 (2007)
47. Y. Sofer, NROM Memory Design, in *ISSCC 2007-Non-Volatile Memory Circuit Design and Technology*, San Francisco, 2007.
48. M. Momodomi, Y. Itoh, R. Shirota, Y. Iwata, R. Nakayama, R. Kirisawa, T. Tanaka, S. Aritome, T. Endoh, K. Ohuchi, F. Masuoka, An experimenatal 4-Mbit CMOS EEPROM with a NAND-structured cell. IEEE J. Solid-State Circuits **24**, 1238–1243 (1989)
49. D. Richter, *Vorlesung Halbleiter Bauelemente–Nichtflüchtige Speicher Titel: NVM Shrink Roadmap 3–4bit/cell NAND–Key Performance Indicator, München: TUM* (Lehrstuhl für Technische Elektronik, Fakultät für Elektro- und Informationstechnik, 2008)
50. T.-S. Jung, Y.-J. Choi, K.-D. Suh, B.-H. Suh, J.-K. Kim, Y.-H. Lim, Y.-N. Koh, J.-W. Park, K.-J. Lee, J.-H. Park, K.-T. Park, J.-R. Kim, J.-H. Yi, H.-K. Lim, A 117-mm2 3,3V only 128 Mb multilevel NAND flash memory for mass storage application. IEEE J. Solid-State Circuits **31**, 1575–1583 (1996)
51. T. Tanzawa, T. Tanaka, K. Takeuchi, H. Nakamura, Circuit techniques for a 1,8V only NAND flash memory. IEEE J. Solid-State Circuits **37**, 84–89 (2002)
52. N. Fujita, N. Tokiwa, Y. Shindo, T. Edahiro, T. Kamei, H. Nasu, M. Iwai, K. Kato, Y. Fukuda, N. Kanagawa, N. Abiko, M. Matsumoto, T. Himeno, T. Hashimoto, Y.-C. Liu, H. Chibvongodze, T. Hori, M. Sakai, A 113mm2 32Gb 3b/cell NAND flash memory, in *ISSCC-Digest of Technical Papers*, pp. 242–233, San Francisco, 2009.
53. E. Harari, NAND at center stage, in *Flash Memory Summit*, Santa Clara, 8 August 2007.

54. M. Bauer, NOR flash memory design-non-volatile memories technology and design, in *ISSCC 2004 Memory Circuit Design Forum*, San Francisico, 2004.
55. K. Tedrow, NOR flash memory design, in *ISSCC 2007-Non-Volatile Memory Design Forum*, San Francisco, 2007.
56. C. Friederich, in *Program and erase of NAND memory arrays*, eds. by R. Micheloni, L. Crippa, A. Marelli, Inside NAND Flash Memories (Springer, Dordrecht, 2010), pp. 75–76.
57. T. Tanaka, A quick Intelligent Page-Programming Architecture and a shileded bitline sensing method for 3 V only NAND Flash Memory. IEEE J. Solid-State Circuits **29**, 1366–1373 (1994)
58. F. Koichi, Y. Watanabe, E. Makino, K. Kawakami, J. Sato, T. Takagiwa, N. Kanagawa, H. Shiga, N. Tokiwa, Y. Shindo, T. Edahiro, T. Ogawa, M. Iwai, O. Nagao, J. Musha, T. Minamoto, K. Yanagidaira, Y. Suzuki, D. Nakamura, Y. Hosomura, A 151mm2 64Gb MLC NAND flash memory in 24 nm CMOS technology, in ISSCC, Digest of Technical Papers, pp. 198–199, San Francisco, 2011.
59. M. Bauer, Multi-level Cell Design for flash memory, in *ISCC 2006 Tutorial T5: MLC Design for Flash Memory* (Feb, San Francisco, 2006).
60. J. Lee, S.-S. Lee, O.-S. Kwon, K.-H. Lee, D.-S. Byeon, I.-Y. Kim, K.-H. Lee, Y.-H. Lim, B.-S. Choi, J.-S. Lee, W.-C. Shin, J.-H. Choi, K.-D. Suh, A 90-nm CMOS 1.8V 2Gb NAND flash memory for Mass storage application. IEEE J Solid-State Circuits 38(11), 1934–1942 2003 (Bde. %1 von %2).
61. T. Tanaka, NAND flash design, in *ISSCC, Non-Volatile Memory Circuit and Technology Tutorial F1*, San Francisco, 2007.
62. T. Hara, K. Fukuda, K. Kanazawa, N. Shibata, K. Hosono, H. Maejima, M. Nakagawa, T. Abe, M. Kojima, M. Fujiu, Y. Takeuchi, K. Amemiya, M. Morooka, T. Kamei, H. Nasu, C.-M. Wang, K. Sakurai, N. Tokiwa, H. Waki, T. Maruyama, S. Yoshikawa, A 146-mm2 8-Gb Multi-level NAND flash memory with 70-nm CMOS technology. IEEE J. Solid-State Circuits **41**, 161–169 (2006)
63. R. Cernea, D. Lee, M. Mofidi, E. Chang, W. Y. Chien, L. Goh, Y. Fong, J. Yuan, G. Samachisa,D. Guterman, S. Mehrotra, K. Sato, H. Onishi, K. Ueda, F. Noro, K. Mijamoto, M. Morita, K. Umeda, K. Kubo, A 34Mb 3.3V Serial Flash EEPROM for solid-state disk applications, in *ISSCC Digest of Tech Papers*, pp. 126–127, San Francisco, 1995.
64. R. Cernea, o. Pham, F. Moogat, S. Chan, B. Le, Y. Li, S. Tsao, T.-Y. Tseng, K. Nguyen, J. Li, J. Hu, J. Park, C. Hsu, F. Zhang, T. Kamei, H. Nasu, P. Kliza, K. Htoo, J. Lutze, und Y. Dong, A 34MB/s-program-throughput 16Gb MLC NAND with all-bitline architecture in 56nm, in *ISSCC Digest of technical Papers*, pp. 420–424, San Francisco, 2008.
65. J.F. Dickson, On-chip high-voltage generation in NMOS integrated circuits using an improved voltage multiplier technique. IEEE J. Solid-State Circuits **11**, 374–378 (1976)
66. C.-C. Wang, J.-C. Wu, Efficiency improvement in charge pump circuits. IEEE J. Solid-State Circuits **32**(6), 852–860 (1998)
67. T. Tanzawa, T. Tanaka, A stable programming pulse generator for single power supply flash memories. IEEE J. Solid-State Circuits **32**(6), 845–851 (1997)
68. F. Pan, T. Samaddar, Charge pump circuit design, 1st edn. (Mcgraw-Hill Professional, New York, 2006).
69. K. Kim, G. Jeong, Memory technologies in the nano-era: challenges and opportunities, in *ISSCC Digest of Technical Papers*, pp. 576–578, San Francsico, 2005.
70. K. Kim, J. Coi, Future outlook of NAND flash technology for 40 nm node and beyond, in *IEEE NVSMW*, pp. 9–11, Monterey, CA, 2006.
71. T. Tanaka, A quick intelligent page-programming architecture and a shielded bitline sensing method for 3 V-Only NAND flash memory. IEEE JSSC Bd. **29**(11), 1366 (1994)
72. C. Kim, J. Ryu, T. Lee, H. Kim, J. Lim, J. Jeong, S. Seo, H. Jeon, B. Kim, I. Lee, D. Lee, P. Kwak, S. Cho, Y. Yim, C. Cho, W. Jeong, K. Park, J.-M. Hạn, D. Song, K. Kyung, A 21 nm High Performance 64 Gb MLC NAND flash memory With 400 MB/s asynchronous toggle DDR interface. IEEE J. Solid-State Circuits **47**, 981–989 (2012)
73. J. Han, B. Lee, J. Han, W. Kwon, C. Chang, S. Sim, C. Park, K. Kim, A critical failure source in 65nm-MLC NOR flash memory incorporating co-salicidation process, in *Integrated Reliability Workshop Final Report*, pp. 80–82, South Lake Tahoe, CA, Sep 2006.

74. A. Bergemont, M.-H. Chi, H. Haggag, Low voltage NVGTM: a new high performance 3 V/5 V flash technology for portable computing and telecommunications applications. IEEE Trans. Electron Devices **43**(9), 1510–1517 (1996)
75. N. Ito, Y. Yamauchi, N. Ueda, K. Yamamoto, Y. Sugita, T. Mineyama, A. Ishihama, K. Moritani, A novel program and read architecture for contact-less virtual ground NOR flash memory for high density application, in *Symposium on VLSI Circuits, Digest of Technical Papers*, pp. 116–117, Honolul, HI, 2006.
76. N. Zous, M. Lee, W. Tsai, A. Kuo, L. Huang, T. Lu, C. Liu, T. Wang, W. Lu, W. Ting, J. Ku, C.-Y. Lu, Lateral migration of trapped holes in a nitride storage flash memory cell and its qualification methodology. IEEE Electron Device Lett. **25**, 649–651 (2004)
77. Y. Roizin, Extending endurance of NROM memories to over 1 million program/erase cycles, in *Proceedings of 21st Non-Volatile Semiconductor Memory, Workshop*, pp. 74–75, Feb 2006.
78. T. Kuo, N. Yang, N. Leong, E. Wang, F. Lai, A. Lee, H. Chen, S. Chandra, Y. Wu, T. Akaogi, A. Melik-Martirosian, A. Pourkeramati, J. Thomas, M. VanBuskirk, Design of 90nm 1Gb ORNAND(TM) flash memory with mirrorBit(TM) technology, in *Symposium on VLSI Circuits Digest of Technical Papers*, pp. 114–115, Honolulu, HI, 2006.
79. N. Shibata, H. Maejima, K. Isobe, K. Iwasa, M. Nakagawa, M. Fujiu, T. Shimizu, M. Honma, S. Hoshi, T. Kawaai, K. Kanebako, S. Yoshikawa, H. Tabata, A. Inoue, T. Takahashi, T. Shano, Y. Komatsu, K. Nagaba, M. Kosakai, N. Motohashi, A 70 nm 16 Gb 16-level-cell NAND flash memory. IEEE J. Solid-State Circuits **43**, 929–937 (April 2008)
80. Y. Li, S. Lee, Y. Fong, F. Pan, T.-C. Kuo, P. J., T. Samaddar, H. T. Nguyen, M. Mui, K. Htoo, T. Kamei, M. Higashitani, E. Yero, G. Kwon, P. Kliza, J. Wan, T. Kaneko, H. Maejima, H. Shiga, M. Hamada und N. Fujita, A 16 Gb 3-bit per cell (X3) NAND flash memory on 56 nm technology with 8 MB/s write rate. IEEE J. Solid-State Circuits 44(1), 195–207 (2009) (Bd. 44).
81. B.T. Park, J.H. Song, E.S. Cho, S.W. Hong, J.Y. Kim, Y.J. Choi, Y.S. Kim, S.J. Lee, C.K. Lee, D.W. Kang, D.J. Lee, B.T. Kim, Y.J. Choi, W.K. Lee, J.-H. Choi, K.-D. Su, 32nm 3-bit 32Gb NAND flash memory with DPT (double patterning technology) process for mass production, in *VLSI Technology (VLSIT)*, pp. 125–126, 2010.
82. N. Giovanni, A 3bit/Cell 32Gb NAND Flash Memory at 34nm with 6MB/s Program Throughput and with Dynamic 2b/Cell Blocks Configuration Mode for a Program Throughput Increase up to 13MB/s, San Francisco, 2010.

Chapter 3
Performance Figures of Non-Volatile Memories

The selection of one memory architecture during the system development process is based on an assessment of cost per bit, scalability, and power efficiency and performance values.

Performance parameters of flash memories are typical values. They are specified within a pre-defined range. Flash cell characteristics and their statistical behaviour have to be taken into account. The description of the non-deterministic behaviour of non-volatile memories is one subject of this chapter.

- The definition of latency, cycle time and read and write data bandwidth is introduced for volatile and non-volatile memories to highlight obvious and hidden differences.
- The write operation of a non-volatile memory can be physically one write sequence or a combination out of two operations—for flash memories erase first and program afterwards.

Two performance parameters are used for a memory performance overview. The program and write data bandwidth/throughput are shown in Fig. 3.1 to illustrate the performance increase over time.

An impact analysis of flash performance parameters on the achievable system performance is done. The dependencies between performance values and durability values are introduced and exemplarily calculated. Durability optimization strategies are introduced.

A non-volatile memory cell is changing its behaviour over lifetime. Due to this fact all flash performance parameters can be dependent on reliability stress parameter. The expected performance degradation or improvement over life time will be discussed in the reliability chapter.

D. Richter, *Flash Memories*, Springer Series in Advanced Microelectronics 40,
DOI: 10.1007/978-94-007-6082-0_3,

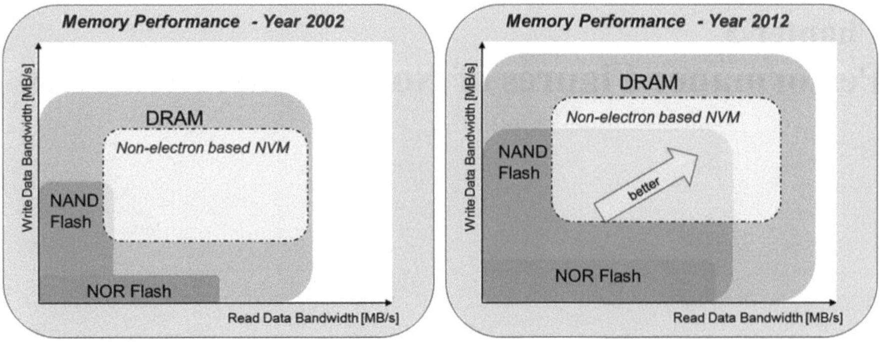

Fig. 3.1 Memory performance overview

3.1 Memory Performance Parameter Definition

The definition of performance parameters is developed for **R**andom **A**ccess **M**emories in general. A performance capability assessment of a system is based on the parameter set known from volatile memories like SRAM and DRAM.

The details behind relevant performance parameters are the basis for the performance indicator methodology developed in this work. Relevant memory performance parameters are introduced in this chapter and the specific behavior of flash memories is illustrated.

3.1.1 Read Performance Parameters: Definition for Volatile Memories

Random Access Memories are characterized by two main operations—read and write—including all corresponding timing and address parameters.

The read operation transports data bits from the target address within the memory array to the data output pins of the memory. The memory design translates a logical address into a physical location—reading from the target row (x) and column (y) address. The access time depends on the physical status of the accessed row. Two read access parameters are defined for memories:

- **Random Read Access latency**—row and column access time.
  - The first parameter RAS—Row Access Strobe—defines the access time to a new row.

    The timing is shown for a DRAM access in Fig. 3.2.

    The row access time is in the range of 50–70 ns;

    The row access time for a DRAM is:

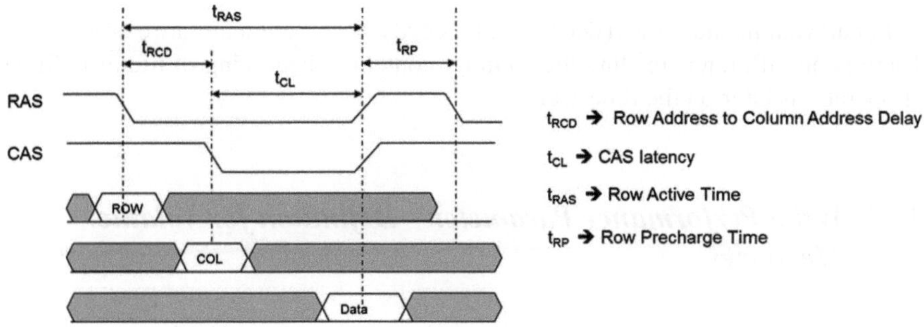

Fig. 3.2 DRAM READ access timing cycle—random read access latency

$$Row\ Access\ Time\,[ns]: \quad t_{RAS} = t_{RCD} + t_{CL}$$

- **Sequential Read Access Latency**—column access time.
  - The second parameter CAS—Column Access Strobe—defines the access time to a new column address within an already open row.
    Sequential burst access is in the range of 5–10 ns
    The parameter is also called CAS Latency (CL) shown in Fig. 3.2.

$$Column\ Access\ Time\,[ns]: \quad t_{CAS} = t_{CL}$$

- The **Read Data Bandwidth** is defined for a sequential burst access to consecutive addresses.
  - The data bandwidth or data throughput or data rate (DR) is based on the bus width of the interface and the clock rate which is typically defined in MB/s (10^6 bytes per second). The burst data rate for a column access is based on these parameters:

$$Burst\ Data\ Rate\left[MB/s\right]: \quad DR_{RD\_Burst} = \frac{Bus\_Width_{Data}}{t_{CL}}$$

  - The read data bandwidth including the row access timing is calculated:

$$Read\ Data\ Bandwidth\left[MB/s\right]: \quad BW_{RD\_DRAM} = \frac{N * Bus\_Width_{Data}}{\left(t_{RCD} + \left[N * t_{CL}\right] + t_{RP}\right)}$$

The burst data rate can be maintained continuously if an internal cache or a memory bank structure is available to hide the row access time to the next row address during the time transferring data.

The timing parameters t_{CL}, t_{RCD}, t_{RP} and t_{RAS} for a DRAM read operation are shown in Fig. 3.2.

For all volatile memories (DRAM and SRAM) these parameters are deterministic. There is no difference reading the memory content at beginning of life or millions of operations later on the time scale.

3.1.2 Write Performance Parameter—Definition for Volatile Memories

The write operation writes physically data to the target row (x) and column (y) address within the array. Data and corresponding address are transported from input pins to the address decoder and write amplifier.

The logical write command can address an already open page within the memory—write burst mode—or a not addressed page. The write operation stores the data—physically putting charge in a DRAM capacitor or flipping a SRAM cell—in the memory cell and closes the corresponding page—finishing the write operation on the corresponding row address. The write cycle has a similar timing compared to the read cycle except that the write data has to be applied together with the addresses shown in Fig. 3.3.

Only one write access parameter is defined for random data and addresses:

- Write cycle time—row and column access time and time to change the cell content
  - Row and column access time to a random address are part of the write cycle. After the correct row address is selected the column address is used to select the correct bit line and data are transported to the corresponding local sense/write amplifier. During the write recovery (t_{WR}) time the local bit lines are changed to the correct voltage value linked to the corresponding write data. The pre-charge cycle of the row (t_{RP}) finalizes the write operation and puts the memory in the initial state.
  - The physical change of the cell content is done during write recovery and pre-charge.
  - The write cycle time is calculated:

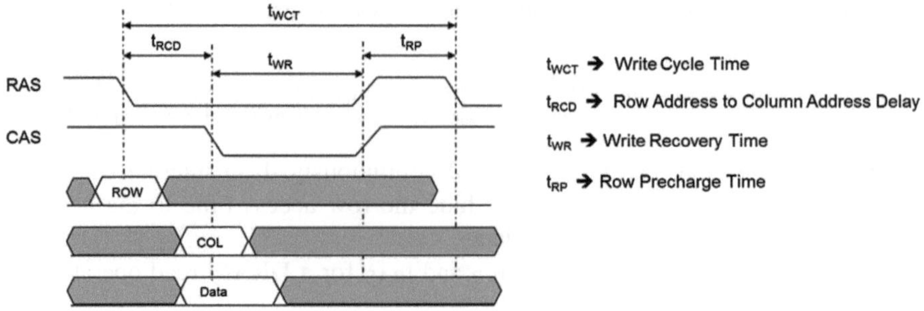

Fig. 3.3 DRAM write timing cycle—WRITE cycle time

$$Write\ Cycle\ Time\,[ns]:\quad t_{WCT} = t_{RCD} + t_{WR} + t_{RP}$$

- **Write data bandwidth** is defined as the write operation data transfer rate to consecutive column addresses.
  - Write data are transported to internal sense amplifier latches, so that this amplifier is forced to apply the desired low or high voltage state on the selected bit line pair.
  - The write burst operation writes continuously data to a complete row.
  - The write data bandwidth can be calculated based on the bus data width and the specified timing shown in the write cycle:

$$Write\ Data\ Bandwidth\,[MB/s]:\quad BW_{WR\_DRAM} = \frac{N * Bus\_Width_{Data}}{(t_{RCD} + [N * t_{WR}] + t_{RP})}$$

The write data bandwidth can be maintained continuously in page mode operation, by using multi-bank memory array architecture to hide first asynchronous row access and row pre-charge time.

For all volatile memories (DRAM and SRAM) the write parameters are **deterministic**. There is no difference writing data to a memory at beginning of life or millions of operations later.

3.1.3 Sequential Data Throughput: Read and Write Bandwidth

The data throughput is defined as the average rate of successfully delivered data bits per second on a data channel. For data storage systems the data throughput is also called the data transfer rate.

The data throughput for a solid state memory depends on or is based on

- read and write data bandwidth of the memory architecture,
- data packages size, and
- sequences of read and write commands.

The data throughput defines how many data packages can be written or read within a given time. The time penalty switching between read and write operations is an important parameter.

The data throughput or the data bandwidth is the key performance parameter used in this work to analyse different memory architecture in depth. Therefore this overview for volatile memories is included here and can be used to compare with the definition for non-volatile memories.

The **sequential data throughput** is defined as the measure to identify the capability of a memory architecture to deliver in a sustained mode data packages in one direction either read or write.

The **read data bandwidth** is the transfer rate on the investigated read channel and the **write data bandwidth** on the write channel.

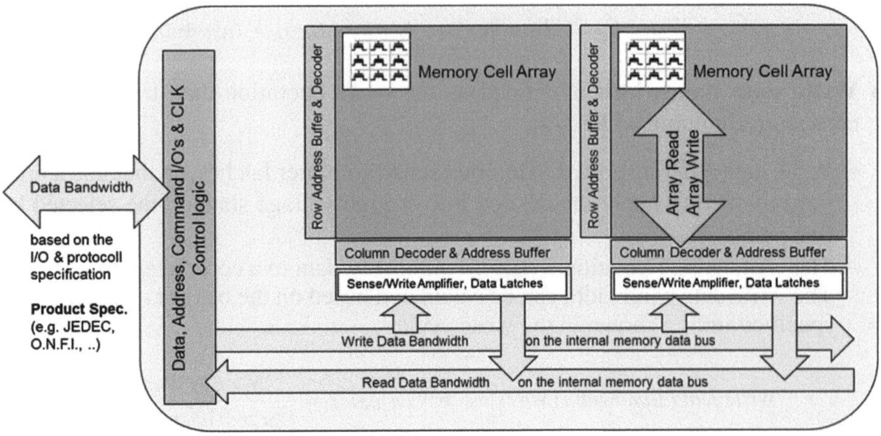

Fig. 3.4 Data bandwidth and data transfer rate on different channels for a generic memory

The read and the write channel have to be divided once more as shown in Fig. 3.4.

For memory architectures with a high number of parallel sense amplifiers we differentiate between these two data transfer channels:

- Memory array data transfer channel—data are moved from cells or into cells.
- Internal memory data bus transfer channel—data are moved from or to data latches.

Performance parameters for volatile random access memories are characterized by:

- Sequential **Read** and **Write Data Throughput** have roughly the same values.
- Switching between read and write is only limited by the data latency of the last read operation, only if the same memory array (memory bank) is assessed by both operations.
- An endless write operation—write data throughput—is allowed and overwrites stored data automatically without any restriction.

Random access memories are optimized for data throughput and for read latency. The read latency is one of the main key performance parameters and determines how fast random data can be accessed and how fast the switching between read and write operation can be maintained.

3.1.4 Random Data Throughput

The random data throughput is the random data transfer rate achievable for randomly distributed data packages transferred in both directions so a mixing of read and write operation. This parameter describes the ideal application requirement for most applications.

Memories enabling the intelligent mixing of read and write supported by special commands like read modify write could be the preferred choice. The random data throughput for volatile memories is limited by the design concept (especially for read latency and write cache capability), the memory array segmentation and the interface architecture.

A worst case calculation for random data throughput is made for flash memories in Sect. 3.2.

3.1.5 Refresh of Memories: First Non-Deterministic Behaviour

The cost effective storage of one bit in a capacitance storage element controlled by a transistor—the Dynamic RAM array architecture—requires a periodical refresh of data stored within the capacitor.

The refresh logic rewrites periodically all data within the memory array, which are not accessed by the application. Row by row is addressed and opened up. The local sense amplifier senses and amplifies the signal on the bit lines and writes the data values back to each cell belonging to this row. This operation has to be repeated typically every 64 ms for each row in a DRAM.

- A DRAM with 8096 rows would require an interrupt of the system access every 7.905 μs executing the refresh of one of the rows, if the time would be equally distributed.
- The DRAM refresh is typically supported by a memory internal refresh counter.

The DRAM refresh enforces a hidden non-deterministic system task responsible to ensure enough spare time to execute fault free DRAM refresh. Large memory systems based on DRAMs consume a lot of power for this simple refresh, which can be saved by non-volatile memories.

The refresh operation turns a DRAM into a very reliable memory assessed for data retention. The retention of each DRAM cell can be tested with a guard band of a factor of two to four. A comparable retention test with guard band in volume production under worst case environmental conditions is not doable per design for SRAMs. Non-volatile memories are tested with high temperature bake procedures; some physical failure modes are covered but the coverage is based on statistical assumptions (not every single defect can be accelerated and detected) [1].

3.1.6 Read Performance Parameters for Non-Volatile Memories

Non-**V**olatile **M**emories are classified in this work into electron based—flash memories—and non-electron based non-volatile **R**andom **A**ccess **M**emories—Fe**RAM** and **MRAM**. Non-volatile RAM-like memories are more similar to volatile memories. Therefore only flash memories are the subject of this non-deterministic performance excursus.

The read performance parameters are the same as for volatile memories. The access to a certain logical address depends on the physical status of the accessed row. The read access can be executed faster in case data are already latched in the sense amplifier belonging to the target row address. Two read access parameters are defined for non-volatile memories:

- **Random Read Access Latency**—row and column access time and sensing time.
  - The first random read access time depends on array and sensing architecture and in case of MLC on the coding scheme.
  - The first read access is often utilized to adapt internal voltage and timing parameter to improve the read accuracy or to find the read Vth window.
  - The first random read latency can differ depending on the operation history.
- **Sequential Read Access Latency**—column access time.
  - The sequential read access latency is defined as access time to an address within an already open row.
- The **read data throughput** is defined by the internal memory bus data bandwidth—number of sense amplifier which can be read in parallel—and the interface performance and width—how many data packages can be transferred.

The read bandwidth can be maintained continuously if an internal cache or a memory bank structure is used to hide the row access.

The following read performance parameter summary is made for non-volatile memories

- The **Read** access has typically the specified value.
- For Multi-level memories dependencies based on address and on coding scheme result into different random read access latencies under different conditions.
- Non-volatile memories using the Ready/Busy signal to indicate read data availability and generate often a flag indicating that the sensed data are correct and non-corrupted.

The read operation of non-volatile memories is typically combined with error detection and correction. Therefore, read access times could vary by a factor, depending on the need for error detection and error correction, and impacts the average data throughput.

3.1.7 Write Performance Parameters for Non-Volatile Memories

The write operation stores the supplied data into a non-volatile storage element at the target row and column address within the memory array.

- A write of a volatile memory is a logical change of a voltage level within a storage capacitor (dynamic storage e.g. DRAM) or a change of the logical state (static storage e.g. SRAM).

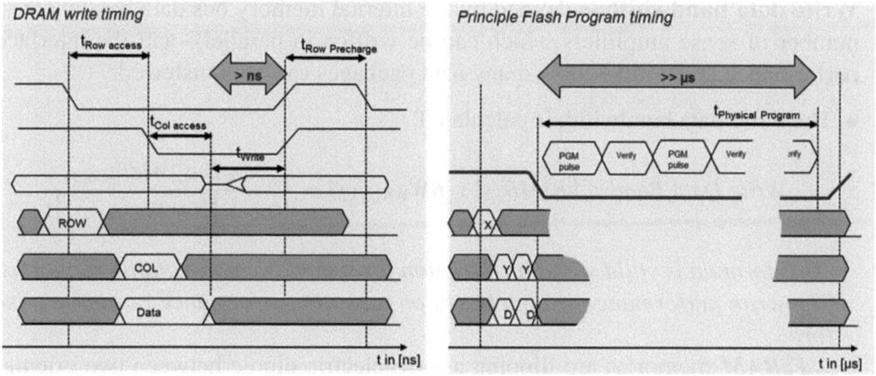

Fig. 3.5 Write performance—the physical write cycle for volatile and non-volatile memories

- The write of a non-volatile memory is a physical change of a storage element, which takes time and requires often an algorithm combining the physical change of the storage element and a verify operation of the value afterwards.

Figure 3.5 illustrates the difference between writing a volatile and a non-volatile memory. The logical write is a small portion of the write cycle of a volatile memory shown on the left side.

The write performance parameter definition is divided into two sections: The classical performance parameter—including the physical storage principle—and the discussion and the implication of the time and data size specified to write logical "1" and logical "0" and to re-write data.

Write performance parameter definition:

Write cycle time—time for data in (t_{DI}) and for physical change of the storage element

- Change the storage element corresponding to logical write data depends on:
  - Time to store a physical "0" and "1"
    Same time for both—symmetric operation—or different
  - Granularity specified for this physical operations
    Bit, Byte, Page or Block (e.g. flash block erase)
- The write cycle time for a symmetric memory is defined by the storage time ($t_{NV\_Storage}$) for the non-volatile physical change and is calculated:

$$Write\ Cycle\ Time\ [ns]: \quad t_{WCT} = t_{DI} + t_{NV\_storage}$$

- The write cycle time for an asymmetric memory with different granularities—segments ($S_{Granularity}$)—writing "0" and "1" and different times for both operations is based on a calculation formula assuming typical or worst case scenario.

$$Write\ Cycle\ Time\ [ns]: \ t_{WCT} = t_{DI} + t_{NV\_storage\_``0"} + (t_{NV_{storage_1}} * \frac{S_{Granularity_0}}{S_{Granularity_1}})$$

Write data bandwidth is defined by the internal memory bus data bandwidth—number of sense amplifiers which can be written in parallel—and the interface performance and width—how many data packages can be transferred.

- The write data bandwidth is calculated:

$$Write\ Data\ Bandwidth\,[MB/s]:\ BW_{WR\_NVM} = \frac{N * Bus\_Width_{Data}}{\left(t_{DI} + \left[N * t_{NV\_storage}\right]\right)}$$

This formula is valid without restriction for symmetric non-volatile memories. The write performance is then based on a single performance parameter like for:

FeRAM memories are flipping a ferro-electric dipole between two orientations.

MRAM memories are changing the magnetic orientation—flipping the spins.

The**Write data bandwidth** calculation for non-volatile memories with asymmetric data operations has to combine options for randomly distributed data sizes and directions of changes. A write of one bit "1" requires a bit, byte, or page—physical—write operation and the same assessment has to be done for the opposite physical data change for a write of one bit "0".

The consequence of cell and block operation is illustrated in Fig. 3.6 and will be calculated in detail for NAND flash memories in Sects. 3.2 and 3.3.

The write data bandwidth includes all operations to allow an endless write to randomly distributed addresses for randomly distributed data packages (4 KByte assumed if no other size specified).

3.2 Performance Parameters for Flash Memories

The links between product specification and cell, array architecture and algorithm are discussed and key performance parameters are derived to describe the performance behaviour of flash memories. The availability of product specification and

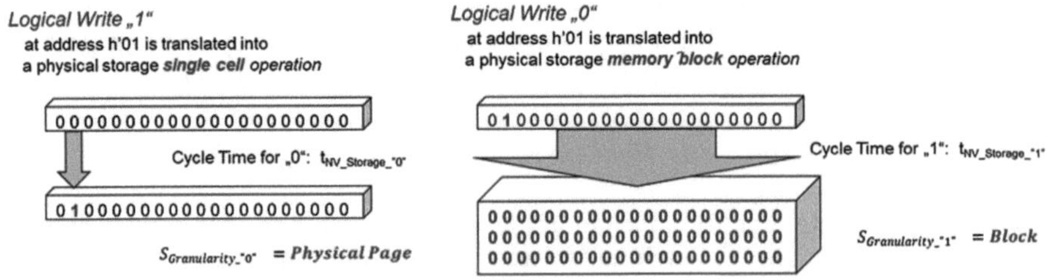

Fig. 3.6 Write performance—asymmetric NVM—cell versus block operation

application notes is qualifying these documents to the initial reference for the system development and optimization process.

3.2.1 *Flash Memory Product Specification*

The knowledge about interferences between cell, array, sensing and technology over lifetime is embedded in the algorithm specification and hidden for the customer. The details of implemented solutions to manage these interferences have to be understood based on available product specification and based on memory characterization. This is one subject of this chapter.

Very often system engineers make decisions based on product specification and available application notes. A strong link between specification, application notes and available reverse engineering data is necessary to understand the hidden behaviour of flash memories for the system development.

Parameters given for NAND and NOR flash selected from product specifications are slightly different compared to the key parameters identified in Chap. 2 of this work. Typical parameters for NAND and VG NOR flash memories are listed and compared in Table 3.1 as an example.

The logical memory address organization is shown Fig. 3.7 and does not fit to the expected 32 word lines belonging to a 32 cell NAND string. The differences are enforced by the logical to physical mapping of MLC NAND pages and the odd and even shielding bit line architecture.

Flash memory specifications define program and erase parameters with the attributes—minimum, typical and maximum. Algorithm duration time required to program or erase a number of cells are based on a statistical behaviour and depend on history, neighbour data and technology variations.

A comprehensive specification should always define all values - typical, minimum and maximum.

The specification normally does not explain under which conditions the min and max values shown in Table 3.2 are occur. A system design should not be based on typical performance values, because the risk is high to detect a weakness in the validation phase during system testing.

The page program time (t_{PROG}) is a key parameter to define the system write performance. The typical value of 600–800 μs derived from the specification is good for an average system write performance calculation. System architecture design has to use array and algorithm knowledge of MLC NAND memories introduced in Figs. 2.86 and 2.92. For logical pages belonging to the LSB (first level program)—short program times are expected—and for MSBMSB (second and third level programming)—longer page program times are expected. These differences are deterministic and can be utilized on system level to optimize the data throughput.

Figure 3.8 shows the logical page to physical word line mapping on the right side and the expected program times for each page for one erase block. The difference in performance indicates that the reliability margin is well balanced for the assumed

Table 3.1 NOR and NAND flash product specification comparison

Specification	Parameter (generic)	NAND (spec)	VG NOR (spec)
Memory	Density [Gbit]	16 Gbit	1 Gbit
	Bits per cell	2 (Multi-level cell)	2 (MirrorBit)
Voltage	Core voltage [V]	3.3 V	3.3 V
	I/O voltage [V]	1.8 V or 3.3 V	3.3 V
Interface	Asynchronous/syn.	Asynchr.	
	SDR/DDR	SDR/DDR	SDR
Array organization	Read unit size	4 kByte	64 word
	Write unit size	4 kByte	64 word
	Page unit size	4 kByte	
	Physical WL unit size	8 kByte	
	Erase Block unit size	512 kByte	64 k word
	ECC unit size per page	224 byte per page	
Performance	Random read	Typ. 50 μs	120 ns
	Data output cycle	25 ns	25 ns
	Data input cycle	25 ns	
	Program (write unit)	Typ. 800 μs	60 μs (single word PGM)
	Program page	Typ. 800 μs	
	Erase time (smallest)	Typ. 2 ms	Typ. 500 ms
	Block erase time	Typ. 2 ms	Typ. 500 ms
Reliability	Endurance	10 k cycles	typ. 100 k cycles per sec
	Data retention	10 years	20 years
Package	Package type	TSOP1-48	56-pin TSOP
	Multi-die package	2x, 4x	MCP

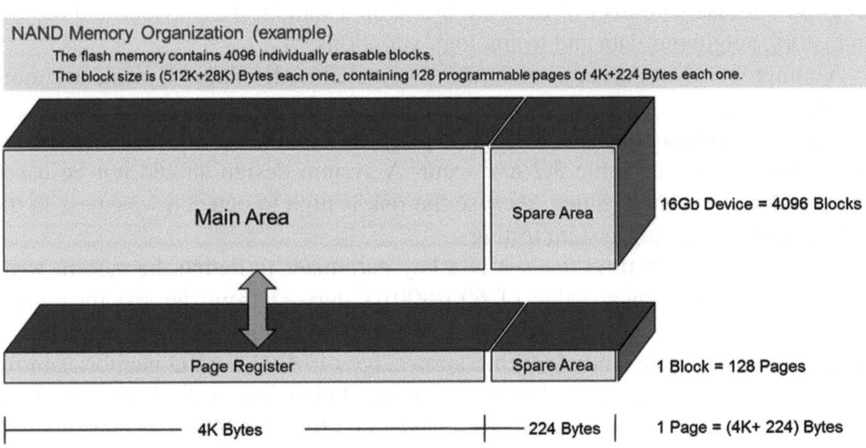

Fig. 3.7 16 Gbit MLC NAND memory organization example

Table 3.2 NAND Product Specification performance differences from vendor to vendor

NAND specification	Page program time (t_{PROG}) (μs)			Block erase time (t_{BERS}) (ms)		
	Min. (μs)	Typ. (μs)	Max. (μs)	Min. (μs)	Typ. (μs)	Max. (μs)
Vendor 1	150	800	1400	1200	2000	6000
Vendor 2	200	600	1250	1800	2000	3800

algorithm. A more homogeneous program time for the same algorithm would indicate a difference in the reliability margin for the first and the second programmed MSB page.

Performance figures of flash memories have to consider all parameter required to derive the overall performance over lifetime. The timing behaviour of a specific flash operation can have variations in the range between the specified minimum and maximum values. The expected variations (typical R/B times, ECC implementation and adaptive techniques) have to be applied in the system performance simulation in the correct order.

3.2.2 Array Architecture Impact on Flash Performance

The strong array architecture impact on performance parameters was introduced in Chap. 2. The major differences impacting all performance figures are summarized within Fig. 3.9.

The DRAM combines a fast localized full page sensing in *folded array architecture* [2] with a fast secondary sense amplifier to transport the data across the chip. This array architecture enables a combination of fast access and fast data throughput for read and write. The FeRAM is based on the same array architecture, which results into the same performance benefits.

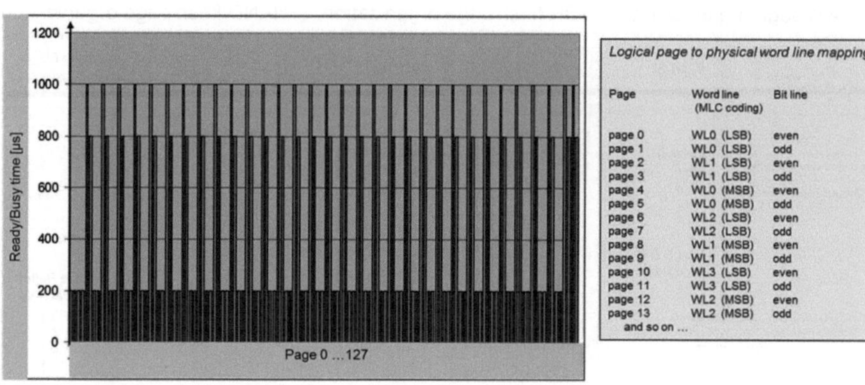

Fig. 3.8 Systematic page program time variations over 128 pages belonging to one erase block

The NOR array with short local bit lines supports a fast read and a fast program operation per cell. The parallel access is limited to 128–256 global sense amplifiers. Therefore the program data throughput is limited, because a real page mode operation comparable to a DRAM is missing.

The NAND array with extremely long bit lines supports slow read and program operations in terms of cycle time. Each operation is executed in parallel on 32000 or more sense amplifiers—page buffers in the NAND design wording. This results into an excellent read data throughput and a reasonably high program data throughput.

The next level of array architecture impact on the performance figures is based on resistance and capacitance values of bit- and word-lines. The time constant τ (tau) is the rise time characterizing the response time at a certain position of the bit line to a change of the bit line voltage at the driver.

$$tau\,[s] = \tau = RC$$

Where R is the resistance in ohms and C is the capacitance in farads.

The cycle time to access and sense a cell is dependent on the tau value of the bit line. Figure 3.10 illustrates the dependency of the achievable accuracy of the voltage applied to a bit line on the wait time—major part of the cycle time of this operation—given as multiple of tau.

The accuracy of the word line voltage is the important parameter for the sense accuracy for flash memories. The designed cycle time is balanced between the tau values of word and bit line to achieve the best balance between performance, accuracy and die size.

Figure 3.10 shows the dependency between tau values and achieved voltage values of the applied voltage which corresponds with the achievable accuracy. A shorter rise time of 2.5 τ instead of 3.0 reduces the margin for reliable memory operation by 5 %. This calculated loss of operational margin can become visible after months or years.

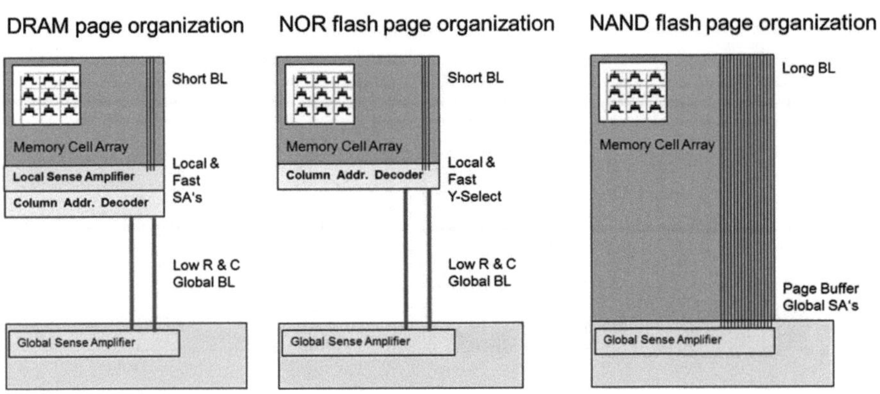

Fig. 3.9 Page array organization impact on timing and performance

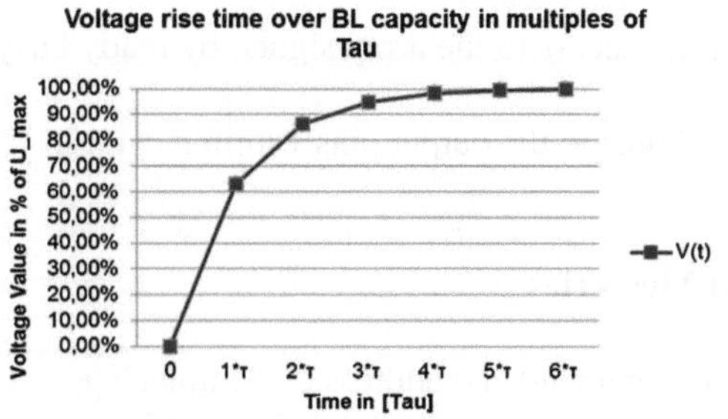

Time [tau]	Voltage Values during BL capacity charging [%]
1 * tau	63,21
2 * tau	86,47
3 * tau	95,02
4 * tau	98,17
5 * tau	99,33

Fig. 3.10 Voltage increase over bit line capacity after driver on—rise and fault time dependency from τ

Pin Names	NAND flash signal name
$\overline{CE}$	Chip enable
CLE	Command latch enable
ALE	Address latch enable
$\overline{WE}$	Write enable
$\overline{RE}$	Read enable
$R/\overline{B}$	Ready/Busy
I/O	Data Input/Outputs

Symbol	NAND timing parameter
t_{WB}	$\overline{WE}$ high to busy
t_R	Array read access to data register
t_{RC}	Read cycle time
t_{RHZ}	$\overline{RE}$ High to Output high impedance

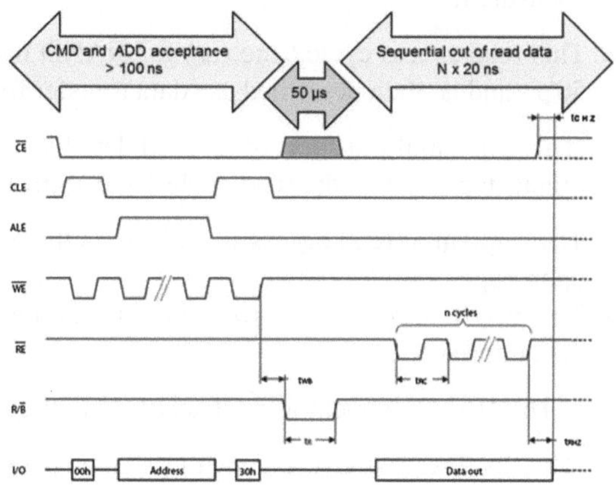

Fig. 3.11 NAND flash read timing—page read timing—asynchronous standard NAND

The voltage and current values have no direct influence on the cycle time. In case the required current supplied to the memory array is higher than the maximum allowed specification values the current specification defines the cycle time and limits a performance increase of the memory.

3.2.3 Read Cycle Time, Ready/Busy Time and Data Throughput

The flash read operation is defined in more detail for a NAND flash memory in this chapter. NOR flash memories have a read performance comparable to DRAM memories and only remarks are included.

The read timing cycle is divided into three sections shown in Fig. 3.11:

- Command Accepting Phase: accept read command and corresponding read address;

- Array Read Access: asynchronous read access to the array signed by ready busy signal;
- Data Out: transport of the sensed read data to the output pins within n cycles.

3.2.3.1 Read Cycle for NAND Flash Memories

The NAND interface timing defines the command and address cycle time. The first five cycles ($5 * t_{RC}$) and the write enable high to busy time (t_{WB}) define the **Command Accepting Phase.**

The **Array Read Access** latency includes the row and column access time and the sensing time.

- This read access cycle time (t_R) for NAND flash is in the range between 20 and 50 μs and is often described as "data transfer from flash array to data register" [3].

The sequential transport of all read data located in the data register to the pins is the final step to finish the read cycle—the **Data Out Phase.**

- The sequential read access is defined as the access time to an address within an open page.
- This access time is synchronized with the read cycle time (t_{RC}) of the external data interface.
  - The NAND sequential read access is in the range of 10–20 ns

The Read Access Cycle Time of one data package can be calculated based on tau values.

$$\textit{Read Access Cycle Time}\ [\mu s]: \quad t_{RACT} = 5 * t_{RC} + t_{WB} + t_R + 2 * t_{RC}$$

The read access time of a flash memory depends on the operation history of the system.

- An erase operation could block a read access to a certain memory bank for the specified erase time up to milliseconds. Functionality like erase suspend can be used to interrupt the erase and shorten the wait corresponding to the read access time.
- The first read operation of the cell content can be combined with ECC and adaptive techniques. Additional logical operations are executed during the read access and therefore this first access time could vary by a factor of two or three.

Remark: NOR memories operate like RAM memories.

- The typical random read access values are between 30 and 90 ns for NOR flash memories.

3.2.3.2 Read Data Throughput for NAND Flash Memories

The calculation of the read data throughput for NAND flash is defined in such a way that all data stored in the page buffer after sensing are transported to the interface pins.

$$Read\ Data\ Throughput\ [MB/s]:\quad DT_{RD} = \frac{Page\ size\ (2048\ Byte)}{10 * t_{RC} + t_R + 2048 * t_{RC}}$$

Table 3.3 shows the improvement potential for the read data throughput for different interface cycle times (t_{RC}) and page sizes. The page sizes are given without spare area, because the spare area can be different between products and does not contribute to the user data throughput.

The importance of the slow read access parameter decreases for larger page sizes and faster interface specifications. The read data throughput can be improved further by parallel read operation using independent planes. The dual plane concept is discussed in Sect. 3.2.5.2.

The traditional asynchronous NAND data interface can be significantly improved applying a source synchronous data interface including a clock (t_{CLK}) and the double data rate concept. The **O**pen **N**AND **F**lash **I**nterface (**ONFI**) working group introduced first the double data rate interface for NAND memories as an open standard. Table 3.4 shows the achievable read data rates for a single and a dual plane configuration. The dual plane configuration hides the data out times completely.

3.2.4 Write Cycle Time, Ready/Busy Time and Data Throughput

The write performance of a flash memory is the combination out of the time to transport the data into the data register and the time to program the flash cells. The program operation is controlled by an incremental step pulse algorithm which executes a certain number of program pulses.

The ***Program time*** can be calculated multiplying the program pulse and verify duration time with the number of program pulses. The number of program pulses

Table 3.3 Read data throughput (MB/s) dependency from NAND page size and interface cycle time

Interface cycle time (t_{RC}) and interface frequency ($f_{IF} = 1/t_{RC}$)	Page size (without spare area) 2048 Byte (MB/s)	4096 Byte (MB/s)	8192 Byte (MB/s)
25 ns (40 MHz)	28.76	33.46	36.44
20 ns (50 MHz)	33.60	40.19	44.56
15 ns (66 MHz)	40.38	50.29	57.33
10 ns (100 MHz)	50.59	67.19	80.38

Table 3.4 Read data throughput (MB/s) for single and dual plane NAND configuration and DDR cycle time

Page size(without spare area) / Interface cycle time (t_{CLK})	4096 Byte single plane (MB/s)	4096 Byte dual plane configuration (MB/s)	Comments
10 ns (100 MHz) DDR	101.19	200	ONFI 2.1
5 ns (200 MHz) DDR	135.45	400	ONFI 3.0

can be estimated theoretically using the erase distribution width well defined by both borders V_{th_LL} and V_{th_LH}.

$$Number\ of\ PGM\ pulses: \quad N_{Theoretical} = \frac{V_{th\_LH} - V_{th\_LL}}{\Delta V_{PP}}$$

The accurate calculation of the number of program pulses is based on the first program pulse distribution width, the guard banding for temperature offset and for all effects increasing the target program distribution width. The difference between the theoretical calculated program pulses—three—and the required – five—is illustrated in Fig. 3.12.

A Write operation—programming—could only be executed fault free in case the selected page or the complete block was erased in advance shown in Fig. 3.6. Two operation strategies are possible and impact the write performance significantly:

- The erase operation has to be executed in front of a program operation per default.
  - This strategy can be recommended only if the erase block size is as small as the typical program unit.
- The erase operation is executed in the background. Therefore a program operation can always be started on an empty page within a spare block.

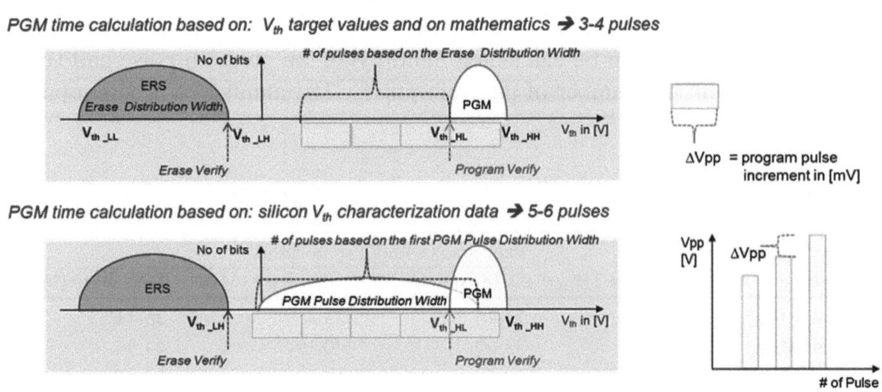

Fig. 3.12 Non-volatile memory—simplified program time calculation

- This strategy requires dedicated software memorizing which logical blocks are physically erased. In case a program is started to a certain logical page address a remapping of physical and logical block and/or page addresses is necessary.

This work assumes the second strategy is applied on system level to hide the erase time.

3.2.4.1 Write Cycle Time for NAND Flash Memories

The write cycle time for a flash memory is defined as the average cycle time required for a complete write operation of a logical data size smaller than page and block sizes to a random address.

- Sequential transport of write data into the data register (NAND page buffer data latches);
- Execution of the program operation—the flash memory is blocked by the ready busy time;
- Add a virtual time adder for the erase time of one page (physical erase time divided by the number of pages per erase block).

$$Write\ Cycle\ [\mu s]\ \ t_{WCT} = t_{DI} + t_{PROG} + \frac{t_{BERS}}{\#\,pages}$$

3.2.4.2 Write Data Throughput for NAND Flash Memories

The calculation of the write data throughput for NAND flash is defined in such a way that all data registers are filled with write data and afterwards the program operation is started.

$$Write\ Data\ Throughput\ [MB/s]\ \ DT_{WR} = \frac{Page\ size\ (2048\ Byte)}{(2048 * t_{WC}) + t_{PROG} + \frac{t_{BERS}}{\#\,pages}}$$

Table 3.5 shows the write data throughput dependency from different interface cycle times and page sizes based on the above described formula for $t_{PROG} = 220\,\mu s$ and $t_{BERS} = 2\,ms$.

The **write data throughput** is analysed assuming reasonable worst case scenarios. The different flash array segment sizes designed for program and erase strongly impact the write data throughput.

- **W**orst **C**ase **E**rase scenario **I** (**WCE_I**):
  - For every program operation the full erase time is added, because the operation has to be executed under worst case conditions in front of the first program. All data of the block to be erased are invalid; the erase could immediately be started.

Table 3.5 Write data throughput (MB/s) dependency from NAND page size and interface cycle time

Interface cycle time (t_{WC}) and interface frequency ($f_{IF} = 1/t_{WC}$)	Page size (without spare area)		
	2048 Byte (MB/s)	4096 Byte (MB/s)	8192 Byte (MB/s)
25 ns (40 MHz)	6.77	11.58	17.96
20 ns (50 MHz)	7.01	12.29	19.74
15 ns (66 MHz)	7.26	13.10	21.90
10 ns (100 MHz)	7.54	14.02	24.59

Table 3.6 Write data throughput (MB/s]) dependency from worst case erase scenario

Scenario and interface frequency (f_{IF})	Page size (without spare area)			Comments
	2048 Byte (MB/s)	4096 Byte (MB/s)	8192 Byte (MB/s)	
Default (100 MHz)	7.54	14.02	24.59	Distributed average erase time
WCE_I (100 MHz)	0.91	1.81	3.56	Full erase time in front
WCE_II (100 MHz)	0.22	0.44	0.88	50 % move data and full erase
WCE_III (100 MHz)	0.15	0.31	0.61	50 % move data, erase fails—marking—erase on next block

- W**orst** C**ase** E**rase scenario** **II** (**WCE_II**):
  - A reasonable number of pages have to be moved (assumption made half of max. number of pages per block) and copied to new locations, before the erase operation on the target block can be started. Afterwards the logical data write could be started.
- W**orst** C**ase** E**rase scenario** **III** (**WCE_III**):
  - A reasonable number of pages are copied and afterwards the erase operation fails on the target block due to reliability issues of flash cells. A second erase operation has to be started on a different block and afterwards the logical data write is started.

Table 3.6 the calculated write data throughput for the four scenarios—the default scenario with average erase times and three worst cases WC_I, WC_II and WC_III.

The write performance is impacted by orders of magnitude if the worst case scenarios will appear in the real application world. The same dramatic reduction is visible for the read data throughput, in case a read operation has to be executed on a memory bank which currently executes an erase.

The impact of the erase operation on the Read and Write Data Throughput is a bottleneck for flash memories and will be covered in the relevant sub-chapters. Dependencies shown in Table 3.6 have to be understood for relevant application cases to be solved on system software level.

The worst case performance assessment has to be made for failing program and erase operation during operational lifetime to enable the corresponding recovery strategy as part of the specified performance behaviour of the non-volatile memory in use.

3.2.5 Program Data Throughput

The programming performance and the threshold voltage operation window of flash memories are the two focuses of this work. Flash memories have two array granularities used for program and erase operation. Consequently two performance parameters are used to specify data throughputs:

- The **program data throughput** is defined by the memory array architecture—the number of sense amplifiers defines the cell program parallelism—and the flash program algorithm—the number of pulses and verify operations required to finish the program operation.
- The **write data throughput** is defined by the sum out of the time to erase the smallest erasable block and the program data throughput for the specified data size and the strategy to hide the erase time in such a way that the assumed average values are guaranteed.

3.2.5.1 Program Data Throughput for NAND Flash Memories

The calculation of program data throughput for NAND flash is the same like for the write data throughput without the erase term.

$$\textit{Program Data Throughput}\ [MB/s]\quad DT_{PGM} = \frac{\textit{Page size}}{t_{PROG} + t_{DI}}$$

Simple assumptions achieve in most cases the best understanding of important dependencies.

The PGM time can be replaced by the number of PGM pulses multiplied by twice the time required for the read operation—first a program pulse with the required accuracy and afterwards a verify operation.

$$\textit{Program Data Throughput}\ [MB/s] = \frac{\textit{Page size}\ (2048\ \textit{Byte})}{\textit{PGM pulses} * (9 * (\tau_{BL} + \tau_{WL})) + (2048 * t_{WC})}$$

Table 3.7 shows the program data throughput for different interface cycle times and page sizes based on the above described formula where t_{PROG} is replaced by $t_{PROG} = 5$ pulses*(9*(2 μs + 3 μs)) = 225 μs.

The data interface time can be hidden by using two flash memories with shared I/O channels in a chip interleaved mode [4].

Figure 3.13 compares graphical the program data throughput for different page sizes. A larger page size improves significantly the program data throughput. Dual plane or dual chip configurations are one possibility to improve the sustainable data rate as shown in Fig. 3.14.

3.2.5.2 Write Data Throughput: Impact of Dual Chip and Dual Plane

A worst case scenario calculation for a two NAND chip memory sub-system architecture shows the benefit of dual plane and dual chip utilization. The first and the second worst case scenario are not visible any longer and can be hided completely. The worst case write data rate is equal to the best case program data rate for a single chip dual plane configuration. Programming at one time only on one plane is assumed; the second plane is only used to hide the data in time.

The decision on system level to use a memory sub-system configuration with two NAND chips in parallel offers the required flexibility to hide the erase and a certain number of worst case write performance scenarios. A memory sub-system based on two packages with each time two NAND dies stacked enables enough flexibility to

Table 3.7 Program data throughput (MB/s) dependency from NAND page size and interface cycle time

Interface cycle time (t_{WC}) and interface frequency (f_{IF})	Page size (without spare area)		
	2048 Byte (MB/s)	4096 Byte (MB/s)	8192 Byte (MB/s)
25 ns (40 MHz)	7.41	12.51	19.06
20 ns (50 MHz)	7.70	13.35	21.07
15 ns (66 MHz)	8.01	14.30	23.55
10 ns (100 MHz)	8.34	15.40	26.69

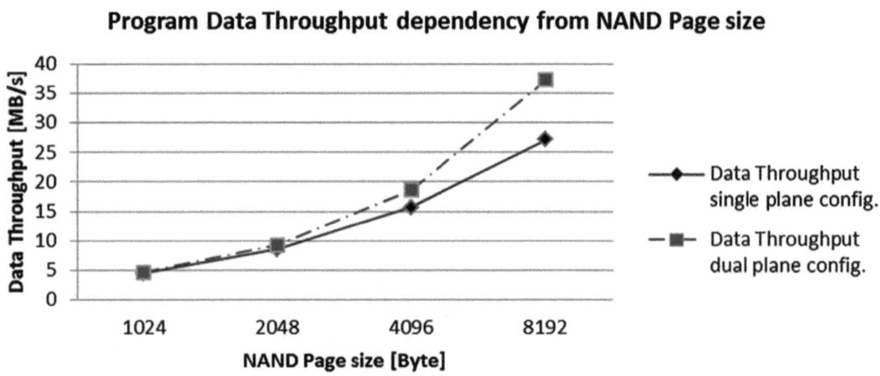

Fig. 3.13 NAND program data throughput for different page sizes with dual plane configuration

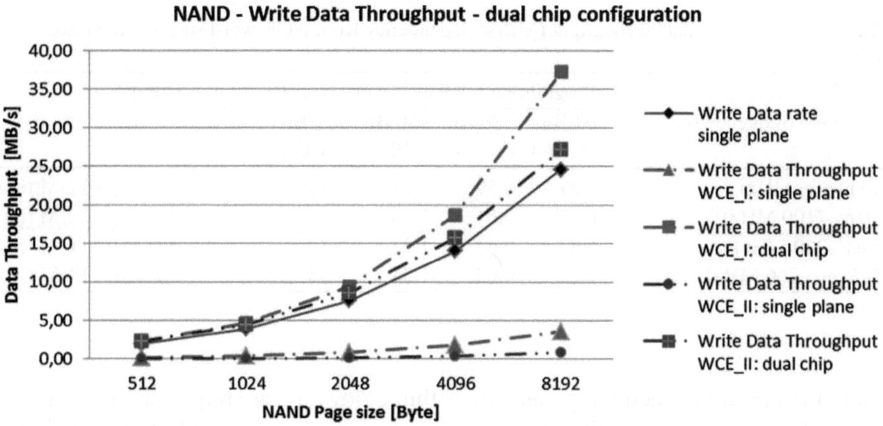

Fig. 3.14 NAND write data rate for dual chip configuration—worst case compared to single plane/chip

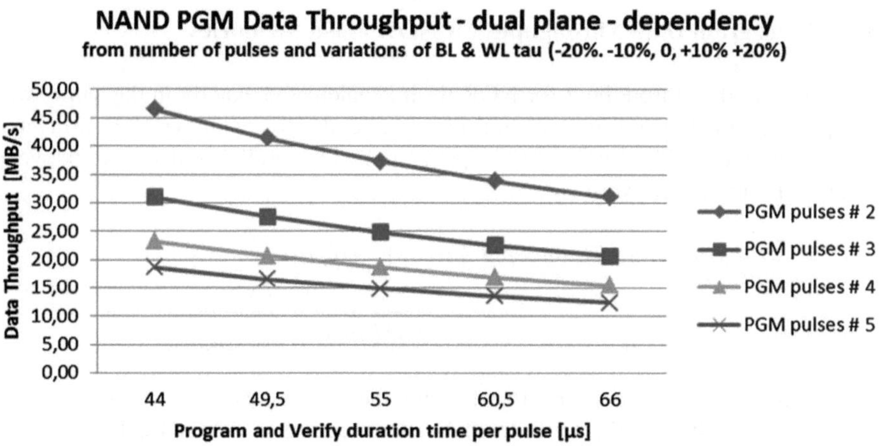

Fig. 3.15 NAND PGM data throughput dependency from number of PGM pulses and BL & WL Tau variations

support a sustainable write performance which should never fall below the single die program data throughput value.

The memory array program data throughput is the dominant performance parameter for a non-volatile memory product assessment and the basis for the performance indicator methodology.

Figure 3.15 illustrates the impact of the parameters introduced in the PGM data throughput formula—the number of program pulses and the impact of the rise time depending on R and C of the bit and word lines within the memory array.

Technology variations will increase the number of required pulses along the shrink roadmap, and the smaller dimensions and smaller distances will increase automati-

Table 3.8 Program data throughput (MB/s) dependency from NOR word size and interface cycle time

Interface cycle time (t_{IF}) and Interface frequency (f_{IF})	Program parallelism—word size (Bits)				Comments
	64 Bit (MB/s)	128 Bit (MB/s)	256 Bit (MB/s)	512 Bit (MB/s)	
15 ns (66 MHz)	0.31	0.59	1.11	1.96	for 25 μs PGM time
10 ns (100 MHz)	0.31	0.61	1.16	2.12	for 25 μs PGM time
5 ns (200 MHz)	0.32	0.62	1.22	2.32	for 25 μs PGM time
3,75 ns (266 MHz)	0.32	0.63	1.23	2.38	for 25 μs PGM time

cally the tau value. Technology and algorithm innovation are required for every new technology node. The quality of these innovations is covered and fully visible in the memory array performance parameter **Program Data Throughput**.

3.2.5.3 Program Data Throughput for NOR Flash Memories

The program data throughput for NOR flash is calculated exactly in the same way. The parallelism of the program operation is smaller due to the known NOR array and sensing limitation. The number of bits programmed in parallel is in a range between 64 and 512 and called word size.

$$Program\ Data\ Throughput\ [MB/s]\ \ DT_{PGM} = \frac{Word\ size}{PGM\ time + \frac{Word\ size}{Bus\_width_{Data}} * t_{IF}}$$

The assumption of program time follows the same rules. The combination out of local and global bit lines and local Y-select requires a detailed calculation for design to be investigated. Therefore no generic formula is introduced for NOR and VG-NOR flash memories.

Table 3.8 shows program data throughput values for NOR flash memories assuming different interface cycle times and different word sizes—the definition for the parallelism of NOR operations.

The physics of the program principle for NOR flash limits the parallelism and a significantly further increase of the program data throughput. The write data throughput is even worse due to the relative long erase times which will be discussed in the next chapter.

3.2.6 Erase Performance and Erase Suspend Commands

The erase performance for flash memories is listed in Table 3.9 and compared with other non-volatile memories. The ratio between erase and program for the same block size indicates the probability to hide the erase cycle time impact on system level.

The erase cycle times from 2.0 ms for NAND up to 600 ms for NOR flash are translated into wait times on system level. The assessment on application level decides if wait times can be accepted.

- The erase wait time is acceptable on application level:
  - A single die solution is possible.
  - Shift the erase operation into sleep modes and power up/down times.
- The erase cycle time is not acceptable on system and application level:
  - Stop the erase if needed ⇒ introduction of an erase suspend command.
  - The erase is hidden and executed as a background operation.

A continuous sustainable write performance of a flash memory based sub-system requires that the ratio between erase and program is below 10 % better below 5 %. A detailed assessment of the ERS/PGM ratio and strategies to handle the mismatch are introduced in the durability chapter.

NOR flash memories introduce an erase suspend command, which will stop the erase operation running on the flash die. The stop of high voltage operations takes typically a couple of μs and reduces the wait time by orders of magnitudes.

Beside the direct impact on the system performance the erase operation defines position and width of the erase distribution. These parameters have an indirect impact on the system performance.

- A well designed erase algorithm can maintain better the number of outlier bits—reduced effort for EDC and ECC and therefore reduced energy consumption—and achieves a longer run time – improves the endurance parameter for flash memories.

Table 3.9 Erase performance—absolute values and ratio between program and erase time

Memory	Erase time per ERS block (ms)	Program time per page/word (μs)	Program time per ERS block (ms)	Ratio ERS/PGM per ERS block
Non-volatile flash memories (electron-based)				
SLC NOR	600	30	30	20
VG NOR	30	30	30	1
SLC NAND	2	200	12.8	0.15
MLC NAND	2	800	102	0.02
Non-volatile emerging memories (non-electron-based)				
FeRAM	Not applicable	100 ns	Not applicable	Not applicable
PCM	Not applicable	100 ns	Not applicable	Not applicable

Table 3.10 Performance Parameters summary for different non-volatile memories

Memory	Read access cycle time	Read data throughput	Program cycle time	Program data throughput	Write data throughput
Non-Volatile Flash Memories (electron-based)					
SLC NOR	Fast	High	Medium	Medium	Slow
FG 1-bit/cell	25 ns	>100 MB/s	10–20 μs	1–10 MB/s	0,1–1 MB/s
VG NOR	Fast	High	Medium (μs)	Medium	Slow
CT 2-bit/cell	25–50 ns	>100 MB/s	60 μs	2–8 MB/s	1–4 MB/s
SLC NAND	Slow	High	Slow	High	High
FG 1-bit/cell	25 μs	>100 MB/s	>200 μs	>100 MB/s	>80 MB/s
MLC NAND	Slow	High	Slow	Medium	Medium
FG 2-bit/cell	50 μs	>100 MB/s	>700 μs	>25 MB/s	>20 MB/s
Non-Volatile Emerging Memories (non-electron-based)					
FeRAM	Fast	High	Fast	High	Not applicable
	25–50 ns	>100 MB/s	100 ns	>100 MB/s	
PCM	Fast	High	Fast	Medium	Not applicable
	25–50 ns	>100 MB/s	100 ns		

- A smaller erase distribution can be translated into one or two program pulses less to finish the program operation. This impact was calculated for NAND flash and shown in Fig. 3.15.

The erase operation improves and maintains the write (program) performance over life time.

3.2.7 *Identification of Key Performance Parameter*

The specified performance parameters for flash memories are analysed including the impact of array and technology parameters and variations. An impact analysis of these independent parameters is done on achievable read and writes system performance values for different configurations.

Different types of flash memories are summarized in Table 3.10 and two non-electron based emerging non-volatile memories are added to complete the overview and highlight the differences.

For flash memories the following conclusions are summarized:

- Array operation cycle times and the corresponding data throughput are not linked;
- Increased bit density on NAND page sizes can be translated into increased data throughput;
- Increased sense amplifier density on NOR can be translated into increased read data throughput, the link is not that strong for program throughput due to current restrictions;

- Sustainable program data throughput has to be achieved by memory-system architecture;
  - NAND erase cycle times for multiple plane and multiple die memory subsystems can be hidden in a way that a reduced program data throughput is guaranteed.

For the development of an independent key performance parameter set the **"Program Data Throughput"** is selected as the first parameter characterizing the cell, array and algorithm combined memory performance.

The second parameter is the **"Read Access Cycle time"** important for the system performance simulation.

The **"Write Data Throughput"** has to be selected if the ratio between erase and program is higher than 10 % instead of the **"Program Data Throughput"** as the main key performance parameter.

The additional impact on the write and read performance due to the finite cycle capability of flash memories and the significant effort to manage the different sizes between program and erase array segments (data granularity) are discussed in the next chapter.

The performance benefits of emerging non-volatile memories are described in Tables 3.9 and 3.10. Both emerging memory examples are leading the performance comparison in all parameter classes. The economic success of NAND flash emphasizes the importance of a combined cost and performance assessment including the improvement potential on memory sub-system level.

3.3 Performance and Durability

The performance parameter analysis is combined with an assessment of the durability of non–volatile memories. We introduce the wording durability at this stage to discuss performance and durability requirements and interferences between both based on the typical wording and understanding developed for volatile memories. This chapter will introduce all technical challenges linked to the limited durability of most non-volatile memories.

3.3.1 Durability Definition for Combined Memory Operation

The durability of a memory operation can be calculated based on the expected operating lifetime of the memory. We calculate the number of possible read or write operations based on the specified cycle time—100 ns—for 10 years lifetime within the specified temperature range.

- A volatile memory guarantees 10^{15} fault free operations per cell.

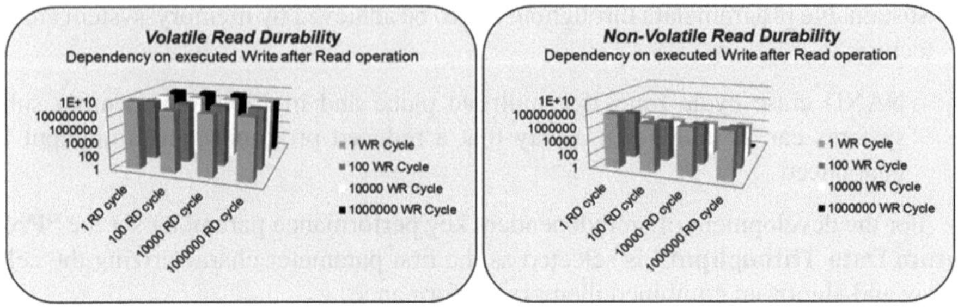

Fig. 3.16 Read durability dependency—compared for volatile and non-volatile memories

  - The specification assumes duty cycles to reduce voltage stress to gate oxides.
- This durability number is guaranteed for any combination of reads and writes
  - We can combine 9×10^{14} writes with 1×10^{14} reads or the opposite.

A non-volatile memory is specified with the same value for the operating lifetime. However, the number of write cycles—also called cycle endurance—is limited to a typical value of 10^5 cycles. The memory specification limits also the read cycles to a value of typically 10^7 due to read disturbances getting stronger during lifetime. We use the same calculation method and combine the read and write cycles:

- This non-volatile memory should be able to guarantee 10^{12} fault free operations assuming a correct sequencing of reads and writes is always maintained over lifetime per cell.
  - Reliability effects will reduce these numbers and are discussed in Chap. 4 in detail.
- Based on this calculation the assumed non-volatile cell guarantees at least 1 h fault free operation per cell multiplied by the density of the memory device divided by the parallel array access to the array which is important for disturbance and cycling stress.
- For a non-volatile memory larger than 4 Gbit based on this calculation a continuously fault free operation should be possible in a time range of 10 years.

The importance of this durability analysis illustrates Fig. 3.16 comparing volatile and non-volatile memories for durability requirements in the operation cycle range of 10^{10}.

The right side of the figure illustrates the intrinsic weakness of most non-volatile cells; the change of the non-volatile storage element introduces a physical stress and change the reliability behaviour.

An effective method making this impact visible is a characterization of the bit failure rate during read.

A reliable SLC NAND flash device from 70 nm node was tested on a bench tester with the described NAND flash characterization flow at room temperature.

The weakness of non-volatile memories described in Fig. 3.16 is now tested with SLC NAND. The NAND functionality is demonstrated for all combination of reads and writes, only the bit failure rate per page increases as shown in Fig. 3.17.

This NAND characterization flow can be extended to more memory devices. Afterwards a model describing the stress dependency shown in Fig. 3.18 can be generated to support the system engineering team to develop a correct durability specification for the memory system.

This behaviour is different for each non-volatile memory. The mathematical dependency will be varying over technology nodes and during production. Design, algorithm and application improvements can extend the durability for a dedicated data size by orders of magnitude.

The additional effort defining the correct durability is now becoming clear and the result depends on performance values of the complete system architecture based on the performance parameters of the flash memory.

3.3.2 Design for Read Durability: Extended Values for Small Data Sizes

Flash memories have a limited durability for read and for write. A design for durability can be based on different approaches. Two examples to extend the read durability are introduced for NAND flash:

- An optimized sequence of read cycles—10^7—and rewrite cycles—10^5—can extend the read durability to 10^{12}. This concept refreshes the disturbed read data by a write operation after the maximum period of allowed reads. The error detection of NAND flash can be used to define a threshold for an acceptable failure rate and then the rewrite is triggered after this threshold is achieved.

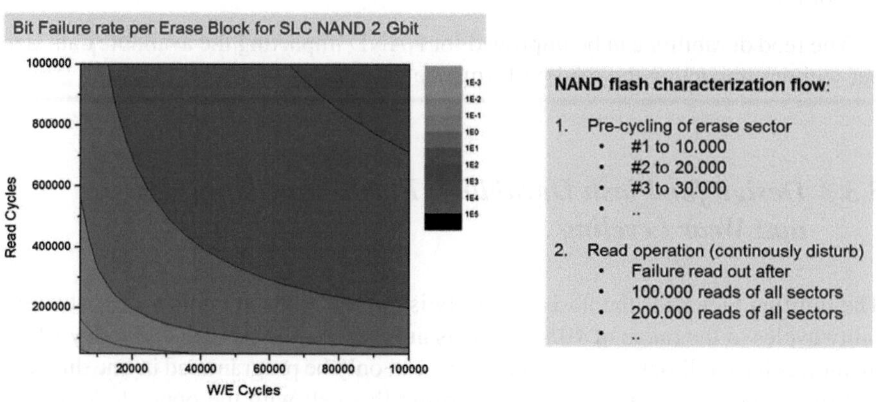

Fig. 3.17 Bit-failure rate per NAND erase block for read and write durability

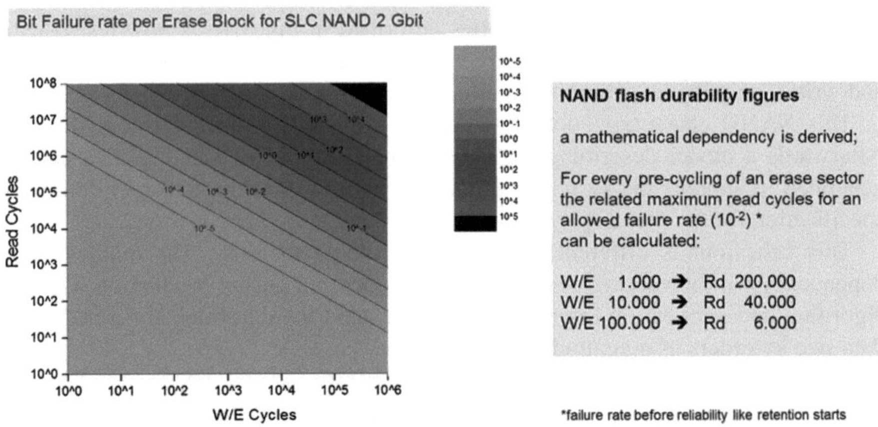

Fig. 3.18 Bit-failure rate on a log scale to estimate combined read and write durability

  - This requires a serious overhead and one issue is not solved, what will be the failure rate in case this sequence is interrupted by a 5 year's retention time.
  - Interference between durability and reliability is covered in Chap. 6 in detail.

- A read operation within a NAND flash array is disturbing the cells along pass word lines by the read pass overdrive voltage (V_{RD_PASS}). The target word line which is repetitive read is not disturbed at all. A system concept can use only one physical word line out of 32 per erase block. This word line guarantees a read durability comparable to values of volatile memories.

  - In case the MLC program algorithm of the flash memory is known in detail the disturb of up to 32 word lines can be improved by overlaying the user data with complementary data to generate two small program distributions which are closer to the V_{RD_PASS} and therefore more robust to read disturbance by orders of magnitude.

The read durability can be improved for NAND impacting the available data size, but without impacting the read performance values.

3.3.3 Design for Flash Durability: Flash Translation Layer and Wear Leveling

The limiting factor for the flash durability is not the reduced number of write durability cycles in the range of 10^4–10^5. This number is valid per cell and today's flash memories have billions of cells. Flash cells can only be programmed in one direction and the granularity is a page size. Rewriting this cell with the opposite logic data requires an erase operation applied to an erase block with a typical size of 1–4 Mbit for NAND flash.

- Design for Flash Durability (DfFD) requires a system—and application—which physically writes data packages with the size of one erase block or a multiple of this size. As an example the memory sub-system has to cache a large number of operations and starts the physical program operation only with a granularity of 1, 2 or 4 Mbit.

We have to highlight this point again to make it very clear: A system "Design for Flash Durability"—*"full utilization of cell endurance values"*—has to support a physical programming of a complete erase sector size. During this time frame the power supply has to be maintained within a tight specification range without voltage spikes and changes.

A system of hardware and software is wrapped around the flash memory and called flash translation layer—FTL. The main tasks of this wrapper are:

- A logical change of data is decoupled by a logical to physical address mapping.
- A change of a Byte within a logical page enforces a physical write operation of one or more physical pages, but not of a complete erase sector.
- The address mapping will fragment the physical address space with non-valid pages—which cannot be used because the erase sector still contains valid physical pages.

At a certain trigger level—depending on the additional spare area of the flash memory—still valid pages have to be copied to another erase block to freeze up the current one for an erase operation. The concept is called "garbage collection" and these additional write operations are reducing the data throughput and the values for the durability and performance calculations.

3.3.3.1 Concept of Wear Leveling

Wear leveling is a method to extend the usable life of block-erasable flash memories. Due to the finite cycling capability of flash memories wear leveling distributes the physically executed erase operation evenly across the whole memory or better across the memory sub-system.

Consider the case without wear leveling—a flash based system changes a data file—256 byte—every 10 ms with a new measurement value. The last value overwrites the old one. After 24 h runtime this specific logical address has to be overwritten 8.640.000 times.

- *A system built with **SRAM** and a buffer battery would have exactly accessed and modified 8.640.000 times the* 256×8 *memory cells.*
- *A system built out of* **NAND** flash would have cycled the addressed erase block *$\gg$100.000 times till end of life—would stop or would cycle the next erase sector till it is failing.*

Wear leveling [5] strategies and special file systems are developed to extend the useable life of flash memories. We compare two strategies for a system with two NAND flash memories. Table 3.11 shows the results for 8.640.000 write operation of 256 byte at the same address.

- **Strategy 1**: Distribute the P/E cycling stress over all empty blocks—static wear level;
- **Strategy 2**: Distribute the cycling over all sectors—move data—dynamic wear level;

The effectiveness of these two wear leveling strategies is disappointing. 256 Byte data write is going to change only 0.48 % of one erase block. Due to the difference in page and block sizes the logical change of 256 Bytes forces an erase operation on 524.288 Bytes.

The wear level strategies reduce the risk to wear out the addressed block. The wear level and garbage collection technique will have an impact on the system performance. This impact can be measured with additional executed program operations to operate a logical write. This factor is called "Write Amplification" [6] and is defined as operated physical writes per requested logical write.

The continuous operation of the selected application example over two years would physically destroy both NAND flash devices within a couple of weeks for the static wear leveling and within a couple of months for the dynamic wear leveling based on a flash translation layer granularity of erase block level. These strategies are typically applied for low cost mobile storage devices.

3.3.3.2 Concept of Data Analysis and Address Mapping

A system using flash memories has to decouple the logical page from the physical page address; otherwise an erase block linked to a logical address would have to be programmed and erased millions of times.

- **Strategy 3**: Remapping of logical page addresses:

Table 3.11 Durability—wear level strategies

8192 erase blocks [EB] in total 8.640.000 times write of 256 byte in 24 h run time	Start condition 4000 empty erase blocks	Start condition 200 empty erase blocks
Static wear leveling	2160 P/E cycles	43200 P/E cycles
– Select the next empty EB and mark the last one as non-valid → empty	*for each empty EB*	*for each empty EB*
Dynamic wear leveling	1054 P/E cycles	1054 P/E cycles
– Distribute the cycling over all EB's	*for each empty EB add 1054 P/E cycles to all EB's*	*for each empty EB add 1054 P/E cycles to all EB's*

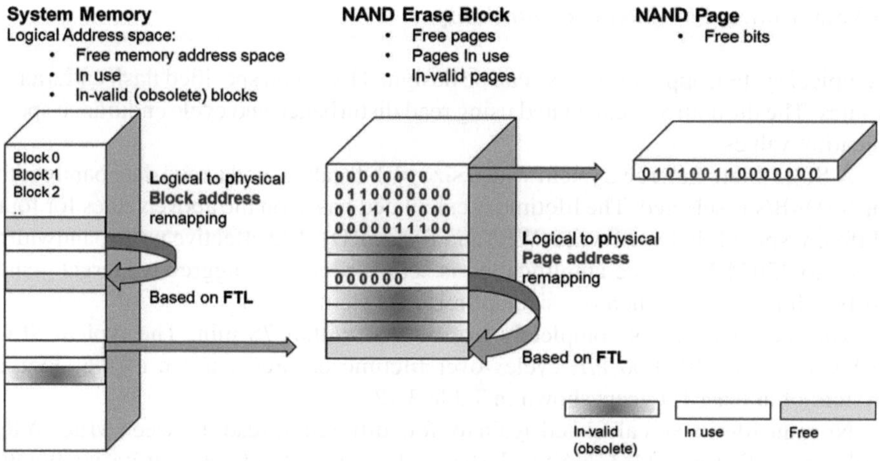

Fig. 3.19 Design for flash durability—address mapping and garbage collection

- Write data—256 bytes—to the next physical page address and remap the address to the logical correct one
  (a) 2 P/E cycle for 4000 "empty" blocks combined with static wear leveling
  (b) 43 P/E cycle for 200 "empty" blocks combined with static wear leveling

- **Strategy 4**: Data analysis:
  - Compare new data first with already stored data by a logical operation on controller level. In case only a small number of bits are changed:

    Only changes and bit position are stored—independent of requested write size.

    Changes are stored in cache and are not released for physical NAND write.

Logical to physical address mapping and data analysis combined with dynamic wear leveling reduce the physical stress to the flash memories by orders of magnitude shown in Fig. 3.19. The difference in size between the smallest unit to program and to erase requires a logical to physical address mapping which results into a data fragmentation of the memory. A garbage collection is required to freeze up erasable blocks with a certain number of in-valid pages. The remaining valid pages are copied to empty erase blocks. This cleaning of the physical address space can be counted as a refresh operation for the system retention calculation.

The performance impact of different strategies and algorithms to improve the flash life cycle are covered by the three performance parameters write amplification, additional physical memory space and cache size available on system level.

3.3.3.3 Concept of Extended Durability

A typical system application case can be designed based on specified flash endurance values. The durability is calculated using read disturbance and cycle endurance specification values.

A Solid-State Storage System with a size of 1 Terabyte and a read data bandwidth of 500 MB/s is selected. The lifetime is calculated based on the above values for four different spreads between write (WR) and Read (RD). The effective write bandwidth is set to 250 MB/s. Write amplification is set to value 2.5—aggressive target—and to 10—for a cost and memory size optimized system.

The memory size is completely overwritten within 78 min. The typical SLC NAND spec of 100.000 P/E cycles over lifetime ensures a usage of the storage system of at least 1.4 years shown in Table 3.12.

Now the lifetime is calculated again for four different spreads between write (WR) and Read (RD) for an MLC NAND flash based system. The effective write bandwidth is set to 125 MB/s.

The memory size is completely overwritten within 155 min. A MLC NAND specification of 10.000 P/E cycles over lifetime ensures a 3 months usage of the storage system according to Table 3.13.

Adding additional physical address space improves durability for SLC and MLC. The right side of both tables achieves promising durability values. The selection of the correct system application case has an essential impact on cost and reliability of flash based systems. The assumed application case values can be exceeded for limited numbers of customer specific cases. Therefore flash based storage system architecture requires a guard band strategy. An extended durability range is defined including the counter measures (reduced retention values, increased ECC coverage, adaptive read algorithm) to maintain the functionality of a storage system even above the assumed spread between write and read operations.

An extended durability (P/E cycles) can be accepted if the overrun is detected and a significantly reduced retention time is acceptable on application level or is guaranteed by the dynamic wear leveling. The detailed dependency between P/E cycles and retention values is analyzed in the reliability chapter.

Table 3.12 System durability values for continuous operation based on SLC NAND Flash

1 TByte SLC NAND Flash 500 MB/s data throughput	WR and RD 90 %/10 %	WR and RD 50 %/50 %	WR and RD 25 %/75 %	WR and RD 1 %/99 %
Complete capacity Fill (min)	77.7 min	140 min	280 min	6990 min
Effective write bandwidth 250 MB/s	1.3 h	2.33 h	4.66 h	116 h
Write amplification of 2.5	5.9 years	10.6 years	21.2 years	500 years
20 % additional physical space				
Ensures 40.000 cycles for application				
Write amplification of 10	1.4 years	2.6 years	5.3 years	133 years
Less than 10 % additional space				
Ensures 10000 cycles for application				

Table 3.13 System durability for continuous operation values based on MLC NAND flash

1 TByte MLC NAND Flash 500 MB/s data throughput	WR and RD 90 %/10 %	WR and RD 50 %/50 %	WR and RD 25 %/75 %	WR & RD 1 %/99 %
Complete capacity Fill (min)	155,3 min	280 min	560 min	13981 min
Effective write bandwidth 125 MB/s	2.6 h	4.66 h	9.32 h	233 h
Write amplification of 2.5	1.18 years	2.12 years	4.25 years	106 years
20 % additional physical space				
Ensures 4.000 cycles for application				
Write amplification of 10	0.29 years	0.53 years	1.0 years	26.6 years
Less than 10 % additional space				
Ensures 1000 cycles for application				

3.3.4 Design for Flash Durability: Utilize Intrinsic Flash Parameters

A system software architecture based on logical to physical address mapping and wear leveling algorithm is wrapped around physical NAND flash devices to guarantee the durability specification over the operational life time. The logical or mathematical optimization is targeted on software and storage publications in depth [6]. The better algorithm requires more overhead and a larger table size to store all the remapping and additional cycle count information.

The dependency between cell degradation and program or erase parameter degradation is discussed in the reliability chapter. This dependency is the concept basis of utilizing intrinsic flash parameters.

The systems can be designed in a way to make use of available flash parameters. The block-based erase operation enforces the addition FTL effort—a significant overhead in terms of cost and especially maintenance over product lifetime—but offers a very strong intrinsic indicator of the quality of each erase block. The link between Flash devices—enable hidden parameters to steer system tasks—and the system architecture—number of channels and software architecture—is often not

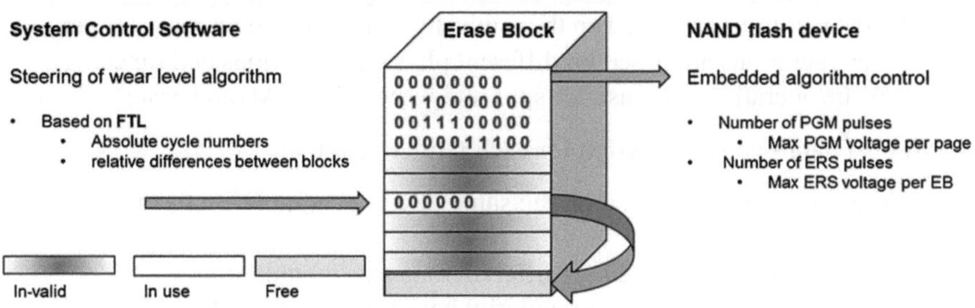

Fig. 3.20 Control of write durability—link between FTL and physical algorithm parameters

analyzed in depth due to the missing understanding of reliability effects and the missing access to these NAND device data.

Variations between erase blocks within a storage system can be one or two orders of magnitude.

- The control of durability, especially the control of the corresponding health level of the erase sector is an important performance parameter to be able to utilize the calculated durability values in the last chapters in Tables 3.12 and 3.13.
- Physical pages and physical erase blocks running out of a certain threshold of operation parameters need a special handling.
  - The intrinsic flash algorithm parameters are the best monitor about the quality of the sector or the page.
  - Bit failure rates can be made available by running EDC/ECC hard- and software and are a second diagnostic channel for the quality of the erase sector.

The link between the logical parameters collected within the FTL and the physical parameters can be established based on statistical relevant characterization data and an excellent understanding of all reliability effects impacting the operational life time of flash memories. Figure 3.20 illustrates this link.

3.4 Performance Parameter and Durability Summary

Performance parameters are defined for volatile and non-volatile memories. Flash memories are used to discuss the non-deterministic performance behaviour. Cell and array parameters are selected to calculate the dependency of cycle time and data throughput for flash memories. The NAND page size is used to estimate the impact of array operation parallelism on performance parameters. This approach is becoming the basis for the memory array model introduced in Chap. 6.

The performance and durability assessment of non-volatile memories focuses on the following additional parameters or properties:

- Cycle time investigation for each physical write operation:
  - Write operations are based on the same physical principle for write "0" and "1".
  - Write operations are based on different physical operations and times.
    Write operations for flash are split into write "0" (PGM) and write "1" (ERS).
- Physical data size investigation for each write operation:
  - Write operations are based on the same data size granularity for write "0" and "1".
  - Write operations are based on different data size granularities.
    An asymmetric data size granularity between write "0" (PGM) and write "1" (ERS) requires a ratio factor in size and in time between both operations.

Table 3.14 Performance parameter classification—impact of asymmetric, non-deterministic and durability

Memory performance	PGM data throughput	Ratio per EB between ERS/PGM	Non-determin. performance	Durability: write amplification	Classification
Non-volatile flash memories (electron-based)					
NOR FG	Fast	10	Erase and PGM	Less important	Asymmetric
VG NOR	Fast	1	Erase and PGM	Less important	Asymmetric
SLC NAND	Fast	0.15	All operations	Yes	Asymmetric
MLC NAND	Fast	0.02	All operations	Yes	Asymmetric
Non-volatile emerging memories (non-electron-based)					
FeRAM	Fast	Not applicable	Deterministic	No	Symmetric
PCM	Fast	Not applicable	Deterministic	No	Asymmetric

- Repetitive write and rewrite (over-write) of data within the PGM granularity investigation.
- Investigation of non-deterministic operation behaviour:
  - Operations are in average deterministic and the values depend on address offsets.
  - Operation times depend on number and value of physical bits (NOR flash). Hidden logical scrambling impacts the performance based on bit changes.
  - Read Access times depend strongly on the history of operation—started before: Program and Erase suspend modes reduce the worst case access time in case a read is issued to a memory running already an erase operation.

The performance assessment of asymmetric non-volatile memories has to include the application case and the Design for Flash Durability concept on system level. Both parameters have a significant impact on performance, cost and reliability targets of the system design.

Electronic systems based on asymmetric flash memories require additional performance parameters, which has to be considered during the system architecture design phase:

- Write amplification factor;
- Additional physical address space—flash memories or flash blocks;
- Link between logical durability steering parameter and physical degradation parameter;
- Additional cache size or equivalent data analysis capability (HW and SW).

The four major performance parameters are classified with the introduced properties in Table 3.14. SLC and MLC NAND flash is the memory type enforcing significant additional effort to execute the system performance and durability optimization.

References

1. Z. Al-Ars, A. van de Goor, J. Braun, D. Richter, Optimizing stresses for testing DRAM cell defects using electrical simulation, in *Design Automation and Test in Europe Conference and Exhibition*, Munich, 2003
2. R.J. Baker, Memory circuits—the folded array, in *CMOS CircuitDesign, Layout, and Simulation* (Wiley-IEEE Press, Piscataway, 2005), pp. 441–446
3. I. Micron Technologies, NAND Flash Memory Specification MT29F16G08FAA, p. 63 (2006)
4. E. Gal, S. Toledo, Algorithms and data structures for flashmemories. ACMComput. Surv. **37**(2), 138–163 (2005)
5. A. Jagmohan, M. Franceschini, L. Lastras, Write amplification reduction in NAND flash through multi-write coding, in *2010 IEEE 26th Symposium on Mass Storage Systems and Technologies (MSST)*, pp. 1–6 (2010)
6. S. Choudhuri, T. Givargis, Deterministic service guarantees for NAND flash using partial block cleaning. J. Softw. **4**(7), 728–737 (2009)

Chapter 4
Fundamentals of Reliability for Flash Memories

The focus of this chapter is the link between reliability specification parameter and the flash behaviour over lifetime. The V_{th} operation window is the tool to introduce reliability understanding and illustrate the difference between application behaviour and product qualification procedures.
System optimization requires the full reliability knowledge of the selected memory type. The reliability behaviour is different comparing different technology nodes and memory designs from different companies even for the same type of flash memory. Therefore reliability parameter assessment and optimization are an important part of the system development and optimization.

The product specification defines two major reliability parameters

- Data integrity [data retention]
- Number of write operations [P/E cycles, endurance]

The reliability parameter are introduced in depth and classified. A V_{th} window margin analysis is described to visualize the differences between reliability effects and cell and array noise effects.

Reliability factors are introduced to prepare the setup of the Key Performance Indicator setup.

4.1 Reliability Parameter Based on V_{th} Window Margin Analysis

The reliability parameter chapter does not start with data retention and cycle endurance. The introduction into the understanding of reliability effects will be discussed with the typical use case of a memory reading data. The interaction between read disturbance—soft programming —and array noise effects for NAND flash memories and the impact on the functionality is the focus.

D. Richter, *Flash Memories*, Springer Series in Advanced Microelectronics 40,
DOI: 10.1007/978-94-007-6082-0_4, © Springer Science+Business Media Dordrecht 2014

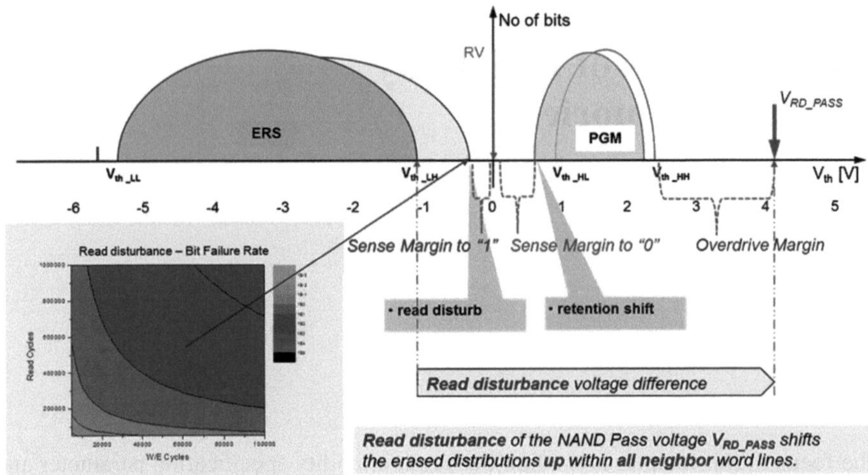

Fig. 4.1 NAND read disturbance reliability—V_{th} window margin read disturbance and retention

4.1.1 NAND Read Disturbance: V_{th} Window Margin Impact

Reading continuously data from a NAND page disturbs all neighbour word lines. The applied pass voltage forces a soft programming to the neighbour—unselected—word lines and will shift the cells up in their V_{th} belonging to the erase distributions. The read disturbance reliability margin will ensure enough sense margin to "1" for a fault free operation within the specified disturbance limits. At the same time the programmed distribution has a tendency to shift down due to data retention.

The amount of the V_{th} shift of disturbed bits depends on the read NAND pass voltage value, the number of read disturbance operations and the status of the endurance stress level—number of pre-cycling—of the erase block. The NAND pass voltage is a kind of soft programming for the erased distribution shown in Fig. 4.1.

The read disturbance increases the number of stored electrons within the non-volatile storage element. A physical change of the storage element is characterized as a reliability effect.

4.1.2 NAND Array Noise Effects: V_{th} Window Margin Impact

NAND array noise effects are impacting the threshold voltage of the cell transistor—enforce a V_{th} shift but do not change the number of stored electrons in the storage element. The important NAND noise effects are listed below and are discussed in detail in the literature [1, 2]:

- Back pattern noise, Source line noise, Well noise,
- Bit line cross coupling, Word line cross coupling,
- Random telegraph noise and Floating Gate cross coupling.

The sense current of a cell within a NAND array is modulated by the string resistance as discussed and introduced in Fig. 2.50. The so called background pattern dependency (BPD) describes the effect that the string current is impacted by data pattern, which influences the series resistance of the string. The first word line sees during programming only erased cells along the string. Programming the following pages (word lines) increases the string resistance and reduces the current capability of the string, which causes an increased V_{th} of the cells programmed on the first word lines.

The background pattern dependency shifts the V_{th} of all cells up shown in Fig. 4.2. The typical application use case is a combination out of page programming and continuous read operation—read disturbance. If the page programming is executed after a long read disturbance time both V_{th} shift values have to be added.

The reliability assessment and optimization is always based on reliability effects physically changing the V_{th} of the cells and array noise effects shifting the V_{th} of the same cells. Both effects depend on the data pattern and the order of execution. The next chapter classifies all required parameters impacting the reliability of a flash memory operated in system applications.

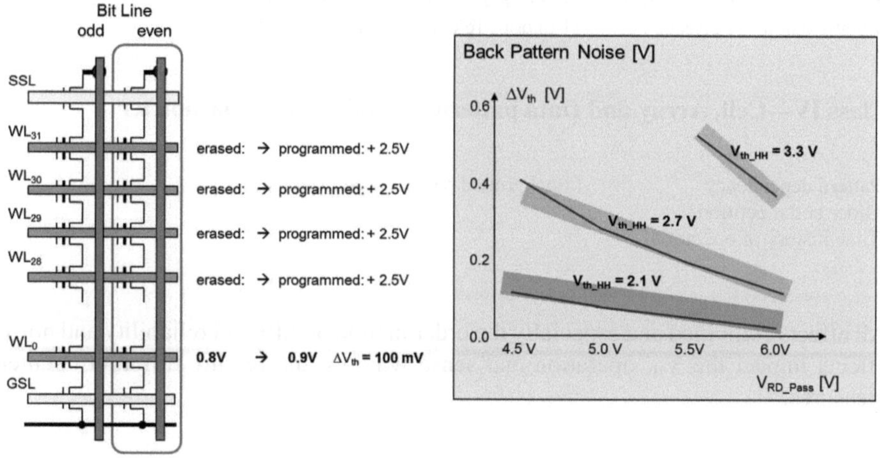

Fig. 4.2 Back pattern noise—delta V_{th} dependency from operation window and read pass voltage

4.1.3 Reliability Parameter Definition

The reliability assessment in this work includes all parameters impacting the V_{th} window margin. The reliability parameters are classified differentiating between cell, array and noise effects.

Class I—Cell based reliability parameter

• Data integrity	[Data retention]
• Number of write operation	[P/E cycles, endurance]
• Number of read operation	[Read disturbance by read cell conditions]

Class II—Array and Cell reliability parameter

• PGM disturbances	[Soft programming, leakages]
• ERASE disturbances	[Neighbour blocks within VG-NOR array]
• Read disturbances	[Neighbour word lines within NAND array]

Class III—Array noise and Cell noise parameter

• Line and well noise effects	[Source line noise, Well noise]
• Line and cell coupling effects	[FG to FG coupling, BL to BL coupling, WL to WL coupling]
• Statistical noise effects	[Random Telegraph Noise, erratic bits]

Class IV—Cell, Array and Data pattern dependent noise parameter

• Pattern dependency	[Background pattern dependency]
• Order and repetition	
• Time history of operation	

All effects combined and especially the order in time of different reliability and noise effects impact the V_{th} operation and sense window and require different counter measures.

4.2 Data Integrity—Data Retention Parameter

4.2.1 Flash Retention Parameter—Cell Retention Behaviour

The data retention is a **class I reliability parameter** defined by the cell concept and the V_{th} window margin defined in the flash product characterization and qualification procedures. The data retention is typically specified over 10 years for a pre-defined data pattern and a storage temperature between 25° and 55 °C.

The physical limitations are defined by the bottom (tunnel) oxide thickness of the cell. Material and thickness combined with the physical stress due to program and erase impact the retention time.

Data retention time of a flash memory is a statistical value and depends on the number of erase operations executed on the flash memory.

- **Temperature:** acceleration of data loss is strongly temperature dependent.
- **P/E cycles:** cycling conditions—voltage values and ramp timing impacts the injected stress

A MLC retention model derived from device characterization is used to illustrate the raw bit failure rate for a continuous P/E cycling shown in Fig. 4.3. This retention model is based on the intrinsic retention behavior affecting all cells, related to material property and visible on small cell entities.

The data integrity has to be guaranteed for the intrinsic and the extrinsic reliability failure types. Extrinsic failures have a typical failure rate of 10^{-6} per MByte and are related to extrinsic defects (point or cluster defects).

4.2.2 *Superposition of Cell Retention and Array Noise Effects*

The flash qualification procedure requires a data pattern with all cells charged—all "0" pattern—to test and stress the highest possible number of cells. The threshold voltage window analysis illustrates the different size of the effective data retention

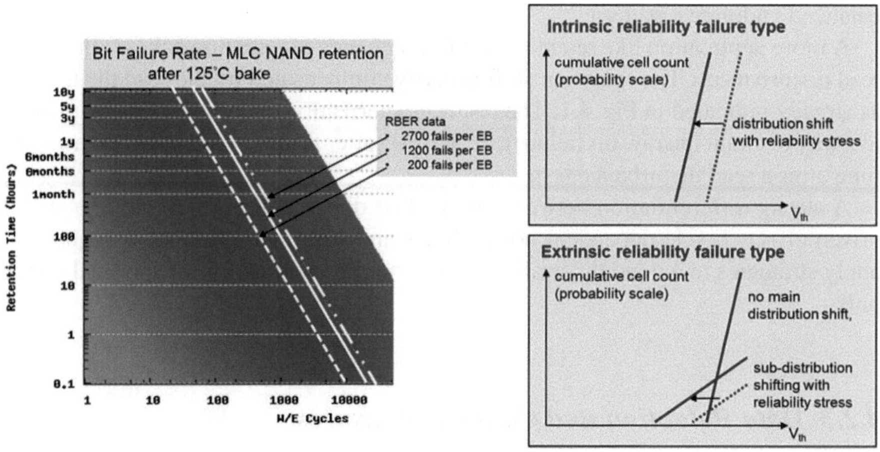

Fig. 4.3 Retention time (BFR) dependency from program/erase cycles—intrinsic versus extrinsic

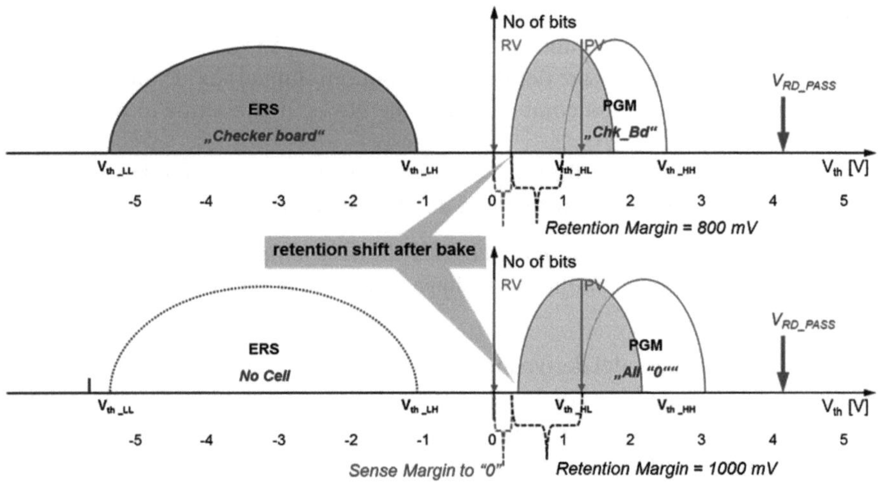

Fig. 4.4 Data retention margin dependency from FG to FG coupling noise effect

V_{th} margin for a "checker board" pattern—low FG to FG coupling—and a "All "0"" pattern—full FG to FG coupling— shown Fig. 4.4.

The figure illustrates an effect often seen during the flash reliability characterization work. The expected worst case reliability stress is not the critical case in the V_{th} window margin analysis due to the superposition with other array and pattern noise effects.

Depending on the specific product program algorithm implementation the reliability margin will be pattern dependent. The second difference is the pre-cycling. A randomly distributed application pattern enforces a different stress compared to a balanced endurance pattern.

A more application like retention test is a sequence of retention bakes followed by read disturbances. The retention shift is partly compensated by the read disturbance as already indicated in Fig. 4.1. The test characterization experiment can be planned in a way to count the raw bit failure rate (RBFR) after the retention bake and a second time after a read disturbance test.

A strong differentiation between charge loss due to cell physics and application driven effects (cycle dependent noise effects and soft programming) is required to apply strategies to extend the available flash margin window for the physical effects only.

4.2.3 Data Retention and Counter Measures

Product data retention characterization has to cover the complete data pattern range and all environmental conditions. The counter measures regarding data retention on

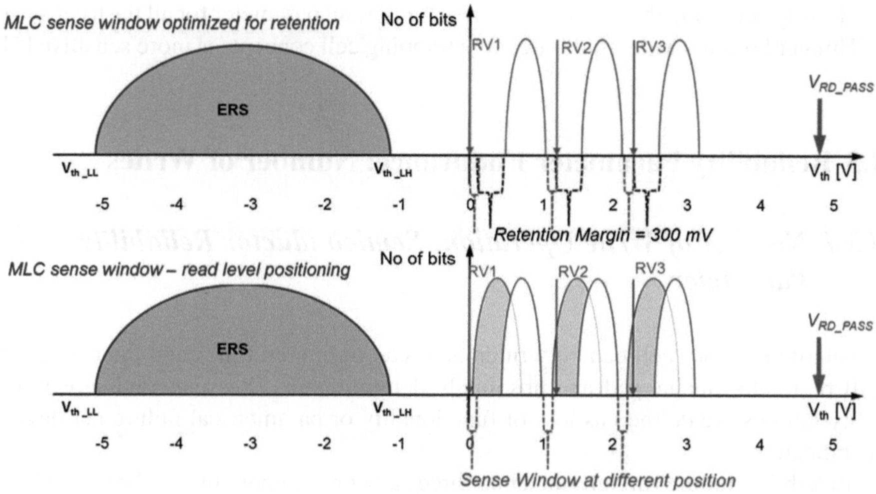

Fig. 4.5 MLC NAND V_{th} window margin—retention shift and read window

product level have to be focused on the physical cell behavior and on the design quality regarding the V_{th} margin definition.

(1) Increase the capability/strength of the ECC coding scheme [3]

- Twice as much ECC capability extends the data retention time for example from half a year into the range of 5 years for a 5k/85 °C MLC retention stress;

(2) Ensure an exact control of max cycling count per erase sector

- Develop a plausibility measure—utilize the erase voltage or the number of erase pulses to guarantee the retention quality of each erase sector;

(3) Design the V_{th} operation window margin with a strong link to the application case

- Design adaptive read techniques to follow the shift of the threshold voltage and ensure the data integrity as shown Fig. 4.5.

The sense window can be influenced directly if the NAND flash offers this functionality or indirectly by an increased read pass voltage. A system optimization strategy has to guarantee enough retention margin or establish advanced read recovery techniques—combine different read parameters with the corresponding ECC algorithm. The charge trapping flash memories have successfully implemented adaptive read techniques with searching algorithm for the window position.

Quality of the bottom oxide is the key material parameter impacting the data retention parameter.

The selection of the storage element—floating gate versus charge trapping—impacts data retention:

- The trap density in the bottom oxide is an important parameter for all flash cells [4];
- Thinner bottom oxide used for charge trapping cell concepts is more sensitive [5].

4.3 Reliability Parameter Endurance: Number of Writes

4.3.1 Number of Write Operation: Semiconductor Reliability Parameter

A volatile memory cell can be written and read unlimited times and the behaviour will be totally unchanged and absolutely deterministic. The semiconductor reliability failures are defined as loss of functionality or parametrical failure per device specifications.

Reliability failures are grouped into three categories depending on the time where the failures occur during the operational lifetime:

- Early failure rate—infant mortality
- Constant failure rate—useful life time
- Wear out failures.

The classical semiconductor failure behaviour and the cumulative failure rate of non-volatile memories over life time are shown in Fig. 4.6.

Every write cycle to a non-volatile memory results into a certain physical stress to a barrier—which has to be tunnelled by electrons or passed by hot electrons—or

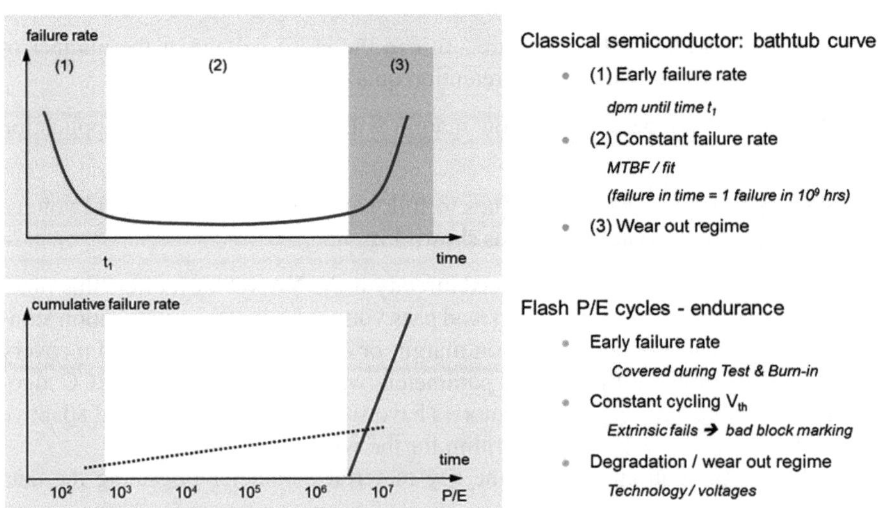

Fig. 4.6 Failure rate during operational life time—bathtub curve for RAM and FLASH

into change of a material—which has to be modified to switch a resistance or other magnetic or ferro-electric parameters.

4.3.2 Program/Erase Cycles: Operational Lifetime Limiter

The physical write and re-write operation will damage parts of the non-volatile cell. Therefore the main research is focused on non-volatile cells with physical write operation which could be applied unlimited times.

Electron based non-volatile cells are based on the physical effects of tunnelling (Band to Band or Fowler-Nordheim) or hot electron or hot hole injection processes. The electrons have to cross an oxide barrier and by crossing the barrier they damage the barrier a little bit. The FN tunnelling operation is based on high voltages and high electrically fields and the damage to the bottom oxide is irreversible.

Hot electron and hot hole injection processes are using lower voltages and the induced damage could be repaired partially by voltage or temperature stress operations.

The typical flash cell degradation behaviour over the number of Program/Erase cycles is shown in Fig. 4.7 for a constant voltage cycling—same program and same erase voltage is applied for 100.000 cycles. The flash cell is degrading over the cycle count. The operational V_{th} window will become smaller and shift up.

The degradation of the V_{th} operation window is only indirectly visible on flash product level. The program and erase algorithms ensure the exact position of the programmed and the erased distributions. The analysis of typical program and erase times of a NAND flash product shows the same effect clearly as shown in Fig. 4.8.

Table 4.1 summarizes NAND flash reliability failure modes linked to single or combined flash operations. Endurance is combined with reliability stresses and a classification into intrinsic or extrinsic failure types is added for each flash operation and failure mode combination [6, 7].

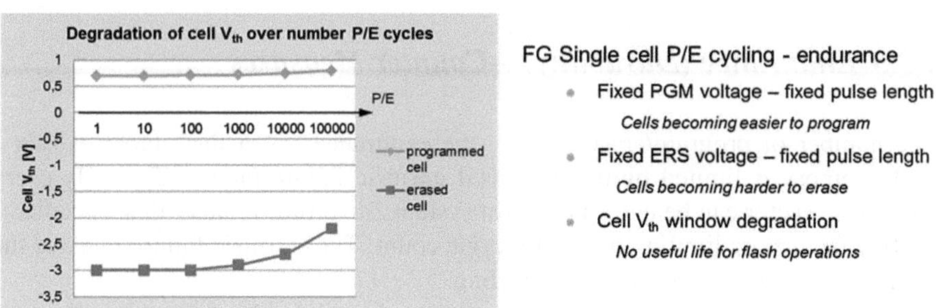

Fig. 4.7 Effect of flash cell cycling on program and erase behavior—cell V_{th} dependency

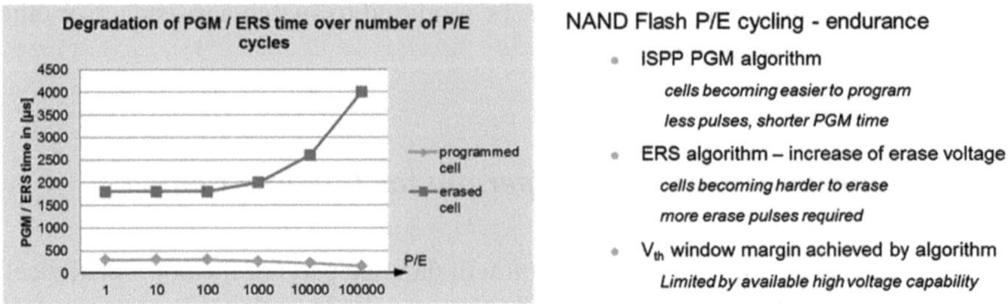

Fig. 4.8 Effect of flash cell cycling on program and erase behavior—PGM/ERS time dependency

Table 4.1 Flash reliability failure modes—classification into intrinsic or extrinsic

Flash operation	Failure description	Physical failure	Classification
Endurance	End of life regime	Oxide defect; wear out	Intrinsic
Endurance	Bad Block Management	Oxide defect	Extrinsic
Endurance and read	Erratic programming	Fast tunneling enhanced by hole cluster	Extrinsic
Endurance and read	Erratic disturbance	Pre-stage of erratic programming	Extrinsic
Endurance combined with data retention	High temperature cell V_{th} shift	De-trapping from oxide	Intrinsic
Endurance combined with data retention	Room temperature cell V_{th} shift	Trap assisted tunneling	Extrinsic
Endurance and read disturb	Read disturbance	Soft programming	"Extrinsic"
Endurance and read	Random telegraph noise	V_{th} fluctuation by trap	"In between"

4.3.3 Endurance (Durability)—Counter Measures

The number of program/erase cycles—the endurance—of a flash block within a flash memory is limited by the described intrinsic failure modes. Typically every flash erase sector can be cycled on a test system 2 to 5 times longer than specified.

Two factors are limiting the useful cycle count: The extrinsic failure rate and the retention dependency from the pre-cycling.

The extrinsic failure rate is strongly influenced by all kinds of array disturbances. The array structure defines the number of gate and drain disturbances every cell sees during a program operation of a complete sector.

- The Virtual-Ground NOR array for large block sizes is a worst case example for a disturbance assessment. Every program pulse to a cell implies a gate disturbance to all cells along the word line and a drain disturb to all cells along the bit line.
- The NAND array applying FN-tunnel programming reduces the number of gate disturbs to the number of program pulses required to program the corresponding word line.

Another aspect for a non-volatile memory is the ***defect propagation*** probability of program and erase failures during lifetime and their impact on other sectors. The NAND specification is prepared for the statistical influence on large memories with billions of cells. Program or erase failure within an erase sector will typically only influence the sector, which will be identified by the internal algorithm control and can be marked outside of the memory as ***bad block***. The technique to identify bad blocks, restore the remaining data and replace the block is called bad block management (BBM).

The retention dependency from pre-cycling was already discussed and illustrated in Fig. 4.3. The following overview includes intrinsic and extrinsic reliability effects for two positions within the normal endurance operation time span shown in Fig. 4.9.

A higher cycling count corresponds to a larger retention shift (and a larger distribution widening) and results into a higher extrinsic failure rate.

4.4 Reliability Margin Optimization on Product Level

The flash reliability margin optimization is done during the flash design and technology development. The array architecture has a major influence on the achievable product reliability parameters which are derived from cell concept physics. Known

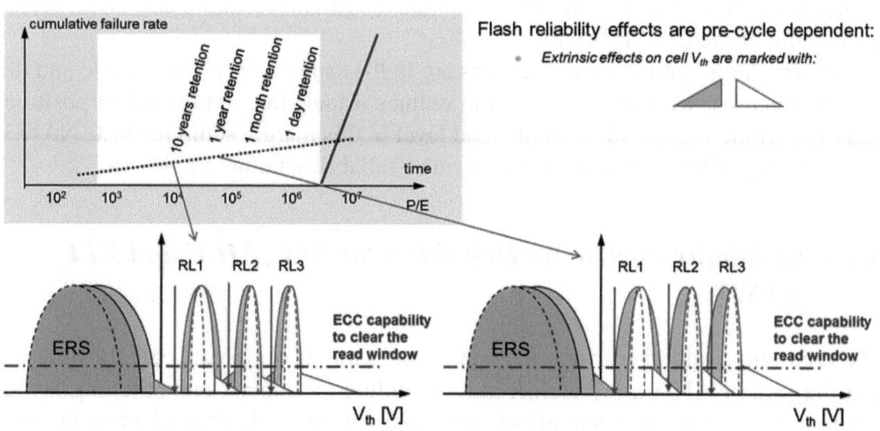

Fig. 4.9 NAND retention dependency from pre-cycling including intrinsic and extrinsic reliability effects

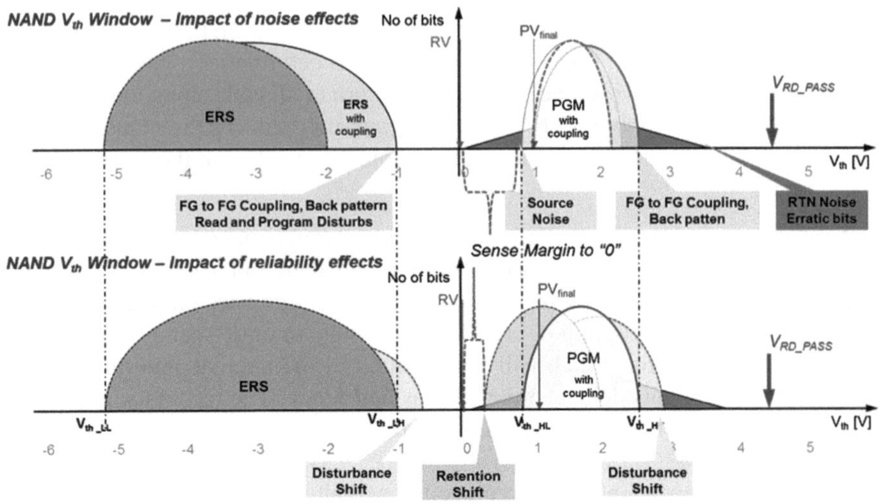

Fig. 4.10 SLC NAND V_{th} window margin –impact of noise and afterwards reliability effects

array weaknesses impacting the reliability parameter and noise effects should be ideally compensated by a robust cell concept for these issues.

4.4.1 Reliability Parameter and Margin Optimization

An ideal programmed distribution is theoretically as small as the program voltage increment. The array and cell noise effect enlarge the programmed and erased distributions already after programming is finished. The programming of the neighbour cell or the complete erase block disturbs the cell further on. Figure 4.10 shows the influence of all combined noise effects on the distribution width. Each effect forces a widening of the final distribution.

All array noise and reliability effects are in the same order of magnitude and the SLC NAND V_{th} window margin setup defines a fault free read window position. Only the erratic bits could cross the read level in this margin setup for SLC NAND. The ECC capability is driven by the extrinsic reliability failure modes.

4.4.2 Reliability Parameter Definition for SLC, MLC and XLC NAND

The reliability parameter example assumes the same floating gate cell construction and the same NAND array architecture including all contact and stitching layout details. All cycling dependent effects and noises are strongly dependent on the max V_{th} shift during programming:

$$\text{deltaV}_{\text{th\_max}} = \text{V}_{\text{th\_HH}} - \text{V}_{\text{th\_LL}}.$$

For a given delta $V_{th\_max} = 8.5$ V and a pre-cycling of 20k the different margin settings are analysed to derive key performance parameters.

The defined SLC NAND margin setup shown in Fig. 4.11 covers all effects and ensures still enough sense margin. A learning about the different reliability effects is not achieved and not necessary.

The MLC NAND margin setup shown in Fig. 4.12 ensures a small sense margin for 400 mV retention shift. The array noise effects have to be eliminated by design, sensing and algorithm measures.

The XLC NAND sense window after a retention shift of 250 mV as shown in Fig. 4.13 is already a superposition of array noise and extrinsic reliability failure modes. The reliability margin optimization is an iterative process based on extensive silicon characterization to define the position of program and read verifies, to reduce the program pulse voltage increment and to maintain or reduce the max V_{th} shift during programming.

Table 4.2 includes a subset of all noise effects and the result for all NAND types is a small sense window. The superposition of all reliability and noise effects can show a negative margin for the sense window too. In this case design and

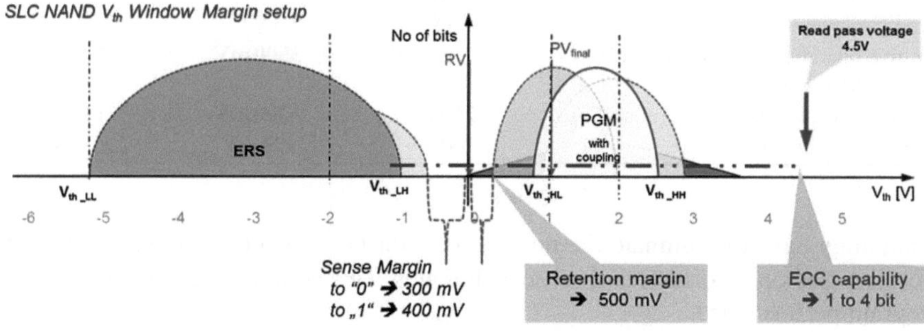

Fig. 4.11 SLC NAND V_{th} window margin setup—retention margin

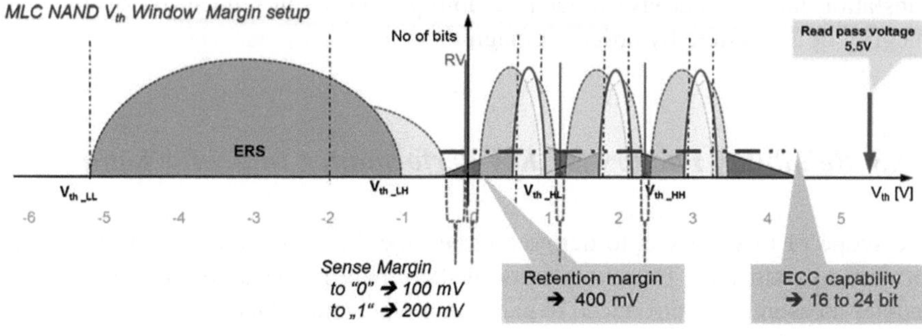

Fig. 4.12 MLC NAND V_{th} window margin setup—retention margin

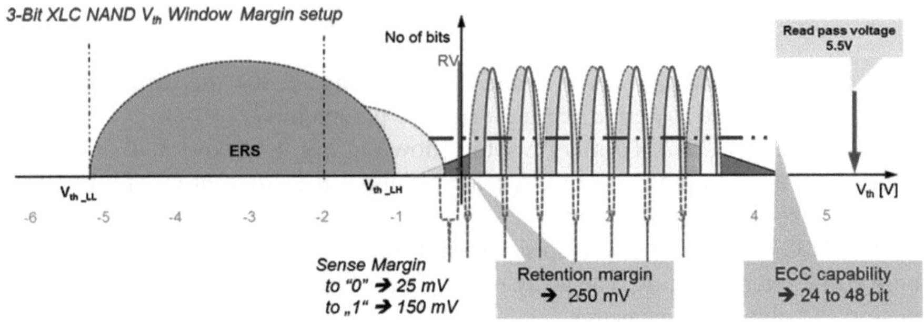

Fig. 4.13 3-bit XLC NAND V_{th} window margin setup—retention margin

Table 4.2 Reliability parameter and array noise effects—V_{th} window margin setup

	SLC NAND 1Bit/cell	MLC NAND 2Bit/cell	XLC NAND 3Bit/cell
Target program distribution width	950	300 mV	100 mV
Program step	900 mV	250 mV	75 mV
FG to FG coupling (for last V_{th} shift)	850 mV	250 mV	25 mV
Random telegraph noise	150 mV	100 mV	50 mV
Back pattern effect	150 mV	50 mV	25 mV
Retention loss => Retention margin	**500 mV**	**400 mV**	**250 mV**
Read pass voltage level	4.5 V	5.3 V	5.5 V
Positive operation window (Read Pass overdrive voltage ~1,8 V)	2700 mV	**3500 mV**	**3700 mV**
– Target width + noise and reliability	2600 mV	**3300 mV**	**3150 mV**
– Sense window margin	100 mV	**<75 mV**	**<75 mV**

technology have to eliminate the noise effects further. The elimination of noise effects is based on algorithm and sense techniques. This additional design effort costs program performance.

The flash V_{th} window margin knowledge supports the right balancing on system level regarding independent hardware channels and threshold levels within the flash translation layer. Otherwise hidden reliability effects can reduce the overall data integrity of the system by orders of magnitude.

4.4.3 Reliability Factors for Key Performance Indicator Setup

The scope of this work is to define a target line for important flash memory performance parameters. The complexity of the reliability behavior and the required counter measures are introduced to gain a complete understanding of the interaction between reliability stresses, noise effects, flash V_{th} operation window settings and the application case.

The performance indicator methodology requires basic quantitative values to compare memory array architectures. Therefore, we combine first the reliability parameter endurance and the density of different multi-level or multi-bit per cell flash types for the typical endurance specification values:

- *SLC factor* = 100.000 * 1 *Gbit* / 100.000 = 1 → $RF_{EnduranceDensity} = 25$
- *MLC factor* = 10.000 * 2 *Gbit* / 100.000 = 0.2 → $RF_{EnduranceDensity} = 5$
- *XLC factor* = 1.000 * 4 *Gbit* / 100.000 = 0.04 → $RF_{EnduranceDensity} = 1$

The following **R**eliability **F**actors (**RF**) have been derived from a detailed V_{th} margin analysis and an exhaustive reliability and noise characterization work on flash memories from two different technology nodes. All reliability factors are normalized for a bit size of 1 F^2.

- The **R**eliability **F**actor based on **Program Throughput** differences is derived from a margin analysis with reduced retention values as shown in Fig. 4.9.
- The **R**eliability **F**actor based on **Endurance** and **Density** combines the specification values is calculated above for 1.000, 10.000 and 100.000 endurance cycles and the corresponding density.
- The **Data Throughput** based **R**eliability **F**actor assumes a sequence of 75 % read and 25 % write operations statistically distributed for the reliability margin analysis.

Table 4.3 compares the NAND endurance specification ratio with the introduced Reliability Factors for NAND flash memories to select the best fitting setup for the performance indicator analysis applicable for the relevant application cases.

Table 4.3 NAND reliability factors—normalized for a bit size of $1F^2$

	SLC NAND 1Bit/cell	MLC NAND 2Bit/cell	XLC NAND 4Bit/cell
NAND Endurance specification ratio	100	10	1
NAND Reliability Factor—$RF_{ProgramThroughput}$	49	7	1
⇨ *The application relevant endurance behaviour is the relevant reliability parameter*—$RF_{ProgramThroughput}$			
NAND Reliability Factor—$RF_{EnduranceDensity}$	25	5	1
NAND Reliability Factor—$RF_{DataThroughput}$	16	4	1

4.5 Flash Memory Reliability Summary

Reliability optimization of non-volatile memories is all about V_{th} window margin optimization. Flash memories have one strong benefit: The non-volatile storage element is embedded in a MOS transistor with an excellent on/off ratio.

Reliability parameters for NAND flash memories are introduced and classified into

- Class I—Cell based reliability parameter —*physical change of cell* V_{th};
- Class II—Array and Cell reliability parameter —*physical change of cell* V_{th};
- Class III—Array noise and Cell noise parameter —*shift of cell* V_{th} *or erratic change of cell* V_{th};
- Class IV—Cell, Array and Data pattern dependent noise parameter.

Reliability stresses and noise effects are in the same order and the static or dynamic—during operation—selection of the correct counter measure for each reliability class is the only way to optimize the reliability behaviour. All reliability stresses and noise effects have to be known (correctly predicted) at begin of the flash memory design development.

ECC and adaptive read techniques are very powerful counter measures, but only for the correct reliability failure class. Reliability optimization is all about understanding and real time prediction of superposition of stress and noise effects.

This short overview is the end of the fundamentals of flash reliability. A more detailed discussion has to be vendor and application specific.

- Data integrity [data retention]
  - Strongly dependent on application/use case.
  - Wear levelling or data remapping on system level can operate as a kind of refresh.
  - Cell physics and cell architecture define the physical reliability limit—e.g. floating gate cells have a very stable V_{th} over time in contrast to charge trapping cells characterized by moving distributions even without disturbances.
- Number of write operation [P/E cycles, endurance]
  - Defines end of operational functionality—Cell V_{th} window closing due to physical cell damage
  - Wear levelling strategy and reduced retention strategy
- Number of read operation [read disturbance, neighbour effects, leakage]
  - The combination out of extended endurance and extensive read disturbance is becoming very critical from node to node, but can be controlled well application case specific.
- Cell and Array interference [program disturbance, coupling effects]
  - Array noise effects are strongly memory array dependent.
  - Physically segmented arrays have a strong benefit in terms of disturbance.

  - The block write operation principle reduces noise effects dramatically, writing a complete erase block within one or concatenated operations.
- Statistical bit effects [Random Noise, erratic bits]
  - Defect and random driven failure events are driven by statistics or by failure density per volume.
  - The cell and technology development has to achieve for these effects values low enough so that they can be handled by error correction.

The reliability assessment of a selected memory is the most difficult task. The effect of assumed innovation on reliability parameters can only be judged by characterization of memories out of volume production. First time published reliability values for emerging memories are based on test chip characterization or low volume sample productions. Years later the product specification values guaranteed for volume production are one or two orders of magnitude lower compared to the originally published values.

Therefore a qualification procedure of a flash memory can cover only a subset of the described complex reliability and noise behaviour. A serious reliability discussion can be made only on statistical data based on high volume memory production and requires continuous reliability test procedures monitoring the production quality.

The architecture optimization of the memory sub-system starts with conservative assumptions based on today's specifications. A thorough analysis in depth creates evidence in the achievable potential of the specific material, design or algorithm innovation. Reliability improvement has to become a part of the system optimization concept.

The **D**esign for **F**lash **D**urability system concept develops software solutions supported by low level flash algorithm parameter. The effective margin loss per page and the physical degradation of the flash erase blocks can be derived based on the introduced V_{th} window margin principles. A reliability guard band strategy can be developed to extend durability values from case to case to optimize the overall performance and reliability of the system.

Reliability parameters will be included based on the introduced **R**eliability **F**actors in the non-volatile memory assessment based on the key performance indicator methodology.

References

1. T. Tanaka, NAND flash design, in *ISSCC, Non-Volatile Memory Circuit and Technology Tutorial F1*, San Francisco, (2007)
2. C. Friederich, Multi level programming scheme with reduced cross coupling, in *Control of harmful effects in program operation of NAND flash*. (Shaker, Aachen, 2011)
3. F. Sun, S. Devarajan, K. Rose und T. Zhang, Design of on-chip error correction systems for multilevel NOR and NAND flash memories, *Circuits, Devices & Systems, IET*, 241–249 June 2007

4. S. Shuto, M. Tanaka, M. Sonoda, T. Idaka, K. Sasaki, S. Mori, Impact of passivation film deposition and post-annealing on the reliability of flash memories, in *35th Annual Proceedings, IEEE, Int. Reliab. Phys. Symp.* (1997)
5. T.-H. Hsu, H.-T. Lue, S.-C. Lai, Y.-C. King, K.-Y. Hsieh, R. Liu, C.-Y. Lu, Reliability of planar and FinFET SONOS devices for NAND flash applications—Field enhancement vs. barrier engineering, in *VLSI Technology, Systems, and Applications, VLSI-TSA*, pp. 154–155. (Hsinchu, 2009)
6. N. Mielke, H. Belgal, I. Kalastirsky, P. Kalavade, A. Kurtz, Q. Meng, N. Righos, J. Wu, Flash EEPROM threshold instabilities due to charge trapping during program/erase cycling. IEEE Trans. Device Mater. Reliab. 4, 335–344 (2004)
7. R. Degraeve, F. Schuler, B. Kaczer, M. Lorenzini, D. Wellekens, P. Hendrickx, M. van Duuren, G. Dormans, J. Van Houdt, L. Haspeslagh, G. Groeseneken, G. Tempel, Analytical percolation model for predicting anomalous charge loss in flash memories. IEEE Trans. Electron Dev. 51, 1392–1400 (2004)

Chapter 5
Memory Based System Development and Optimization

The availability of multi-core microprocessors, fast embedded RAM's and large non-volatile solid-state memories with high data bandwidth influence the system architecture. The principles of optimization of high performance microprocessors and electronic systems are well described in the literature [1–3]. The established hardware-software co-design has to be improved by a concurrent design approach of the system and the application specification and software [4].

The end of the microprocessor GHz race has enforced the adaption of multi-core microprocessor architectures. The need of energy optimized systems continuously forces architectures with distributed calculation power surrounded by embedded memories. The calculation power has to be increased generation by generation which is ensured by multi-core microprocessor architectures.

The semiconductor memories were developed as commodity parts fulfilling system requirement specifications in the past. The full system optimization potential of solid-state non-volatile memories can be utilized if the system architecture including the multi-core microprocessor is designed to offer the best support for flash memories. The issue is how to reduce the complexity of flash based memory sub-systems to derive cost, durability and performance optimized decisions on system architecture level based on quantitative values.

The subject of this work is to introduce a model-based quantitative performance indicator methodology applicable for performance, cost and reliability optimization of non-volatile memories and memory-centric systems. This chapter describes the current separated development processes, introduces a set of efficiency parameter and describes the non-volatile complexity dilemma.

5.1 Memory-Centric System Specification

The semiconductor industry has executed over decades the CMOS shrink roadmap. The increased transistor switching speed has enhanced the performance of existing system architectures automatically. The gap between microprocessor calculation

D. Richter, *Flash Memories*, Springer Series in Advanced Microelectronics 40, DOI: 10.1007/978-94-007-6082-0_5, © Springer Science+Business Media Dordrecht 2014

power and memory access times and bandwidth especially for non-volatile memory sub-systems—hard disc drives—became bigger year by year. The target to optimize system performance and energy consumption was emerging with the memory–centric mobile applications growth.

Year by year upcoming applications require a continuous increase of system performance. The system architecture is becoming more data centric and the processor architecture is adapting the concept of parallelism. Multi-core processor architectures restart the concept phase for data path optimization and memory partitioning. Non-volatile memories with high data bandwidth are suitable to replace large volatile memories like DRAM due to their better energy balance.

A memory sub-system can be developed based on a single memory type, a combination out of two or three memory types within a package or a system with a dedicated memory controller and different types of memories on board. Data path optimization requires fast read access combined with highest data bandwidth which requires memory architecture optimization into two contradicting directions for all cost-competitive memories. The multi-core microprocessor architecture is adding on this contradicting requirement the need for a distributed data space.

A requirement based memory sub-system development process is introduced step by step.

5.1.1 System Performance Values: Memory Sub-System Requirements

5.1.1.1 System Requirement Specification: Performance Values

System architecture and concept decisions are based on a requirement driven development process. The requirement specification for the memory sub-system is derived from performance modelling based on memory specification and memory product roadmap data.

The requirement specification defines five groups of parameters for a memory sub-system:

- **Memory Density per volume** (per package)
  - The system dimensions are fixed. The available space for the memory sub-system is a key criterion for the selection of a memory type.
  - A memory density increase roadmap is expected:
    Roadmap: 2012: 4 GB; 2013: 8 GB; 2014: 16 GB; 2015: 24 GB per package
- **Cost per GBit** / Cost per Bit
  - The **B**ill **o**f **M**aterial of a memory sub-system decides about memory size and type. The BOM of mobile applications is dominated by memory costs in contrast

to stand-alone computer applications, which are dominated by processor and electronic parts.
  - A memory cost decrease roadmap is expected:
    Roadmap: 2012: 2$/Gbit; 2013: 1,3$/Gbit; 2014: 0,88$/Gbit

- **Data Throughput**
  - **Read Data Rate** / Read data Throughput 100 .. X00 MB/s
  - **Write Data Rate** / Write data Throughput 100 .. X00 MB/s
  - **Access Cycle Time** 10 .. X00 ns
  - The values are typically derived from high level system simulations assuming symmetric and deterministic memory sub-system behaviour.

- **Durability and Reliability values**
  - Successful executed read access operations 10^{12}
  - Successful executed write access operations 10^9
  - 100 % data integrity over operating life time 5... 10 years
  - Non-volatile retention time without power 2... 10 years
    Typical and worst application cases are specified.

- **Energy consumption**
  - Data operation / average RMS current values 25... 50 mA
    Energy needed for each operation;
  - Standby / standby current 100... μA
    Energy needed to keep data information, during the time no access is executed to the memory (e.g. energy required for wear leveling).

All values are carefully defined and an overall system cost and performance balancing is done. Afterwards the requirement specification is finalized and reviewed and the memory selection process is going to start.

The ideal target configuration of a memory sub-system would be based on one single memory type, which fulfils the following **memory performance parameters**:

- **short read access time**—within 5 to 10 processor clock cycles
- **high data bandwidth**—support bandwidth requirements of one microprocessor-core
- **bit wise read, write and re-write capability**
- **non-volatile**—support application targets in the mobile and server market.

Energy consumption of a memory sub-system over lifetime is becoming a key decision parameter. For mobile application energy consumption is directly linked to battery lifetime. For data servers energy consumption increases the cost of operation and becomes a serious cost factor larger than the cost per bit parameter. A lot of energy is wasted to sustain large amounts of data either in huge DRAM caches or on ultra-fast HDD's [5].

5.1.1.2 Flash Memory Specification: Performance Values

Flash memory specifications are developed based on a memory roadmap process, inputs from standardization (JEDEC), customer feedback and the most important point for commodity products the compatibility in terms of interface, commands and package to the last generation.

NOR and NAND flash specification values are introduced in detail in the performance chapter. Flash memory performance parameters are clustered into groups comparable to the system requirement specification.

- **Memory Density per package**
  - The memory density for a given package type (TSOP or BGA) defines the maximum capacity of a mobile system generation.
  - Die size [mm^2]
- **Cost per bit**
  - Bits per cell—the SLC, MLC and XLC product roadmap decreases the cost per bit.
  - Technology node [nm]
  - Normalized cell efficiency [%]

  Additional spare area and related ECC capability
- **Data Throughput** (typical values specified)
  - **Read** Data Throughput 100 .. X00 MB/s
  - **Program** Data Throughput 40 .. X00 MB/s
    Program Page Size 4 KB; 8 KB; ..
  - **Write** Data Throughput 20 .. X00 MB/s
    Erase Block Size 512 KB; 1 MB; ..
  - **Random Access** Cycle Time 10 .. X00 μs
- **Energy consumption**

Data operation/average RMS current values 20 .. mA
Standby/standby current 20 .. μA

- **Reliability values**

P/E cycles/endurance 10^4 (MLC); 10^5 (SLC)
100 % data integrity/retention 5 .. 10 years

The cell, array and bit efficiency parameters and the link between technologies shrink roadmap and memory product design is introduced in Sect. 5.2 in detail.

The final memory specification is reviewed. The cost, performance and innovation balancing is assessed per product design and compared with competitor devices. Afterwards the target specification is released and the memory development process starts.

5.1.1.3 Design for Flash Durability: Link Between Application and Flash

System performance, density, cost and energy requirements are mapped to the flash memory specification and the system architecture is developed. The system durability parameter cannot be mapped to a flash memory. Chapters 3 and 4 have introduced reliability and noise effects as well as hardware and software based wrapper solutions required to match system and flash memory parameters best. This is an iterative process which impacts all components of the system architecture including the memory array.

The Design for Flash Durability parameters are summarized:

- **Memory Density increase**—hidden physical address space
  - Additional memory density of 5–20 % of the total density;
    Write Amplification factor depends strongly on the additional spare area;
- **Durability target** based on the **Write Amplification factor** and the **System memory size**
  - Allowed and restricted combination of concatenated reads and writes;
  - Extended durability concept;
  - Parallel execution of application data transfer and wear level data transfer;
    energy equivalent for reliability enhancement;
- **Data Retention** in combination with reliability enhancement features
- Additional **volatile or non-volatile cache size** enhancement.

An example for a combined reliability and energy assessment parameter set for a flash based memory sub-system is listed below:

Average Write energy—typical block size (512 kByte or 1 MByte)—including wear leveling;
Average Read energy—typical block size (512 kByte or 1 MByte)—including wear leveling;
Restricted combinations for the concatenated duration of read and write operations
Refresh time interval to sustain data—typical medium size 4 GByte.

Detection and sustaining time intervals for data retention and array disturbance have to be carefully developed using the application environmental conditions. The dependency of these intervals from the data size of the memory sub-system is an important parameter. The energy consumption of a memory sub system over lifetime has to include the effort for the reliability enhancement.

The application case—the statistical operation time of read and write—and the impact of this concatenation on reliability and durability values is the key performance parameter for a reliability driven system optimization.

5.1.2 Application Specific Requirements: Impact of Multi-Core Microprocessor

Multi-core microprocessor based systems are designed for parallel execution of different system tasks. Each task can have different requirements to the memory subsystem in terms of access frequency and order of read and write operations. The system architecture defines the dedicated memory size per core and the size of the shared memory for all cores. The cache sizes of multi-core microprocessors are standardized for high volume and low-cost production.

The definition and development of the memory sub-system surrounding the microprocessor is an application specific task, which can have a significant impact on the overall achievable application performance. Figure 5.1 illustrates the performance gap between the external volatile memories assumed with 8.5 Gbyte/s data bandwidth and two storage systems (left HDD and right SDD) within a multi-core microprocessor based application example.

The system specification can achieve an alignment of the internal cache page sizes of the micro-processor and the page sizes of a low latency non-volatile memory. This page size alignment enables the possibility for a fast direct channel between the second level cache and a fast low latency non-volatile memory solution.

Low latency solid state storage based on fast NAND or enhanced VG NOR flash can improve the overall performance and energy efficiency at the same time. Instead of sending always all required data through all DRAM and cache memory hierarchies the system can use an application specific distributed data storage architecture as shown in Fig. 5.3 on the next page.

The page size of the memory sub-system has to be specified in such a way that it fits or is a multiple of the processor cache page size. A Solid-State Disk (SSD) is equipped with more than 64 NAND memory devices. This multiple device architecture offers the capability for a speculative hidden read of the predicted next page address on 128 NAND page sizes in parallel which results into 512 kByte to 1 MByte cached data within the page buffer circuits of all NAND flash devices.

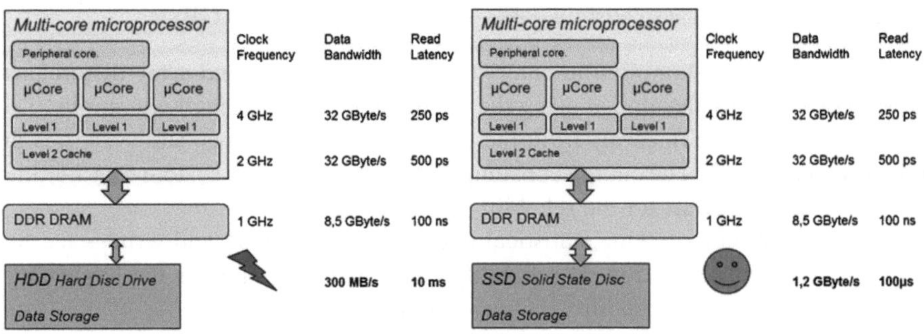

Fig. 5.1 Memory hierarchy optimized for a multi-core microprocessor based system

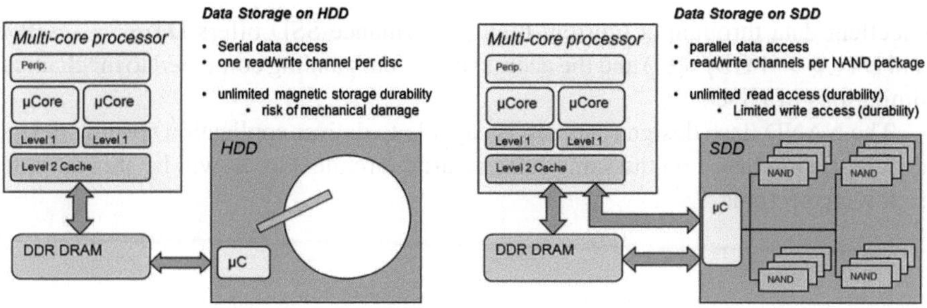

Fig. 5.2 Memory hierarchy based on a parallel access SSD versus a serial access HDD storage system

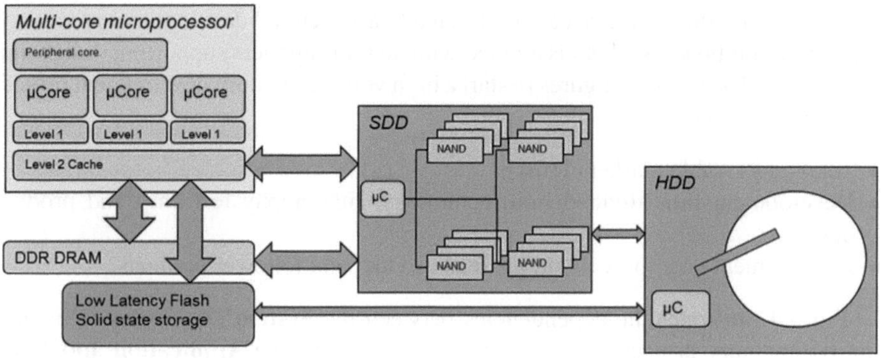

Fig. 5.3 Optimized Memory sub-system for multi-core microprocessor applications

The sequential page access time to a NAND is comparable to the synchronous read cycle to a DRAM. The energy to copy the expected data always from HDD to main DRAM is completely saved. Figure 5.2 illustrates the example to bypass the DRAM and load the data directly from NAND flash.

The typical multi-purpose multi-core system requires a fast read access and a fast read throughput to the non-volatile storage medium. The growth of the required DRAM memory size slows down and the size of the flash based memory sub-systems will increase driven by the cost per bit reduction of the MLC NAND flash memory roadmap from generation to generation.

A dedicated non-volatile memory sub-system for multi-core processor architectures enables fast access to distributed memories and could be based on different application specific adapted non-volatile memory configurations as shown in Fig. 5.3.

The application example shown in Fig. 5.3 indicates different application specific requirements to cost, energy, reliability and performance optimized non-volatile storage systems. The low latency flash solid state storage memory sub-system has to compete on data rate and on read access cycle time with volatile memories (DR > 8 GB/s and $t_{RACT} < 1\,\mu s$). The Solid State Disk (SDD) memory sub-system delivers an excellent data throughput (current high performance SSD offers $DT_{RD} > 2\,GB/s$ and $DT_{WR} > 1\,GB/s$ [6]) and the access time is outstanding compared to mechanical limitations of HDD.

The NAND flash design roadmap is capable to deliver application specific design differentiation based on the same cell and array architecture as will be shown in the following sections.

5.1.3 Memory Technology Roadmap: Impact on System Requirement

Technology roadmaps are necessary to enable a structured development process of electronic components which is aligned with all tool suppliers supporting design and technology. The business figures to start a high volume memory development project have to be worked out:

- Total addressable market (TAM);
- Development time divided into technology, lithography test chip and product design;
- Development cost for technology, lead product and follower products.

Fig. 5.4 illustrates the dependencies between the **System Development** including the memory project cost and the project timeline, the **Application and Market Requirement**—the key system performance parameters and the **Technology Roadmap**.

The product development life cycle has to balance these areas. The system development has to fulfil the market requirements based on the product and technology roadmaps. This work will put the focus on the key enabler features which have to be derived based on the pre-selected technology roadmap.

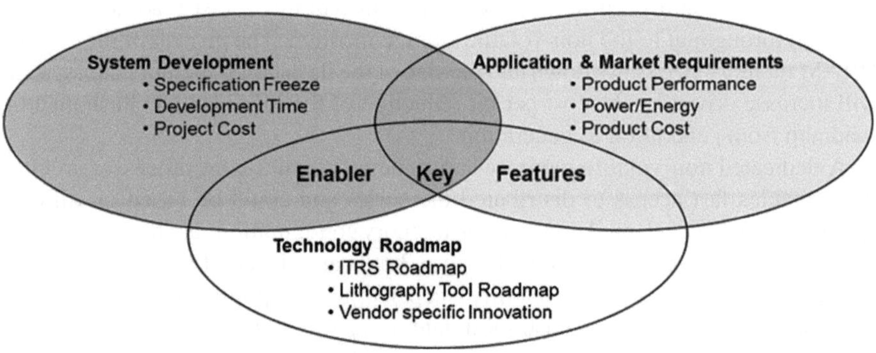

Fig. 5.4 System development—Matching market requirements with rechnology roadmaps

The classical system development starts the architecture definition with a selection of the memory concept. This development approach is not going to have the capability to fully utilize the innovative potential of the memory optimization.

A system definition and optimization process based on the fundamentals of the selected memory and the predicted technology and innovation roadmap has to be used in the future. The system development is becoming more memory centric than in the past and technology innovations are driving the success of application as described in the following citation:

"The principal applications of any sufficiently new and innovative technology always have been—and will continue to be—applications created by that technology."—Herbert Kroemer [7]

Due to the fact that the development time is in the range of years and the system lifetime too, the decisions made during the architecture definition and optimization process influence the competitiveness of the system five or more years ahead. The target of this work is to develop tools and methodologies to support a fast and structured identification process of key benefits of non-volatile memory architectures.

5.2 Memory Efficiency Parameters for Competitiveness

The memory design and technology development is a restricted process based on a structured development handbook. Technology and architecture decisions are made under the focus to ramp within the targeted time window a competitive product. A serious effort is spent to develop memory design and technology innovations. The decision to use these innovations is based on the achievable product competitiveness and the impact on the memory product development timeline.

The memory development is based on a couple of important efficiency parameters. These memory specific efficiency parameters are introduced in the following sections.

5.2.1 Memory Density Parameter

The achievable density of a memory is a generic key parameter. The criterion for the selection of the technology node is the maximum allowed die size, which would fit into the selected package type. The package type for high volume memories is defined by standardization working groups. JEDEC takes the ownership for the DRAM standardization and ONFI is driving the NAND standardization.

Memory publications covering innovation and new architectures are linked to diagrams with unit die size per MB or Bit Density per cm^2. One example is given in Fig. 5.5.

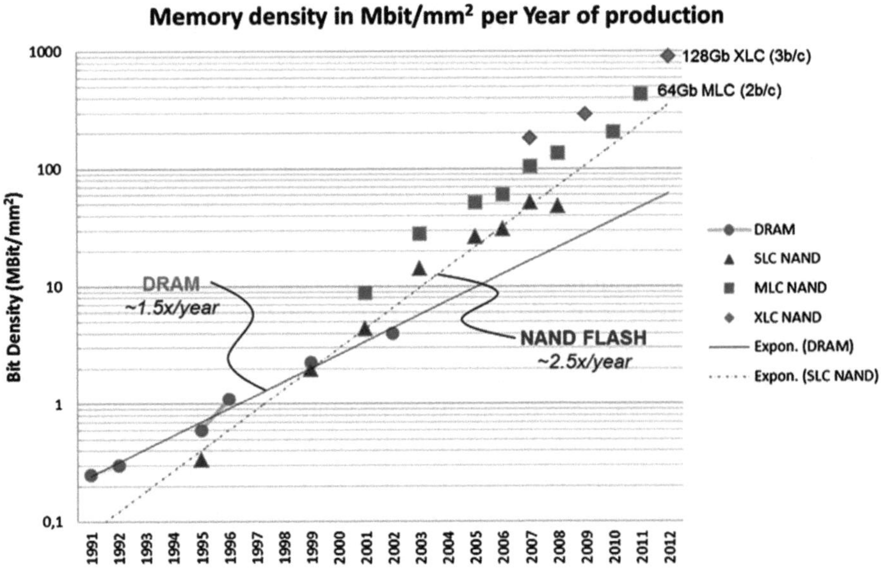

Fig. 5.5 Bit density overview of high volume memories—DRAM and NAND—over time [8]

The memory development process for next generation design concepts can be described by the following rules, which normally are applied step by step with a limited amount of iterations:

- Selection of the largest technology node in nm to achieve the specified memory density—for example 64 GBit—with a die size fitting into the standardized and required package type.
- Generate solid feedback for the critical lithography spaces and support the selection process of the final technology node mainly driven by availability of the lithography within the required time window including cooperation with the tool vendors.
  - Successful applied approaches with larger technology nodes and larger die sizes are less economical, but this drawback is compensated by technology ramp experience. Frequent introductions of new technology nodes result into a fast learning cycle covering all technology effects and fast feedback for design innovations.
- Support the innovation roadmap—design or technology improvements—to fix known serious issues of the next lithography node due to all effects responsible for margin losses affecting memory operations.
- Achieve the memory density with enough manufacturing margin in die size so that 4× or 8× stacking of dies in a MCP package is possible without significant yield loss due to the package manufacturing process.

The company specific design decision is translated into a bit density per area. This parameter is different for every memory product and density. A direct assessment of memory technology and design competitiveness is not supported by the bit density per area. Therefore, four different efficiency parameters are defined in the following subsections.

5.2.2 Array, Cell and Bit Efficiency Parameters

Memory products are characterized by a structured design, a cell array surrounded by word line and bit line decoder structures and read and write amplifier circuits.

Beside the performance values memory products are characterized by the achieved bit density per square mm. NOR and NAND chip layouts are different in terms of array, array segmentation, sensing area and total logic overhead. Quality of design and technology is measured by efficiency parameters.

The definition of the design efficiency is given as the ratio between the die area consumed by cells compared to total die size. NOR and NAND product designs are compared in Fig. 5.6 to illustrate the obvious difference in visible memory area. Both designs are publishing 60 % cell efficiency.

A set of **E**fficiency **P**arameters is introduced to differentiate the impact of technology, design, array and cell architecture on memory product cost.

5.2.2.1 Effective Cell size in F^2

The first parameter is the **size of the basic cell**. The size is defined by construction and based on the selected technology and is normally expressed in N times F^2, in which F is the minimum feature size of the selected technology node.

- The cell principle—storage of electrons within a floating gate—is the same for NOR and NAND memories. The effective cell size defined by the combination out

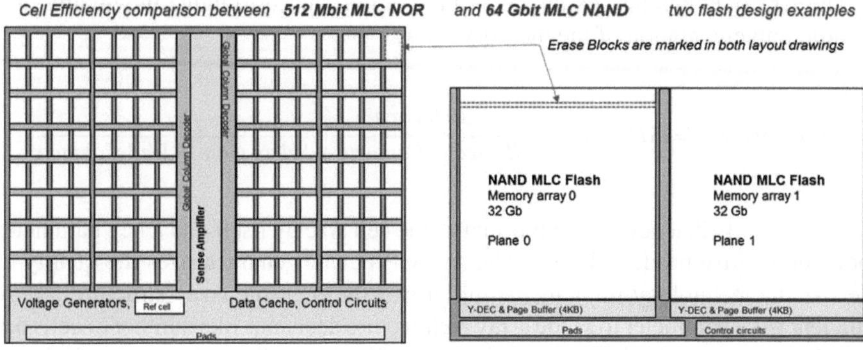

Fig. 5.6 MLC NOR and MLC NAND layout drawings—Cell efficiency comparison

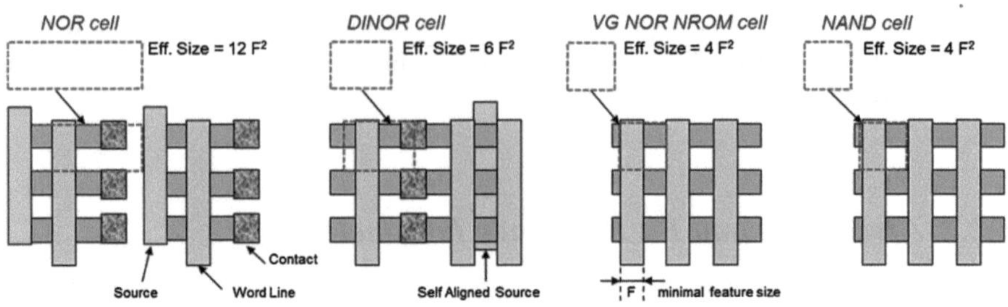

Fig. 5.7 Effective flash cell sizes in F^2 (planar MBC NROM limited in channel length shrink ability)

of cell architecture, program and erase principles, array and technology is different for different concepts.

- The typical NOR cell size is in the range of 8–10 F^2 compared to 4–5 F^2 for NAND flash.
- The number of bits is not the focus of this parameter, because the effective cell size is a technology efficiency parameter characterizing the cell size within a memory array.

A contactless basic array has a smaller cell size as shown in Fig. 5.7 for VG-NOR and NAND. The planar NOR type cells are limited by physics due to CHE programming in channel length reduction below 100 nm.

5.2.2.2 Array Efficiency Parameter

The implementation of the memory array technology—characterized with a specific ***effective cell size***—defines the complete overhead to contact all cells, all select gates and all wells of the array. The grounding and stitching concept of wells and select lines can add a significant portion of additional die size.

The **Array Efficiency** Parameter—**AE**—is introduced to judge the quality of a specific implementation of the memory array.

$$Array\ Efficency = AE_{Array} = \frac{Cell\ Area_{(die\ size\ consumed\ by\ all\ cells)}}{Cell\ Area + Contacts + Select\ Gate + Well\ Contacts}$$

The Array Efficiency is as important as the cell size by itself and a key parameter for a competitive product design. The Array Efficiency characterizes the quality of the specific technology implementation of exactly the same array structure. Array Efficiency is a parameter to guide array architecture decisions like number of cells per NAND string, size of select gates, and length of buried bit lines. The array efficiency is a basic figure handed over to the design team, which could not be improved by design innovations in the later steps of the product development process.

5.2.2.3 Cell Efficiency Parameter

The Cell Efficiency is often used as the target parameter to compare memory designs. The minimization of logic blocks surrounding the memory array, the segmentation and the overhead required for the reference and sense architecture mainly define the design overhead required to develop the specified memory functionality.

Under the assumption of an equal Array Efficiency the **Cell Efficiency Parameter—CE**—characterizes the quality of the implemented design.

$$Cell\ Efficency = CE_{Die} = \frac{Cell\ Area_{(all\ cells)}}{Total\ Die\ Size} = \frac{Cell\ Area_{(all\ cells)}}{Cell\ Area_{(all\ cells)} + Area_{(logic\ blocks)}}$$

The cell efficiency parameter cannot be a fixed target like 60 %, because the logic building blocks have a typical chip size based on the memory segmentation and the high voltage implementation. Therefore the cell efficiency for a defined memory density depends on two parameters:

- Technology node in [nm] selected for the memory density target;
- Array efficiency based on the above selected technology implementation.

The design overhead for the logic blocks in mm^2 is approximately the same even for a smaller technology node. For a given memory density a die with a larger cell area and larger total die size achieves a better cell efficiency. Figure 5.8 illustrates this dependency based on 16 GBit, 32 GBit and 64 GBit MLC NAND memory design model die size calculations.

The model based cell efficiency trend lines are compared with real silicon figures from four NAND flash vendors to prove the model assumptions.

The Cell Efficiency parameter is an excellent parameter to compare competitiveness between the same array and technology concepts. This parameter could become misleading if different array efficiencies and technology nodes are mixed in the assessment.

5.2.2.4 Bit Efficiency Parameter

Array and Cell Efficiency are the parameters of choice to compare the array implementation of different competitors and the design quality of different nodes versus an efficiency target. This target is based on the cell size as the atomic parameter. The application is interested in bits and bytes and in cost figures for bits and bytes.

Therefore a bit efficiency parameter has to be introduced. The reason that bit efficiency is not that widely used is the issue with the normalization factor and the expected low values. The bit efficiency parameter is given in percentage, but should not become higher than 100 %. Therefore, we introduce this normalization factor based on an assumption that 1 bit is the expected bit density out of 1 F^2 for 2D solid-state non-volatile memories.

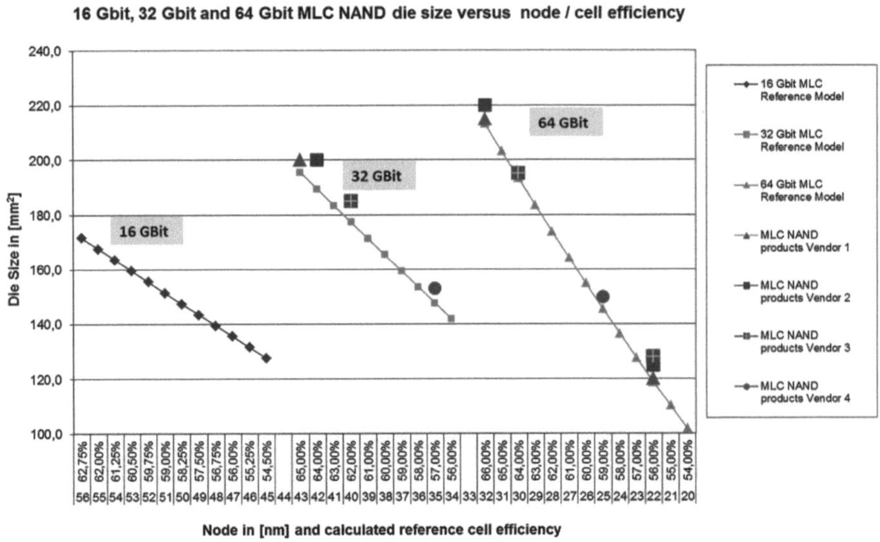

Fig. 5.8 Dependency of cell efficiency from die size and technology node

The **Bit Efficiency Parameter—BE**–characterizes the cost competitiveness of the memory product. This efficiency parameter includes both parameters the number of bits per cell and the size of the cell in F^2. N denotes the number of F^2 for the effective cell size.

$$Bit\ Efficiency = BE_{Die} = \frac{Bits_{(per\ cell)} * Cell\_Area_{(Consumed\ by\ all\ cells)}}{N_{Cell\ size\ in\ F^2} * Total\ Die\ Size}$$

The bit efficiency parameter is introduced to efficiently support the assessment of different array and cell architectures including all kinds of multi-bit and multi-level cells.

5.2.3 Memory Technology and Product Roadmap

A high volume memory product development has to execute a predefined strategy. This strategy includes the short, mid and long term project planning including technology and product innovation. "A technology roadmap is a plan that matches short-term and long-term goals with specific technology solutions to help meet those goals" is an excellent definition given by Garcia and Bray [9].

The benefit of a structured roadmap process for the semiconductor industry is that the ITRS roadmap provides aligned information to support technology investment decisions. Critical technologies are identified and especially within the area

Table 5.1 ITRS Roadmap example—Lithography Technology Requirements

Year of Production	2005	2006	2007	2008	2009	2010	2011	2012	2013
DRAM 1/2 pitch [nm]	80	70	65	57	50	45	40	36	32
DRAM & Flash									
DRAM 1/2 pitch [nm]	80	70	65	57	50	45	40	36	32
Flash 1/2 pitch [nm]	76	64	57	51	45	40	36	32	28

Manufacturable solutions exist and are being optimized
Manufacturable solutions are known
Manufacturable solutions are not known

of lithography ways can be identified to leverage the huge R&D investment among alliances between companies and research institutes.

5.2.3.1 ITRS Roadmap

The reduction of cost by shrinking the design rules—the smallest manufacturable distance between lines and contacts—requires the development of smaller lithography nodes. The International Technology Roadmap for Semiconductors, known as the ITRS roadmap, defines all major technology parameter for certain lithography nodes [10].

Table 5.1 indicates the principle idea of the ITRS roadmap in which the lithography technology requirements—defined in 2005—are summarized. The lithography node and the resulting half pitch dimension for each product group are given with an indication for manufacturability.

The flash memory pitch size is the most aggressive number in this roadmap, stretching the lithography capability always by 10 % compared to the pitch size for DRAM memories.

The ITRS roadmap is defined by semiconductor companies in a structured process to predict the required steps to fulfil Moore's law and achieve a common sense on a conservative judgement. Table 5.2 compares the prediction of the ITRS roadmap from 2005 and 2007 with products on the market in the predicted time line.

The non-volatile semiconductor industry develops technology nodes faster than predicted by the ITRS roadmap. For technology nodes smaller than 70 nm the achievements reported by the vendors at the International Solid State Circuit Conference differ by 1, 2 or more nm compared to the ITRS roadmap. The ITRS roadmap is a conservative prediction of the future and can deliver a rough orientation for the system architecture development process.

Table 5.2 Inaccuracy of ITRS data prediction—roadmap and silicon data comparison [ISSCC publications]

Year of production	Source	2007	2008	2009	2010
Flash ITRS roadmap data					
Flash 1/2 pitch (nm)	ITRS 2005	57	51	45	40
Flash 1/2 pitch (nm)	ITRS 2007	51	45	40	38
NAND Flash silicon data					
Vendor A (nm)	ISSCC publ.	56	43	32	22
Vendor B (nm)	ISSCC publ.	51	42	34	22
Vendor C (nm)	ISSCC publ.	50	34		24
Comparison					
Delta flash silicon data vs. ITRS (nm)	ITRS 2005	7	15	13	18
Delta flash silicon data vs. ITRS (nm)	ITRS 2007	1	9	8	16

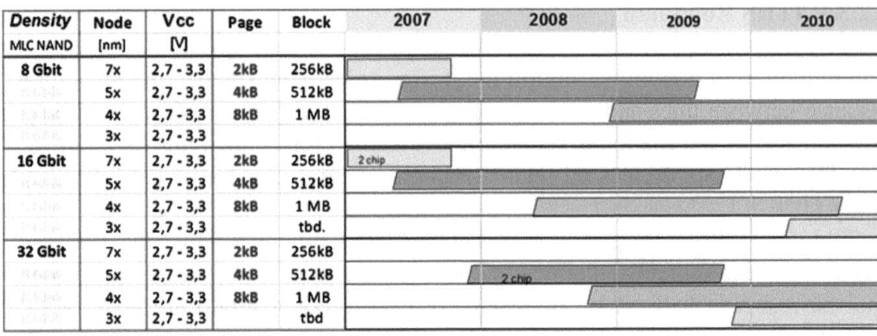

Fig. 5.9 Memory product roadmap—example for MLC NAND flash product roadmap

5.2.3.2 Memory Product Roadmap: Company Specific

Memory companies are presenting their own product roadmaps on memory specific workshops and conferences. A MLC NAND flash memory roadmap example is shown in Fig. 5.9.

The NAND flash specification parameters "Page size" and "Block size" are selected to analyse the product roadmap example. The physical die size is typically known for each technology node from conference publications. The length of the physical word line combined with the bit line array architecture (shielded or all bit line) results into the logical pages size. The length of the NAND string combined with the number and size of the pages results into the erase block size.

A company specific memory product roadmap example is shown in Fig. 5.9. This roadmap example is based on shielded bit line array architecture with 32 cells per NAND string.

The analysis of the NAND page size development trend over time and technology nodes is shown in Fig. 5.10. A link between logical page sizes on system level—

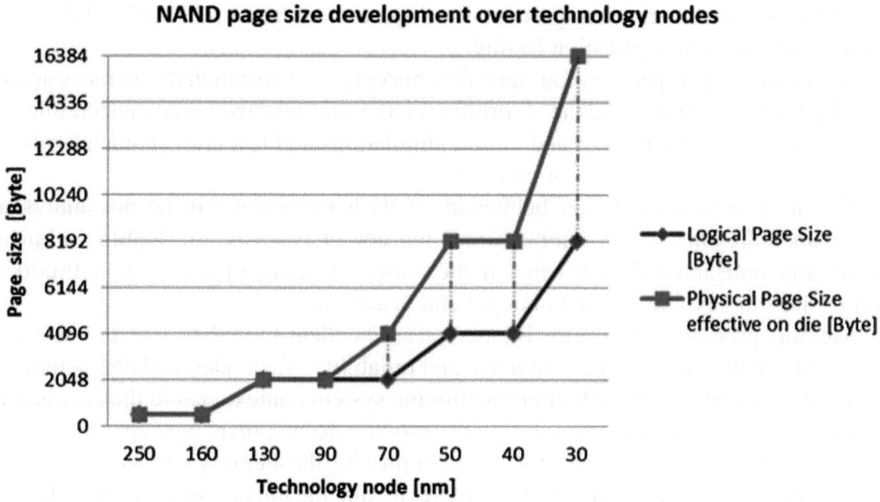

Fig. 5.10 Historical trend analysis of NAND page size over technology nodes

e.g. 512 Byte—and the NAND page size is visible till 160 nm. The shrink roadmap enables a physical page size increase faster than the logical page size increase.

Logical page sizes are coupled to system standards and increase slowly and in steps (512 Byte, 2 KByte, 4 kByte). The page size increase automatically enforces a larger erase block size.

- The mismatch between the logical data size on system level (aligned to the physical flash page size) and the physical erase block size is becoming bigger.

The NAND plane concept [11] offers a smart decoupling of logical and physical page sizes and was applied first time at 2 kByte page size at 70 nm node. A segmented array offers a physical page size aligned to the logical one—the blue line in Fig. 5.10—and operating all planes in parallel the full program and read performance is achieved.

The memory optimization potential and the hidden innovative features are not always assessable for the system architect based on product roadmaps. The next chapter will introduce potential opportunities to optimize and innovate on flash memory design and technology as well on system level.

5.3 Memory System Optimization

The memory development is company specific divided into sub-phases which are executed according to a development process handbook. Semiconductor development processes are in most cases a combination out of waterfall and V-model: product definition, concept definition, product development, test and optimization followed by qualification and application testing.

A memory target specification starts the concept phase, in which the memory array architecture, the sensing and the algorithm features are selected based on technology and lithography data. Design and circuit simulations and test chip characterization data are used to make a proof of concept.

The long term degradation behaviour of flash memories can be not analysed in advance. Testing of all combinations and corner cases is not doable within a reasonable time and with confidence in the statistical quality of the test data. Product innovations have to be verified in depth and in advance.

Memory product optimization is based on an excellent and exhaustive understanding of each cell, array and pattern noise and reliability effect. Detailed characterizations and simulations of each effect within the system context enable the capability to predict and prove innovation during the product development concept phase.

Opportunities to optimize each element improving the memory performance parameters are discussed to highlight the complexity and the impact on the system development roadmap.

5.3.1 Cell Driven Optimization

The cell architecture, the specific cell materials and the cell operation conditions influence strongly the memory performance as well as the durability and reliability parameter.

5.3.1.1 Read Operation: Disturbance, Accuracy and Access Time

A system requirement like continuously repetitive read access to the same address creates specific requirements to the memory sub-system. Unlimited read access capability can be achieved by a non-volatile memory with no restriction regarding read durability or by a read cache architecture. The cache size is balanced between cost for the cache memory and improved performance and reliability based on statistical models for the expected application data access distribution.

A flash memory offering a fast read access and a high read robustness is a cost optimized solution. Read disturbance robustness is derived from cell physics and read operation voltage and time conditions.

The issue of NOR flash cells is a positive low-voltage gate stress and an unwanted CHE injection depending on drain voltage conditions during read. Figure 5.11 illustrates the effect and the solution. The sense operation conditions are optimized by reducing the drain voltage. The gate voltage (V_{GS}) has to follow the cell V_{th} operation window.

The threshold voltage of each cell varies over time—programming is a statistical injection process of charge—and the transconductance (g_m) of the cell

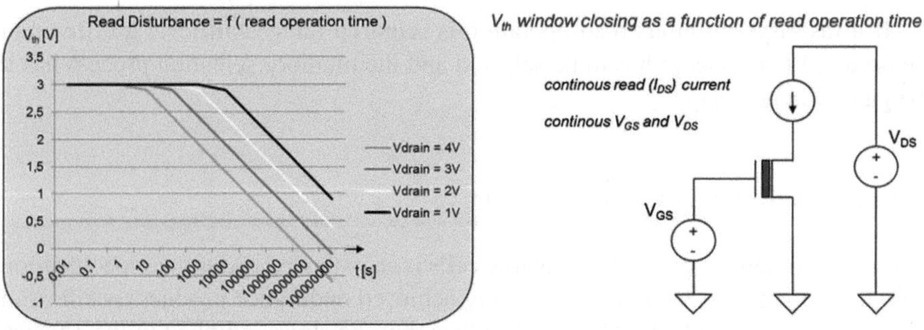

Fig. 5.11 Read disturbance—impact of cell voltage conditions

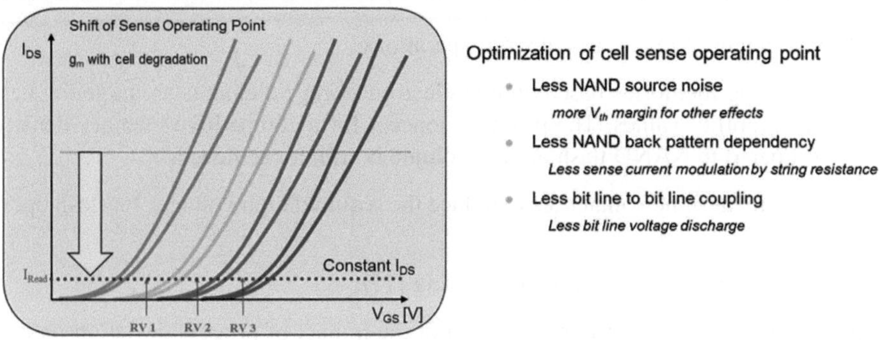

Fig. 5.12 Sense operating point optimization—to compensate cell V_{th} shift and g_m variations

transistor—the slope of the curve V_{GS} versus I_{DS}—degrades over time. Figure 5.12 illustrates possible variations of I/V curves of millions of cells and one possible design counter measure to overcome this degradation effect.

The stability of the sense window is improved by shifting the sense operating point to low sense currents. This sense optimization is in conflict with the requirement of low latency read operation.

Adaptive read techniques have the potential to fulfil the continuous system read requirements. Read operations can be developed which consecutively change the sense operating point in a predefined order. The reduced read disturbance results unfortunately in a larger read latency.

The following questions help to focus during the optimization process:

- How is a repetitive read access to a logical address translated to an array read operation?
- What is the worst case access time to the physical array (cell) of the non-volatile memory?

If a fast asynchronous read operation is required on system level a differential sense amplifier concept has to be selected and the memory selection process has to be guided in this way.

5.3.1.2 Program and Erase Operation

The development target for non-volatile cells is an extended endurance specification. The first target is to maintain the already achieved endurance product specification along the shrink roadmap. Cell sizes reduce, but the stress is linked to the physical principles applied for program and erase.

The endurance optimization on cell level is clustered into the following directions:

- Development of less stressful write operations
  - Emerging memories based on non-electron storage elements are targeting better durability values. The proof of concept for a competitive memory density compared to NAND flash in high volume is still not achieved.
- Development of new materials to reduce the required high voltages for flash operations;
  - For example tunnel barrier engineering [12];
- Optimization of cell architecture to limit the impact of process deviation on cell operation;
  - The geometry of the cell has to be optimized including all statistical relevant effects. A stochastic model for cell–system interaction (MCSI) [13] [FHK+08] allows an accurate prediction of cell geometry changes onto program algorithm and program performance parameters.

NAND cell program and erase operations induce physical stress due to FN tunnelling physics and will degrade the tunnel oxide. This degradation cannot be recovered and defines a hard limit for the flash endurance. The memory sub-system has to detect the specific signatures in advance that cells, pages and blocks have reached the limit of acceptable physical stress.

Hot electron programming and hot hole erase flash cell operations create a different stress level. The bottom oxide is influenced or damaged by these lower voltage operations too. The endurance limits for lateral multi-bit charge storage in charge trapping cells is based on spatial separation of the injected charge quantities. The program and erase efficiencies are changed over the cycling count. For these multi-bit charge storage cells—like NROM—methods are published to refresh the cell or better to clean the storage layer [14], [15]. The lifetime of charge trapping based flash cells could be extended up to millions of cycles.

The system architecture requires performance and durability values which can be fulfilled by a couple of non-volatile memories. Available design and technology data are reflecting a snapshot of the long term memory development roadmap.

- How to balance cell size versus durability?
- How to judge recovery and refresh methods versus robust and simple cell architectures?

5.3.1.3 Bits Per Cell and V_{th} Window Margin

The cell driven optimization is targeting the size of the V_{th} operation window. The number of bits per cell and the cell size in F^2 are the key parameter in terms of cost optimization.

A hidden and indirect optimization strategy can focus on gaining read and write durability by not utilizing the available threshold voltage window. Define a smaller V_{th} operation window and reduce the overall stress for an SLC device by applying the MLC operating conditions.

The SLC NAND product specification defines one order of magnitude better endurance and retention values compared to MLC NAND as introduced in the reliability chapter. Figure 5.13 illustrates the optimization direction to increase the read pass voltage only by 500 mV and accepting less read overdrive to limit read disturbance on the erased bits.

This optimization example illustrates the balancing between reliability stress and array noise. The distribution width for single level cell flash is wider, programming is faster and the effective read margin could become comparably small. The flash design V_{th} window margin and algorithm setup based on optimized cell geometry defines the achievable performance parameter set.

The cell optimization assessment would propose a single level cell not applying FN tunnelling due to high physical stress. The target of high disturbance immunity would enforce the selection of multi-bit instead of multi-level cell architectures.

- The cell driven optimization proposals are in contrast to the commercially most successful non-volatile memory architecture → Multi-level cell NAND flash based on FN tunnelling.

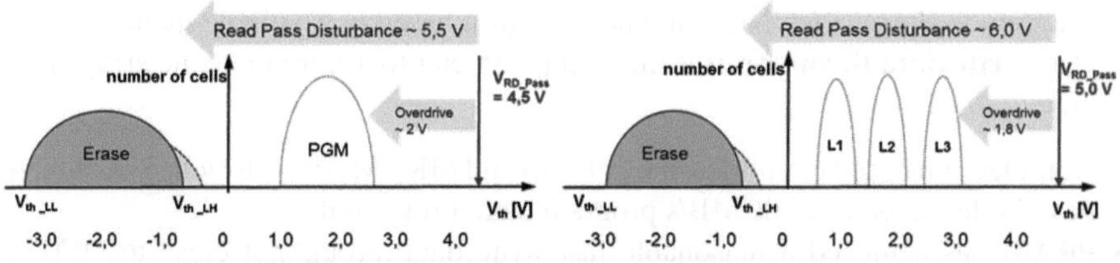

Fig. 5.13 Flash cell margin—read disturbance impact on SLC and MLC cell

5.3.2 Memory Array Driven Optimization

The type of the array has a major impact on the achievable performance of the memory design. The cell efficiency of a memory design combined with the array efficiency of the technology defines the final product cost of the non-volatile memory design.

5.3.2.1 Array Selection Process

The selection process is guided by key performance parameters.

The random **read access** time is the first performance parameter:

- NOR flash memories are characterized by read access times between 30 and 90 ns.
- NAND flash requires read access times between 20 and 50 μs, to sense the selected cell within the NAND string.
- Non-electron based memories have typically a fast read access time.

The sustainable **read data throughput** is the second parameter to identify strength or weaknesses:

- NAND flash memories are doubling the read data throughput generation by generation. The typical values are between 40 and 80 MB/s. Latest generation of double data rate NAND flash designs achieve 200–400 MB/s read throughput.
- NOR like memories are limited by the number of parallel activated sense amplifiers—driven by the array concept or by a cell efficiency optimized array and sense amplifier concept. The achievable read throughput is in the range of 50–100 MB/s.
- Non-electron based memories offer a data read throughput in the same range as flash based memories, based on published data at ISSCC and ASSCC [16].

The read performance parameters identify differences between different array options. The read data throughput is comparable fast; NAND has recovered and achieves today the highest values. The first random access of NAND is three orders of magnitude slower than NOR and non-electron based non-volatile memories.

The **write data throughput** is the next parameter to differentiate the array architecture.

- Non-electron based memories like FeRAM and MRAM compete with SLC NAND flash in the range of >100 MB/s program data throughput
- PRAM has achieved a reasonable fast write data throughput close to 10 MB/s competing with NOR based flash memories.

NOR—direct cell access—array types are preferred for fast access and for randomly mixed operation sequences of read and write.

NAND—indirect cell access—array types are cost competitive and achieve a high data throughput.

5.3.2.2 Array Optimization: Segmentation

Array and cell type define the write data throughput. The achievable program performance values are defined by array segmentation. Segmentation results into shorter bit- and word lines. Cutting the bit line length by a factor of two results into a potential performance increase by a factor of four based on the bit line relevant rise time improvement.

$$tau\,[sec] = \tau = \frac{1}{2}R * \frac{1}{2}C = \frac{1}{4}RC$$

R is the resistance in Ohms and C is the capacitance in Farads.

Array segmentation accelerates the flash operation within the smaller segment and increases the performance on chip by operating more segments in parallel. Operating more segments at the same time depends on the energy consumption of the cell operation and the total array capacity of all segments to be charged and discharged.

NAND array segmentation options are shown in Fig. 5.14:

- one array with long bit lines represents a typical 90 nm NAND die with 512 Byte page size;
- two arrays per die with 2x shorter bit lines result into an optimized two plane NAND die;
- four arrays with 4x shorter bit lines and 2x shorter word lines represents a performance optimized NAND die.

The following parameters can be improved by array segmentation

- Program and Read cycle time—for all rise time (tau) based performance parameters;
- Program parallelism—max cell number per operation;

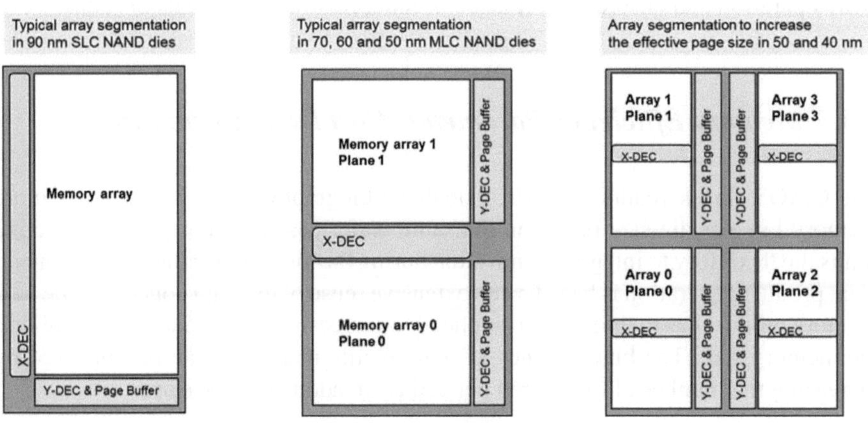

Fig. 5.14 NAND array segmentation—different options

- Energy consumption per bit—shorter bit and word lines reduce the energy required for charge and discharge of the lines;
- Array energy consumption per selected unit (word, page, block).

Array driven energy calculations for program and erase operation set a clear target to choose voltage driven operation—e.g. FN tunnelling. A current in bit line direction limits the parallelism of program and erase and forces the requirement for a low resistive bit line material in combination with smaller memory segments.

The memory array and an optimized partitioning per technology node in combination with lowest possible resistance and capacitance values define the product performance values—both the maximal achievable data bandwidth and the first access time.

5.3.2.3 Array Optimization: Design and Technology Innovation

Design and technology innovation are hard to predict from outside. Memory companies have on-going research activities on all array relevant fields to have the right innovation prepared at time. The developed **P**erformance **I**ndicator **M**ethodology is supporting visibility for upcoming innovations.

Array optimization makes a balance between cell efficiency and array performance. Circuit options like single sided versus double sided driver and decoder are combined with innovative sense and charge concepts.

The dynamic array behaviour can be optimized utilizing self-boosting concepts. Reuse and redirect charge instead of charge and discharge hundreds of times ensures the continuous usage of large NAND flash arrays. A research target is the adiabatic reuse of energy required for array operations.

Technology innovation in material, contact schemes and metal process impacts the performance and the efficiency. Reduction of line resistance, improved shielding capacities and optimized coupling capacities strongly influence the robustness of the array operation and ensure performance or reliability improvements or both.

5.3.3 Memory Efficiency Parameter: Cost Driven Optimization

The CMOS shrink roadmap ensures smaller lithography nodes and enables more memory bits per die size based on the same wafer costs. A short project timeline limits the flexibility to integrate innovation during the memory product development. Next product generation is based on an extensive reuse of existing concepts. The new memory product has to fit in the old application increasing performance and doubling the memory size. The bit efficiency of non-volatile memories can be improved by increasing the number of bits stored per cell as an additional measure.

- Memory efficiency parameter optimization based on multiple bits per cell contradicts the development timeline target to ramp as fast as possible the next generation.

- A memory architecture supporting SLC and MLC with a small additional design overhead allows a fast ramp of SLC products first, followed by a cost competitive volume production of MLC products at a second step.

This bit increase per cell doubles the memory density faster than expected from the technology shrink roadmap. The target for efficiency optimization is a memory bit size set to 1 F^2. NAND and VG NOR array architectures have a 4 F^2 cell size and can achieve a cost competitive memory density.

The 4-bit per cell efficiency target is used to illustrate the optimization process. The program data throughput is not in the focus for the following system architecture decision example. The two array and cell concepts are selected which have demonstrated a bit size of 1 F^2:

- 4-bit per cell MLC NAND and
- 4-bit per cell MBC and MLC VG NOR.

A typical text book flash architecture comparison is shown in Table 5.3.

Figure 5.15 compares the array architectures and the V_{th} operation window for programmed levels.

A very dense NAND memory array type with 100 % voltage controlled memory operation based on a floating gate cell is compared with a fast access NOR memory array type with voltage controlled but current consuming cell operations based on charge trapping multi-bit cell.

4-bit per cell **MBC & MLC** VG NOR achieves short and very competitive cycle times for read and program. The reliability is the development challenge due charge trapping based cells.

4-bit per cell **MLC NAND** achieves a reasonable high data throughput based on highest parallelism of the NAND array architecture. Fifteen different levels have to be programmed within a V_{th} window and the overlap of programmed V_{th} distributions needs a reliable detection and correction technique.

The performance parameter comparison indicates benefits for both concepts shown in Table 5.4. A serious decision is again hard to make based on a snapshot of data for one or two technology nodes.

Additional remarks for multiple-bit per cell flash products are summarized below:

- Multiple bits per cell concepts are strongly linked to the cell threshold voltage stability over time. A distribution movement can be covered by special error correction and data scrambling technologies, but a fast distribution widening is hard to control.
- Multiple bits per cell concepts require much longer program times and sequences including imprint, rough and fine programming phases. Array architectures with a large number of sense amplifiers are preferred due to the periodically required data stored at the right place.
- A higher program parallelism has to be achieved without increasing the energy consumption, which requires a voltage driven change of the non-volatile storage element—for flash a FN based programming behavior—and an array concept which is based on capacitive coupling to achieve the required voltage levels.

Table 5.3 4-Bit per cell comparison of MLC NAND with MLC and MBC VG NOR

	4-bit per cell MLC NAND based on floating gate cell	4-bit per cell MLC and MBC VG NOR based on charge trapping cell
Array architecture	NAND strings with 32 or 64 cells	VG-NOR direct cell access
	In-Y-direction isolation with string select gates (SSG)	In X-direction STI slice isolation;In Y-direction isolation with select gates
Array and cell	PGM Inhibit & Pass voltage disturb	Sharing of phys. WL and phys. BL
Disturbance	Read Pass voltage disturb	Weak ERS and PGM Gate disturbance
		Weak ERS and PGM Drain disturbance
Sense concept	Slow serial high parallel sensing	Fast parallel AC sensing in time domain
	• No reference cells	• Global reference cells
Program algorithm	Incremental step pulse program	Incremental step pulse program
	• Fowler-Nordheim Tunnelling	• Channel Hot Electron Injection
Erase algorithm	FN Pulse 1–2 ms with PAE	PBE, Erase and PAE sequence
	Fowler-Nordheim Tunnelling	Hot Hole Injection
Read algorithm	Parallel sensing of	Adaptive read window search of a pre-defined chunk of data
	• all bit lines—ABL concept	
	• half bit lines—shielded BL	Read latency varies

- A higher accuracy of the cell threshold voltage program sequence requires a real time algorithm which is capable to compensate all technology and interference effects. This is theoretically only possible if all cells are moving together into their target positions.

The Performance Indicator Methodology is applied to derive answers to the following questions:

- Is the increase in array efficiency (4-bit per cell) visible in product cost reduction and not overcompensated by increase in design overhead in time and die size—cell efficiency?
- Will the expected reduction of the reliability be accepted in the market?
- Which performance reduction will be accepted in the market?

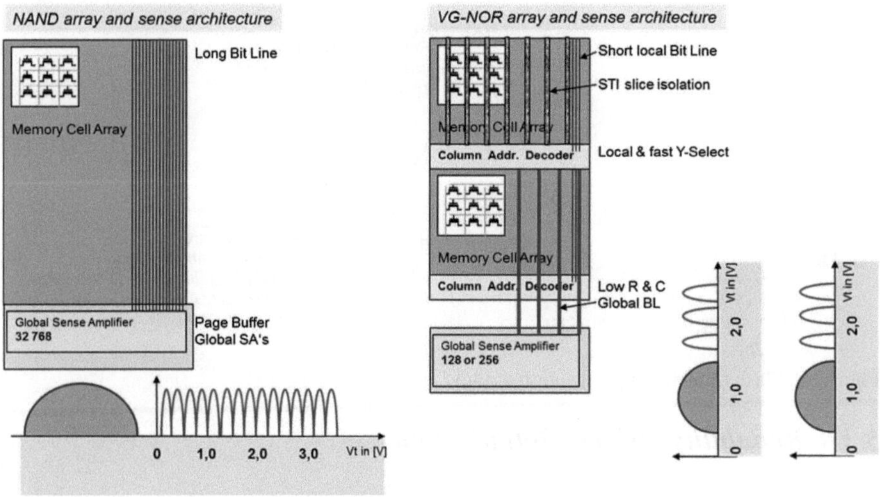

Fig. 5.15 4-Bit per cell comparison of MLC NAND with MLC and MBC VG NOR architecture

The memory efficiency optimization based on 2-bit, 3-bit and 4-bit per cell MLC NAND is discussed by applying performance indicator trend lines focusing on the challenge reliability versus cost in Sect. 6.4.3. The Performance Indicator calculated for a 4-bit per cell design is a quantitative measure and supports the decision process including the questions highlighted before.

Table 5.4 4-bit per cell—Performance parameter comparison based on technology nodes between 50 and 60 nm

	4-bit per cell XLC NAND	4-Bit per cell MLC and MBC VG NOR	Comments
Read Access Cycle time	100 μs	50 μs	(+) for VG NOR
Page (4 kB) PGM time	>2500 μs	>950 μs	(+) for MBC VG NOR
PGM Data Throughput	2–4 MB/s	8–10 MB/s	(+) for VG NOR
Write Data Throughput	2–4 MB/s	2–3 MB/s	(+) for XLC NAND
• System durability behavior is modulated by ERS/PGM size factor and ERS/PGM time ratio			
ERS/PGM size factor	256	16	(+) for VG NOR
Ratio ERS/PGM time per EB	0.01	0.35	(++) for XLC NAND
Write Durability (endurance)	<1000	<100	(++) for XLC NAND
Write Energy μJoule	<75	>175	(++) for XLC NAND

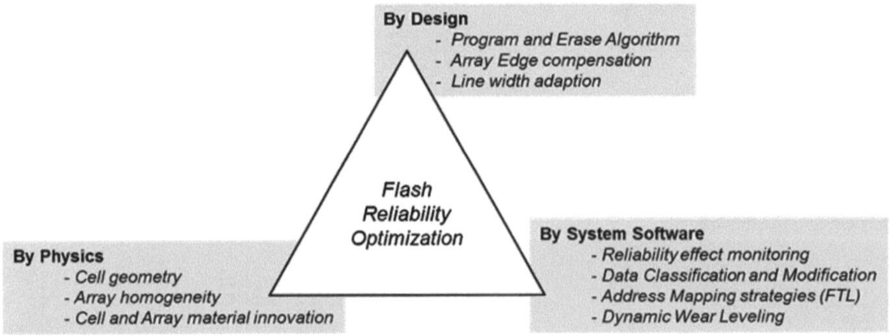

Fig. 5.16 Flash reliability optimization triangle

5.3.4 Reliability Driven Optimization

Reliability driven optimization is linked to the cost of reliability and is achievable by **physics** on cell & array level, by **design** on memory design level and by **software** on memory sub-system level shown in Fig. 5.16.

The research focus of the last years was a universal memory as fast as SRAM and DRAM combined with an excellent reliability and durability behaviour. New cell materials and cell architectures are investigated to find a non-volatile cell, which improves all reliability and durability parameter significantly.

Non-electron based non-volatile memories are reported as the next step to improve performance, cost and reliability since more than 10 years. Orders of magnitude better cycle behaviour was published for FeRAM, MRAM or PCM cells as shown in Table 2.3. Product reliability data based on volume production for emerging memories with a density larger than 256 Mbit are similar to known values from high reliability flash products. There is no single memory to replace NAND and NOR flash as well DRAM and SRAM [17].

An assessment of a non-volatile memory concept has to be based on parameters derived from high density memories in volume production. This role has to be marked as the most important one for non-volatile product and technology development.

A ***technical assessment*** of the development of emerging memories is summarized as follows:

- Improved specific reliability parameter for emerging memories compared to flash has shown no influence to the memory market. As a result emerging memories could not penetrate the key volume markets. They are often used in niche markets where the specific strength is becoming a key decision factor.
- The cost comparison is still the main decision factor in all volume markets. Emerging memories are behind the aggressive shrink roadmaps of the technology driver NAND flash.
- Reduced reliability values of Multi-Level-Cell NAND flash products are accepted on application level and reliability targets are achieved by software on system level.

A ***business assessment*** ends up with a simple conclusion: Reliability optimization is becoming the best cost driven optimization. Every data loss on customer side due to reliability weaknesses can create much more damage to the brand name and additional cost compared to the effort to achieve the expected application targets.

- High volume production experience based on results of investigated application issues improves the reliability of a flash memory product in the fastest possible way.

The flash reliability optimization triangle is used to illustrate the idea of how reliability driven optimization can be implemented in particular for the memory array.

(a) Optimization of the memory array:

Achieve a very homogeneous and robust memory array, which results into reliability figures, which are up to one order of magnitude better than standard products.

(b) Compensation of the memory array inhomogeneities by algorithm and design:

Compensate all inhomogeneous array parts by special design features—dedicated voltages or changed line width—and algorithmic adaptation. The right compensation methodology has to be implemented on design level, but the reliability figures could be significantly improved compared to a standard product in the same technology.

(c) Utilization of the memory array selective by—-low level—system software:

Logical and physical addresses are mapped and classified with reliability attributes. Therefore the storage of the data can be classified. Data with highest safety class are stored only in physical memory areas with highest reliability classification.

The edge word and bit lines have a slightly different behaviour and require a dedicated assessment. The above described methods can be applied to compensate the higher failure rate. The bit failure rate is substantially reduced by adding a dummy word line which is shown Fig. 5.17.

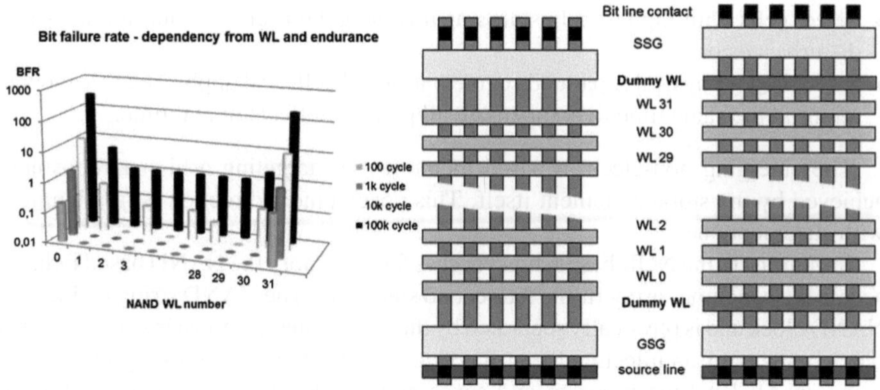

Fig. 5.17 NAND reliability optimization—edge word line

Other optimization methods to compensate the edge effect are a different algorithm applied for edge word lines, an improved error correction code or the storage of a reduced data density.

The time line, the effort and the development cost is different for the different reliability optimization strategies—by physics on cell and array, by design on array and sensing and by software on system and on data density and data scrambling. During the development and optimization process the following questions have to be answered:

- Are the optimizations strategies independently achievable?
- Can all three strategies achieve the same fundamental improvement?
- Does the expected cost or performance impact will be paid and accepted in the market?

The experience has shown that each of the three methods has weaknesses and a serious reliability optimization strategy focus on the combination of all aspects.

A successful reliability optimization strategy includes the following items:

- Focus as early as possible on all expected technology, cell and array inhomogeneities, select always the best technical solution in this context, and never accept the second best solution.
- Focus on the right amount of flexibility in the design architecture. Flexibility is not always given by embedded software; on hardware level flexibility means the design capability to generate enough different voltage levels at the same time and to switch these high voltages to the required word lines.
- Focus on an intensive worst case memory characterization to generate solid statistical data for the system software development.
- Describe all dependencies as simple as possible and clearly defined, so that the software development team can find the best adaption and can implement it correctly.

On a higher abstraction level the reliability optimization of flash memories can be summarized:

- Select cells with highest robustness against neighbour effects, interferences and disturbances or
- Select larger physical separated segmentations of cells and apply the critical operations at the same time to compensate all parasitic effects in real time.

The emerging non-electron based memories are targeting a higher robustness achieved by the storage element itself. This development direction is in alignment with statement one.

In contrast to the NOR based direct access flash memory the NAND based indirect access flash memory array fulfils the second statement. The NAND string defines the NAND block and is physically separated by the select gates as shown in Fig. 5.18. This block based array architecture has still potential to be improved. Sensing innovations make an All Bit Line program and sense architecture doable, which enables the

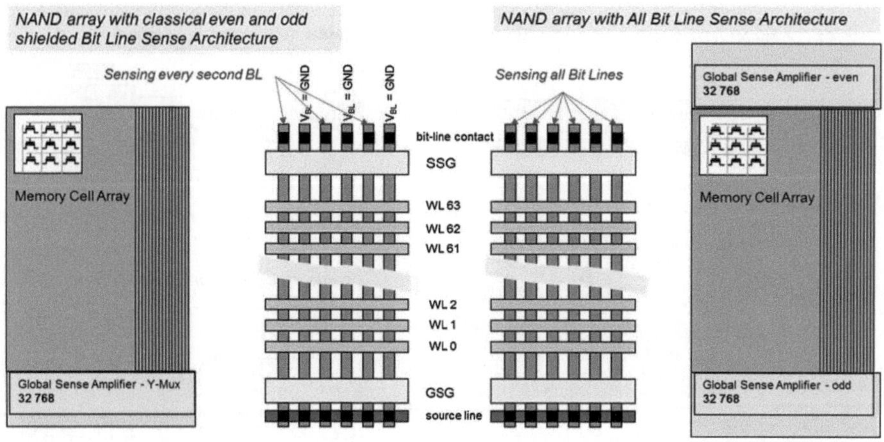

Fig. 5.18 NAND reliability optimization—all BL architecture

possibility to move all cells at the same time to compensate array effects in real time along a word line.

An All Bit Line NAND flash architecture—all cells of a physically separated block are moved in the same moment of time—is compensating all noise effects created by interferences between cells and lines. The influence between cells is becoming stronger (worse) along the shrink roadmap and the All Bit Line NAND array combined with the imprint algorithm is utilizing the stronger cell interferences to make the overall reliability more and more robust.

The reliability driven All Bit Line NAND array innovation improves at the same moment of time performance and reliability and enables the capability to store 3- or 4-bit per cell.

5.3.5 System (Application) Driven Optimization

The reliability driven optimization chapter has already included the system aspect especially the interaction between file translation layer and different wear level software strategies.

The potential of the system architecture has to be analysed in-depth during the optimization process. The number of channels, the number of flash devices sharing the same I/O channel, the number of independent error detection and correction circuits (EDC and ECC granularity) strongly impact the required flash performance, reliability and durability parameters to achieve the specified system target values.

The system architecture enables a high capability to improve flash memory parameters by impacting the design of application (specified (allowed or recommended) application cases).

System configuration impacts flash parameter requirement

System Optimization	Target	System I	System II
		4 Channel	2 Channel
MLC NAND 200 MHz DDR			
System Bandwidth [MB/s]	500		
Read NAND max. [MB/s]	500	793	471
Read NAND min. [MB/s]	250	198	235
Write NAND max. [MB/s]	500	440	466
Write NAND min. [MB/s]	250	220	233
Read latency [µs]	200		
Read IOPS (4k)	50000	39627	21960
Write IOPS (4k)	10000	4444	2222
System Durability		10^9	10^9
NAND single die endurance		25.000	5.000

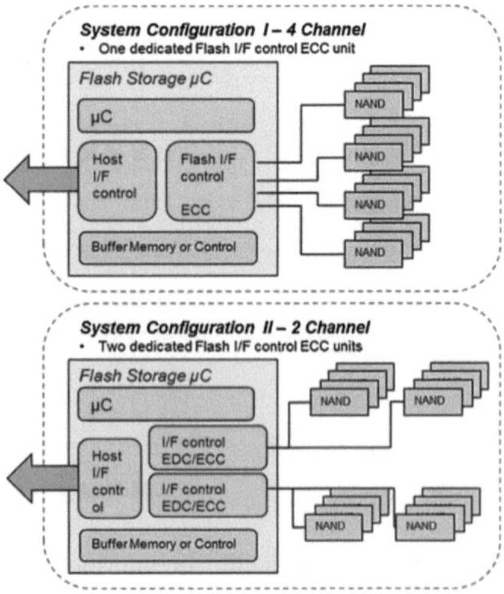

Fig. 5.19 System configuration impact on performance and durability and on flash reliability

The system architecture configuration defines the achievable performance—more channels increase typically the write and the random IOPS values and the energy consumptions.

A smart balancing of parallel channels and optimized and dedicated flash reliability control units (including advanced ECC) can achieve a competitive performance with significantly lower reliability requirements to NAND flash devices. These two contradicting development strategies are illustrated in Fig. 5.19.

System configuration II is used as an example to illustrate the power of the system architecture for an overall performance, cost and reliability optimization of flash memory based systems. System configuration I would be the preferred architecture for SLC NAND flash types due to the higher single die endurance requirements.

The system architecture has to be selected and designed in such a way that an optimized usage of different density types of flash memories can be configured during the system volume production.

5.4 Summary: Flash Memory Based System Optimization

High volume memory design and technology development is based on an on-time project execution combined with cost driven optimization strategy. A delayed production release of next generation products has a serious financial impact on the complete memory company.

Mobile applications are becoming non-volatile memory centric devices. Flash memories are a dominant part of the bill of material—BOM—and a key success factor for a specific system design. The selection process of the memory sub-system and the optimization of the system architecture are important to create a system with unique features for the customer.

The memory selection process is driven by system requirements introduced in Sect. 5.1.

- Target density and granularity to upgrade the memory density (roadmap including page, block and ECC sizes);
- Memory performance parameter (data throughput, cycle time)
- Interface parameter and the interleave capacity
- Energy consumption (active and standby)
- Durability parameter (linked to SLC or MLC flash solutions)
- Cost per bit roadmap for the selected architecture (e.g. 2-bit, 3-bit or 4-bit per cell)

The matching of available memory technologies with required system parameters is done by using spider charts shown in Fig. 5.20. An excellent technology overview [18] is achieved and the decision is based on the best fit of the most important performance parameters.

The memory optimization along the shrink roadmap improves seriously a couple of performance and reliability parameter. The memory selection process has to be based on a pre-assessment followed by an assessment of the innovation and optimization potential of the selected cell, array and technology concept. The innovation potential can be a decision point for memory architecture.

In contrast to the above statement the focus for high volume memories has to be on reuse and optimization to keep the development risk as low as possible and ensure the aggressive development mile stones along the shrink roadmap.

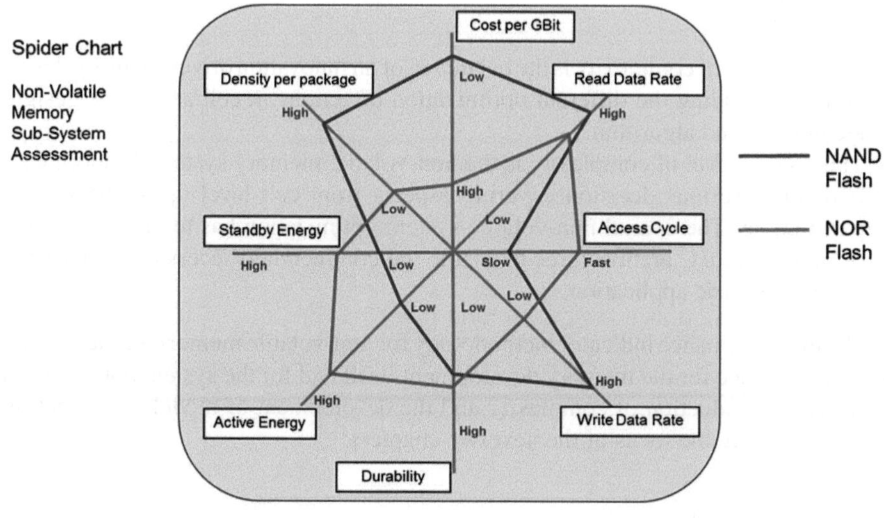

Fig. 5.20 System performance requirements—Comparison of NOR and NAND Flash memories [18]

The result of the concept decisions can be compared with competitor devices and emerging memories using spider diagrams. The judgment is application specific and a balancing of the importance of parameter and between parameters starts:

- Reliability versus Density
- Density versus Performance
- Performance versus Energy consumption
- Energy consumption versus Reliability.

An application specific analysis is made and additional tables and detailed charts are typically generated. The technology and product roadmap of the selected memory concept is analysed in-depth. The company product roadmaps are compared with the ITRS roadmap and a final decision is made. The expected innovation potential along the product roadmap is typically not considered.

The optimization potential for each selected non-volatile memory influence the system performance

- cell and array,
- multiple bits per cell,
- algorithm and V_{th} window margin and
- system architecture and durability optimization.

The design of a flash memory can be optimized to achieve the required V_{th} window margin. The wear level strategy and the FTL can be optimized for a defined flash V_{th} window margin setting to compensate asymmetry and margin weaknesses. The system architecture can be tailored for an assumed application case to enable a certain flexibility to balance performance versus reliability.

The non-volatile memory optimization dilemma is the overall complexity management:

- The first level of complexity is the balancing of an innovation driven memory development including the different optimization directions in cell and array, design, technology and algorithm.
- The second level of complexity is the non-volatile memory system development. It requires serious decision covering aspects from cell level up to the system partitioning. The overall non-volatile system optimization has to drive the usage of multi-core μC architectures to enable the commercial success of the targeted memory-centric application.

A key performance indicator methodology for non-volatile memories is developed to give guidance for the memory development itself and for the system optimization process. The reduction of complexity and the development of models representing the roadmap are the focus of the next two chapters.

References

1. B. Shriver, B. Smith, *The anatomy of a High-Performance Microprocessor A Systems Perspective(IEEE Computer Society Press* (Los Alamitos, California, 1998)
2. V.G. Oklobdzija, *High-Performance System Design: Circuits and Logic: Design Methods and-Circuit Techniques* (Wiley IEEE Press, New York, 1999)
3. D.A. Patterson, J.L. Hennessy, *Computer Organization and Design—The Hardware/Software Interface* (Morgan Kaufmann, Burlington, 2009)
4. V. Getov, A. Hoisie, H.J. Wassermann, Codesign for systems and applications: charting the path to exascale computing. Computer **44**, 19–21 (2011)
5. D. Baral, Life cycle power consumption HDD vs. SSD. http://www.bswd.com/FMS09/FMS09-101-Baral.pdf. Flash Memory Summit, 2009
6. IntelPR, Inter Newsroom, IntelPR, 12 April 2012. http://newsroom.intel.com/community/intel_newsroom/blog/2012/04/12/intel-solid-state-drive-910-series-delivers-high-performance-endurance-and-reliability-for-data-center-tiering-and-caching. [Zugriff am18 November 2012]
7. H. Kroemer, 2000 Nobel Physica Laureate, in *Lex Prix Nobel* (2000).
8. D.-S. Byeon, S.-S. Lee, Y.-H. Lim, D. Kang, W.-K. Han, D.-H. Kim und K.-D. Suh, A comparison between 63nm 8Gb and 90nm 4Gb multi-level cell NAND flash memory for mass storage application, in *Asean Solid-State Circuits Conference*, pp. 13–16, 2005.
9. M.L. Garcia, O.H. Bray, *Fundamentals of Technology Roadmapping* (Sandia National Lab, Albuquerque, 1997), pp. 87185–1378
10. L. Wilson, International technology roadmap for semiconductors, 2009. 2010 International Technology Roadmap for Semiconductors. This page was last updated on 21 Jan 2011.
11. T. Hara, K. Fukuda, K. Kanazawa, N. Shibata, K. Hosono, H. Maejima, M. Nakagawa, T. Abe, M. Kojima, M. Fujiu, Y. Takeuchi, K. Amemiya, M. Morooka, T. Kamei, H. Nasu, C.-M. Wang, K. Sakurai, N. Tokiwa, H. Waki, T. Maruyama, S. Yoshikawa, A 146-mm2 8-Gb Multi-level NAND flashmemory with 70-nm CMOS technology. IEEE J. Solid-State Circuits **41**(161–169), 953 (2006)
12. S. Verma, *Tunnel Barrier Engineering for flash memory technology* (STANFORD UNIVERSITY, Stanford, CA, May, 2010)
13. C. Friederich, J. Hayek, A. Kux, T. Muller, N. Chan, G. Kobernik, M. Specht, D. Richter, D. Schmitt- Landsiedel, Novel model for cell-system interaction (MCSI) in NAND Flash, in *IEDM Technical Digest*, pp. 1–4, Washington, 2008.
14. Y. Roizin, Extending Endurance of NROM Memories to over 1 million program/erase cycles, *Proceedings of 21st Non-Volatile Semiconductor Memory Workshop*, pp. 74–75, Feb 2006.
15. W. von Emden, W. Krautschneider, G. Tempel, R. Hagenbeck, M. Beug, A modified constant field charge pumping method for sensitive profiling of near-junction charges, in *37th European Solid State Device Research Conference, ESSDERC*, pp. 279–282, Munich, 2007.
16. K.C. Smith, A. Wang, L.C. Fiujino, Through the looking glass-trend trecking for ISSCC 2012. IEEE Solid-State Circuits Mag. 4(1), 4–20 (Winter 2012).
17. M. LaPedus, Semiconductor manufacturing & design community, 20 Sep 2012. http://semimd.com/blog/2012/09/20/universal-memories-fall-back-to-earth/. Zugriff am 28 0kt 2012
18. Toshiba America Electronic Components, Inc., NAND vs. NOR Flash Memory Technology Overview, 2006. http://umcs.maine.edu/~cmeadow/courses/cos335/Toshiba%20NAND_vs_NOR_Flash_Memory_Technology_Overviewt.pdf. Zugriff am 10 2012

Chapter 6
Memory Optimization: Key Performance Indicator Methodology

This chapter introduces the model-based quantitative performance indicator methodology applicable for performance, cost and reliability optimization of non-volatile memories. The complex example of NAND flash memories is used to develop the methodology based on a benchmarking of NAND flash product innovations along the CMOS shrink roadmap. A performance and array model is introduced and a set of performance indicators characterizing architecture, cost and durability is defined.

The performance indicator methodology is applied to NAND flash memories to quantify design and technology innovations. A graphical representation based on trend lines is introduced to support a requirement based product development process. The strengths of the methodology is demonstrated by a combination of application trend lines with performance indicators to visualize the application potential of high density multi-level cell NAND flash designs.

The performance indicator methodology is applied to demonstrate the importance of hidden memory parameters for a successful product and system development roadmap.

6.1 Performance, Cost Per Bit, and Reliability Optimization

6.1.1 Flash Memory Complexity Figure

Non-volatile memories are offering a wide space of alternative solutions and optimization directions. An overview focusing on flash memories was introduced and dependencies between performance parameters linked to other parameters of a memory design are discussed in this work. A non-volatile memory sub-system based on solid-state memories can be seen as an excellent example for a complex system. The target is to resolve the complexity and to derive a methodology that supports decisions along the development process.

D. Richter, *Flash Memories*, Springer Series in Advanced Microelectronics 40,
DOI: 10.1007/978-94-007-6082-0_6,

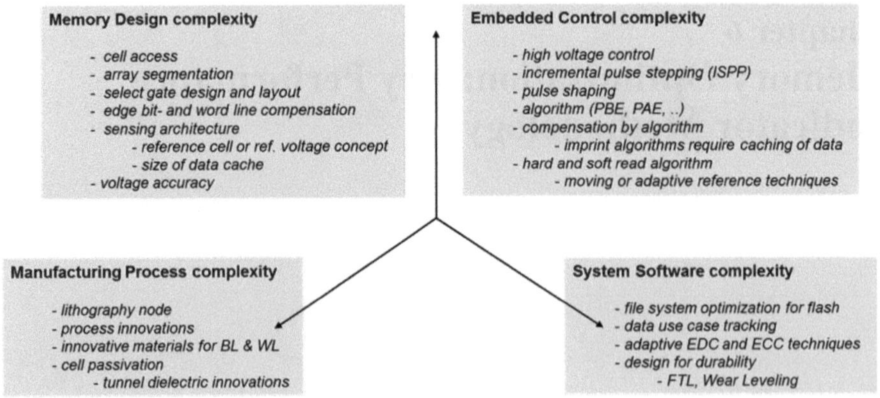

Fig. 6.1 The initial flash complexity figure

The optimization and development target of such a memory system can be defined as an overall efficiency including all parameters covering cost, performance, reliability, energy and availability in high volume production. A successful development strategy has to narrow down the complete complexity figure to focus on the main objective.

This work will start with the non-volatile memory and develop a methodology using flash memories. An initial flash complexity figure is used to divide the flash parameter space into four complexity levels, which are dominated by manufacturing, memory design, embedded control and system software. Cost, performance, reliability and energy are considered as a result based on decisions made within this flash complexity space.

Figure 6.1 shows the four complexity levels and the specific parameters defining each level in more detail. Each of the parameters—e.g. sensing architecture—is linked again to parameters listed at the other two sides.

A known way to resolve this complexity is to start from the result or better from the application target. The key parameters cost, performance, reliability and energy are used to develop a methodology to navigate within the flash complexity figure and achieve the required results.

6.1.2 Flash Memory Parameter Selection

The introduced performance parameters have a substantial impact on performance, reliability and energy consumption of flash memories. Memory product cost is linked to efficiency parameters-cell efficiency is an established key parameter. The flash complexity has to be narrowed down by a careful selection of memory performance parameters for the following analysis process.

Flash memory performance parameters are selected and prioritized to develop the key performance indicator methodology:

- **Program Data Throughput—Priority 1**
  - Comment: includes the non-volatile physics (cell), the program and verify algorithm (sensing and algorithm), the parallelism of the array and the capability to cache data (array and page buffer architecture);
  - All sense and program operations are included;
- **Write Data Throughput—Priority 2**
  - Comment: includes the above described PGM data throughput, the Erase performance, the interface specification and cycle time as well as the used data packet size
- **Read Access Cycle Time—Priority 1**
  - Comments: includes the random access time to read a randomly selected address;
- **IOPS—Priority 2**
  - Comment: includes both the read and write cycle time and the impact of the data package size on cycle times (including program times)

Flash memory durability and reliability parameters representing the flash complexity figure are selected and prioritized:

- **Read Durability—Priority 1** *[Read is assumed as operation with highest use percentage]*
  - Comment: includes the non-volatile behavior (cell), the ***read disturbance*** and the verify algorithm (ECC), the capability to cache read data and to hide the read access to the memory array;
- **Write Durability—Priority 2**
  - Comment: includes the non-volatile program behavior (cell), the ***program disturbance*** and the algorithm efficiency to extend the lifetime of the cells;
- **Data retention—Priority 2**
  - Comment: includes the non-volatile retention behavior (cell) and all disturbance and margin effects;

Flash memory energy parameters representing the flash complexity figure are selected and prioritized:

- **Write Energy—Priority 1**
  - Comment: includes the energy required for program and erase as well as for charging and discharging of the corresponding array parts;

- **Read Energy—Priority 2**
  - Comment: includes the energy required for read including charging and discharging of the corresponding bit- and word lines;
- **Standby Energy—Priority 2**
  - Comment: includes the energy required for background operations designed for durability and reliability improvements;

A limited number of flash memory parameters is selected to prepare the next step reduction of complexity based on the key performance indicator methodology.

6.2 Definition of Performance Indicators

A known strategy to reduce complexity and prepare decisions is the usage of key performance indicators. This chapter develops the idea to define at least one performance indicator per target—performance, reliability, energy and cost—which incorporates all major dependencies between different sub-parameters listed in Fig. 6.1 for design, technology and embedded control. The definition of specific performance indicators is driven by memory architecture and application inputs.

6.2.1 Performance Indicator: Focus on Memory Architecture

A subset representing the memory cell and array dependencies is defined to reduce the available parameter space for non-volatile memories.

The performance indicator methodology is based on technology lithography nodes. The square of the technology node value defines the smallest area which could be printed by lithography and is equal to the introduced unit of measure to characterize the memory cell size—F^2.

The cell efficiency parameter includes the array density and the overhead required for the memory functionality according to product specification.

The memory performance parameters are focused on the data throughput to characterize the cell and array combination including the read and write operations.

The performance indicator generic building rule is shown in the first formula:

$$Performance\ Indicator_{NVM} = \frac{Data\_Throughput * Efficiency_{Design}}{Technology\_Node^2}$$

An increase of data throughput and an increase of memory efficiency (bits per cell) accelerate the performance indicator value. The memory design efficiency has a strong link to the technology node and therefore the square of the node (equal to the cell size for 1 F^2) is included as base line.

Read data throughput and read latency—first random read access—are combined with a normalized density figure based on the cell efficiency for the specific technology node.

The **P**erformance **I**ndicator characterizing the memory **A**rchitecture targeting the **R**ead **A**ccess—**PI_A_RA**—is defined:

$$PI\_A\_RA_{Architecture} = \frac{RD\_Throughput * Cell\_Efficiency_{Die}}{RD\_Cycle * Technology\_Node^2}$$

Non-volatile memories are characterized by the physical storage principle. The physical storage operation defines the cell, the array and the way how to change or program the data. The program throughput incorporates the cell physics, the programming sequence and the V_{th} window margin between the distributions. Design, array and algorithm innovations are automatically included in the program throughput parameter.

The **P**erformance **I**ndicator characterizing the memory **A**rchitecture targeting the **P**rogram **T**hroughput—**PI_A_PT**—is defined:

$$PI\_A\_PT_{Architecture} = \frac{PGM\_Throughput * Cell\_Efficiency_{Die}}{Technology\_Node^2}$$

The performance indicator values are calculated for specific NAND flash memory designs from two memory vendors. The Memory Architecture Model (MAM) for 90 nm NAND flash is used as a reference. The MAM is based on the memory design reference model used in Chap. 5.2.2.3 to calculate the expected cell efficiency.

The performance indicator values are calculated for SLC NAND for four technology nodes. The PI_A values are calculated once per node with the 90 nm NAND performance values. The results are shown in Table 6.1 below each new technology node in the row called "MAM ref_90" to include the PI_A values based on the density increase by the CMOS shrink factor alone and the achieved cell efficiency.

The performance indicator values are normalized at 90 nm for SLC NAND. This normalization factor is applied to all corresponding NAND performance indicators. The results are shown in Table 6.1.

The following conclusion are the results of an assessment of SLC NAND designs for 90, 70, 63 and 51 nm technology nodes using the performance indicator (PI_A_RA and PI_A_PT) values:

- The performance increase of the memory architecture is fully visible for program and read indicators. A linear dependency for NAND flash between density increase and performance improvement is not visible for all nodes and for all designs. The 60 nm technology node does not fit into a density and performance correlation which is possible for 70 and 50 nm.
- Flash memory designs from two selected companies are comparable based on PI_A, nearly independent from their differences in technology node, die size and performance values.

Table 6.1 Performance indicator calculation for SLC NAND devices and technology nodes

Flash memory product	Density Spec (Gbit)	Litho node (nm)	Die size (mm^2)	Cell Efficiency (%)	Program throughput (MB/s)	Read throughput (MB/s)	Read access cycle time (ns)	PI_A_PT normalized	PI_A_RA normalized
Vendor 1: SLC NAND	2	90	138	56	10	20	30000	0.89	0.77
Vendor 2: SLC NAND	2	90	144	56	10	20	30000	0.88	0.77
MAM Model	2	90	118	60	10	20	25000	1	1
Vendor 1: SLC NAND	4	70	145.5	58	20	30	30000	3.03	2
Vendor 2: SLC NAND	4	73	156	59	20	30	30000	2.97	1.86
MAM ref_90	4	70	143	60	10	20	25000	1.65	1.65
Vendor 2: SLC NAND	4	63	131	54	20	40	25000	3.65	3.65
MAM ref_90	4	63	116	60	10	20	25000	2.04	2.04
Vendor 2: SLC NAND	8	51	157	58	40	80	25000	12.04	12.04
MAM ref_90	8	51	152	60	10	20	25000	3.11	3.11

The performance indicators can be utilized to specify the expected performance and density figure for a memory design.

6.2.2 Performance Indicator: Validation with MLC NAND Flash

In the next step the performance indicator based assessment is applied to MLC NAND flash designs from different companies [1–3]. In contrast to Table 6.1 the Memory Architecture Model for MLC NAND is adapted for each technology node in Table 6.2, e.g. MAM_ref_60 for 60 nm node.

The performance indicator values PI_A_PT and PI_A_RA for MLC NAND are expected to be lower than for the corresponding SLC NAND devices. MLC program throughput is significantly reduced and first read access is increased. The density increase is not reflected in this performance indicator characterizing the architecture because the cell efficiency is used as product, design and array efficiency figure. A "cut down" memory product is added to 8 GBit MLC NAND designs and an innovative All Bit Line MLC NAND is added to 16 Gbit MLC NAND designs to enlarge the scope of Table 6.2 for the following assessment.

The assessment of different MLC NAND designs for 90, 70, 60 and 51 nm technology nodes results into more fundamental conclusions obtained by applying performance indicator values:

- The performance increase of MLC NAND combined with the density increase results into a performance indicator value for MLC as large as for the last SLC NAND technology node in average. Both performance indicators [PI_A_PT, PI_A_RA] are for 70 nm MLC NAND designs better than the 90 nm SLC NAND product performance indicator values.
- The physical accessible page size shows a fundamental impact on the PI_A_PT values. The additional MLC NAND designs (8 GBit cut down and All Bit Line MLC NAND) have both an effective page size that is twice as large as the reference designs in the same density class.
- MLC flash memory designs from two different companies are again comparable based on the PI_A nearly independent from their differences in technology node, die size and performance values.

The performance indicator targeting the program throughput is selected for the development of the methodology and the memory array models. The performance indicator targeting the read access is an excellent parameter including the impact of the memory interface.

Table 6.2 Performance indicator calculation for MLC NAND devices and technology nodes

Flash memory product	Density spec (Gbit)	Litho node (nm)	Die size (mm^2)	Cell efficiency (%)	Program through-put (MB/s)	Read through-put (MB/s)	Read access cycle time (ns)	PI_A_PT normal-ized	PI_A_RA normal-ized
Vendor 1: MLC NAND	4	90	138	56	3	20	30000	0.28	0.77
Vendor 2: MLC NAND	4	90	154	52	3	20	30000	0.26	0.72
MAM Model	4	90	118	60	3	20	30000	0.3	0.83
Vendor 1: MLC NAND	8	70	145,5	60	6	30	40000	0.993	1.55
Vendor 2: MLC NAND	8	63	133	64	4.6	30	40000	1.002	2.04
Vendor 1: MLC NAND	8	56	99	52	10	40	45000	2.24	2.49
MAM ref_60	8	60	106	60	5	30	40000	1.13	2.11
Vendor 1: MLC NAND	16	56	173	66	10	40	50000	2.84	2.84
Vendor 2: MLC NAND	16	51	157	60	9	40	40000	2.8	3.89
All bit line MLC NAND	16	56	182	59	34	40	50000	8.64	2.54
MAM ref_50	16	51	152	60	10	40	50000	3.12	3.11

6.2.3 Performance and Cost Indicator

The cell efficiency does not incorporate the improved density achieved by multi-bit and multi-level cells. New performance and cost indicators are required to incorporate the bit density correctly.

The cell efficiency defines mainly the efficiency of the memory design. The cell efficiency of DRAM and Flash memories is comparable in a range between 55 and 65 %. The typically assumed cell sizes are different, typically values are for DRAM: $8F^2$ or $6F^2$, for NOR: $8F^2$ and for NAND: $4F^2$.

Replacing the cell efficiency by the bit efficiency transforms the indicator PI_A_PT into a performance and cost indicator. An effective cost figure is now included and the $Gbit/mm^2$ increase based on more bits per cell is reflected by the performance and cost indicator.

The **P**erformance **I**ndicator characterizing memory **A**rchitecture and **C**ost targeting the **P**rogram **T**hroughput—**PI_AC_PT**—is defined:

$$PI\_AC\_PT_{Architecture\ and\ Cost} = \frac{PGM\_Throughput * Bit\_Efficiency_{Die}}{Technology\_Node^2}$$

The performance indicator characterizing architecture and cost is calculated for three technology nodes based on the modified **M**emory **A**rray **M**odel—SLC and MLC die sizes are different—and compared for SLC and MLC. The more detailed cell and bit efficiency calculation introduced in Fig. 5.8 is applied for the MAM. All performance indicators (for 3 MAM nodes) are shown in Table 6.3.

The performance and cost indicator is dominated by the bit density. The next technology node doubles the bit density and impacts performance and cost stronger than the memory architecture. The performance indicator benefit of SLC over MLC NAND is reduced per node for PI_AC versus PI_A.

A side conclusion can be made based on the performance indicator PI_A_RA. The read performance does not accelerate due to interface limitations. A DDR NAND interface is required from 50 nm on.

6.2.4 Performance Indicator Summary

The proposed solution of combining density, efficiency, performance and cost into an indicator is introduced. A validation was done with 10 different NAND designs out of 4 technology nodes.

Applying performance and cost indicators concludes into the following statements:

- Commercial successful flash memory product designs from different companies are fully comparable based on the introduced new performance indicators. This

Table 6.3 Performance and cost indicator based on the memory array model

Flash memory product	Density spec (Gbit)	Litho node (nm)	Die size (mm^2)	Cell efficiency (%)	Program through. (MB/s)	Read through. (MB/s)	Read acc. cycle time (ns)	Bit effic.(%)	PI_AC_PT normalized	PI_A_PT normalized	PI_A_RA normalized
SLC NAND MAM ref 90	2	90	118	60	10	20	25000	15	0.25	1	1
MLC NAND MAM ref_90	4	90	119.5	60	3	20	30000	30	0.15	0.3	0.83
SLC NAND MAM ref 70	4	70	143	60	20	40	25000	15	0.83	3.3	3.3
MLC NAND MAM ref_70	8	70	144.5	60	5	30	40000	30	0.42	0.83	1.55
SLC NAND MAM ref 54	8	54	170	60	36	40	25000	15	2.50	10.01	5.55
MLC NAND MAM ref_54	16	54	172	60	10	40	50000	30	1.39	2.78	2.78

holds true even if the products differ in technology node by 4 to 6 nm, in program and read performance by 10 to 25 % and in die size.

- Successful memory products have to achieve the same target PI_AC_PT, which results automatically into a competitive product in terms of cost and performance.
- Performance targets for memory design can be assessed best applying PI_A_PT.
- Cost decisions targeting bill of material (BOM) and total cost of ownership (TCO) have to be based on PI_AC_PT assessments.

The following conclusions are made comparing different technology nodes:

- Memory designs with same density produced in smaller technology node have significantly increased performance indicators driven by the larger physical page size and the increased program throughput.
- The memory design with the highest PI_A_PT and PI_AC_PT is cost-wise always the better choice. This conclusion is mainly driven by the cell efficiency target to become larger than 60 %.

The assessment comparing different memory densities results into the conclusion:

- MLC NAND of the technology node /n/—the same node is selected by Memory Array Model – outperforms SLC NAND of the technology node /n-1/ based on both performance and cost indicators.
- The performance capability of memory cell and array architecture is compressed into an indicator value (derived quantity). The performance indicator for read access PI_A_RA is highlighting an expected gap between read throughput capability of NAND and limitations of the asynchronous NAND interface (vendor specific) in the 60 and 50 nm nodes.

All major performance dependencies discussed in Chap. 5 are covered in the first performance and cost indicator assessment. Every memory design available or in development can be characterized with unique performance indicator values to judge competitiveness and usability for an application.

The 5x nm compared to the 7x nm technology node doubles the memory density and doubles the program throughput which is transformed into an increase of the PI_A_PT by a factor of three.

The idea and the process of the performance indicator methodology are now introduced starting with the memory array model. Afterwards performance indicators can be added focusing on durability and reliability parameter.

6.3 Definition of a Performance Indicator Model

The complex example of NAND flash memories is used to develop a performance indicator model which is the basis for the performance indicator methodology. The principle NAND structures of cell, string and array were not changed over generations from 180 nm down to 20 nm. In the same timeframe a significant amount of technology, design and algorithm innovations have solved known issues. These product innovations have ensured 2- and 3-bit per cell MLC NAND volume products.

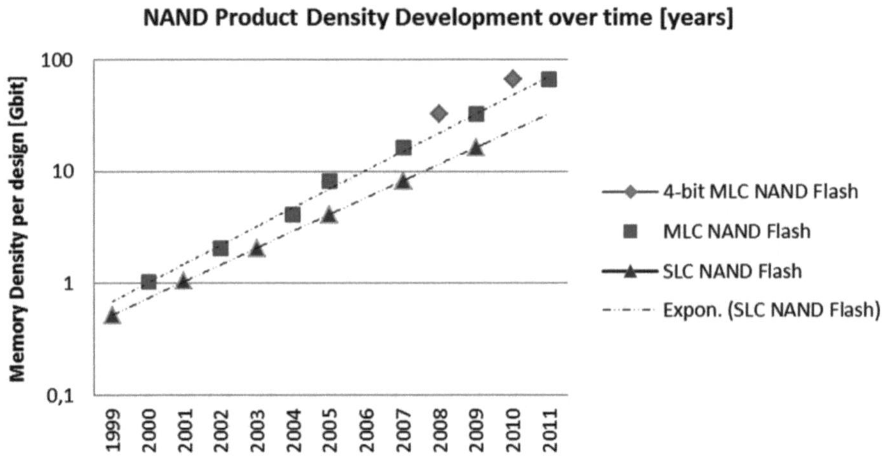

Fig. 6.2 NAND trend chart: density over time

6.3.1 Memory Architecture and Application Trend Analysis

The assessment of known trends and roadmaps on memory and application side will define the setup for the model-based trend analysis.

Figure 6.2 shows the trend of memory density per design. The SLC, MLC and 4-bit per cell MLC NAND designs are based on figures of ISSCC publications and presentations. The trend is well known, every new announced NAND product design doubles the density. The selected SLC and MLC designs follow Hwang's Law and outperform the ITRS roadmap which is predicting slow down. The increase in terms of bit density based on 3- and 4-bit per cell NAND designs outperforms the expectations.

The product density does not include performance and reliability values. Significant differences between flash memory products based on a different bit count per cell are not expressed.

The performance parameters of flash memories define a major part of the success of a mobile memory centric application. The performance requirements defined by the usability of the application are as important as the size of the memory.

The importance of the memory performance is investigated for a portable memory stick to generate an application trend.

A 64 MByte USB stick (based on a 512 Mbit flash die) had a program throughput of approximately 1 MB/s in 2002. A 1 GByte stick had an improved program throughput of 16 MB/s. Flash density and program performance have doubled generation by generation. Today's 16 GByte sticks have program performance values in the range of 16 to 40 MB/s.

Applications entering the market are characterized by density and performance. The density and performance increase along the shrink roadmap is executed up

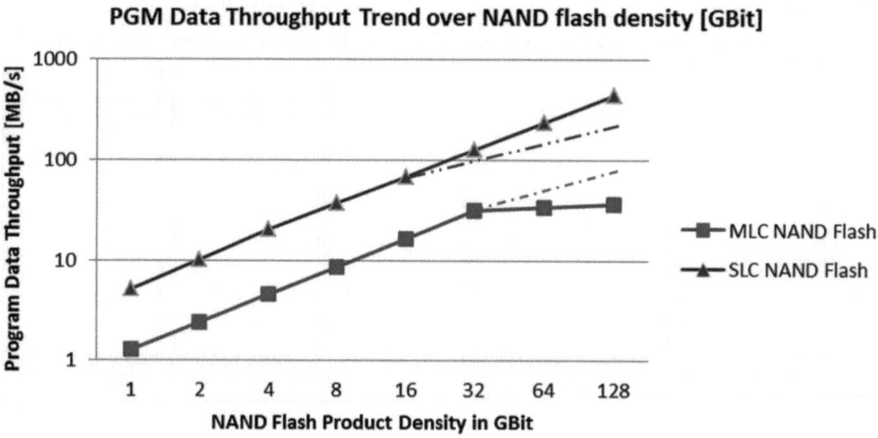

Fig. 6.3 NAND Program throughput over product density

to a performance value, which covers all major applications. Further performance improvement would not substantially increase the usability as shown in Fig. 6.3 for the MLC NAND PGM throughput trend line between 25 and 35 MB/s assumed as an example for a NAND flash product density value of 32 GBit.

The SLC NAND program performance trend line does not saturate in Fig. 6.3. The assumption behind is that the Server Solid State Disc market is predicted as performance demanding application.

These two major trends "doubling the density" and "doubling the program performance" are well aligned with the assessment shown in Table 6.3 for 90, 70 and 50 nm technology nodes.

- A first performance requirement to establish a new memory-centric application successful in the market is defined: ***The time to fill a mobile storage medium has to stay in the same range independent of the size of the medium.***
- A second performance requirement to maintain the commercial growth of a memory-centric application is defined: ***The application (expected file size) specific data throughput has to be fulfilled, but any additional performance increase does not improve the price of the product or the market share of the vendor specific memory design.***

The NAND product roadmap analysis in Chap. 5.2.3.2 has documented both cases, a NAND density and page size doubling and a NAND density doubling without page size increase. The page size doubling is applied on every second technology node. A cost driven development focuses on cell efficiency—smallest possible die size - and results into a page size dependency shown in Fig. 6.4.

The NAND flash operation principle limits the high voltage operation on the erase block, which is very small in size (0.048 %) compared to the total NAND array. The FN tunneling based operation principle enables a performance and cost optimized development roadmap increasing both density and page size.

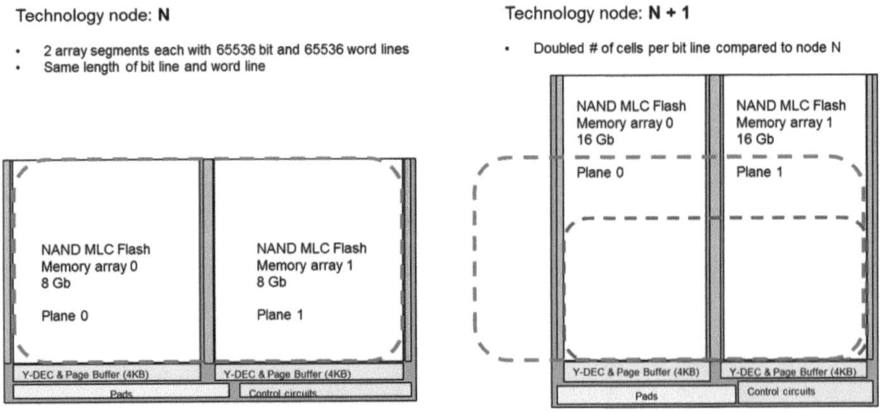

Fig. 6.4 NAND Page size and die size dependency from node to node

MLC and XLC NAND designs will gain significantly from larger physical page sizes. The longer program and read cycle times are over compensated by the page size doubling and the read and program throughput growth is maintained.

NAND became the leading semiconductor memory technology latest with the 90 nm technology node. The model-based approach merges the specific application requirements with the memory density and architecture trend analysis. The effective available page size based on cost optimized memory array segmentation is selected as the lead parameter to develop the memory array model.

6.3.2 Performance Indicator Model Definition and Development

The performance indicator model represents the expected development trend of all parameters required for the calculation of the performance and cost indicators. The performance indicator model developed in this chapter is not linked to a timeline. The unique target of this model based analysis approach is the selection of the best fitting architecture and the judgement of innovation along the shrink and product roadmap.

The time line is a very important factor for project-specific decisions. There are established methods to investigate the project risk and effort for each decision. Memory architecture decisions which enable the success of a memory centric application should not be driven by time constraints.

The second best strategic memory concept decision can never be compensated by faster execution of the memory project time line. Do it first time right, is the important role for memory concept development and assessment.

6.3.2.1 Memory Array Model setup

The memory array model setup follows a couple of strict rules. The model represents an assumed technology development for a known and simplified memory array architecture based on the introduced efficiency parameter.

The detailed knowledge of the memory design constraints and the trend analysis of technology nodes in volume production define the model setup. The application and market requirements influence the model setup with the same weight as technology and design trends to put enough strength in the model to incorporate innovation, which could not be predicted from history analysis.

The Memory Array Model is based on the following setup:

- Array efficiency of 80 % is targeted for NAND flash
  - redundancy and spare blocks according to specification;
  - number of cells per string fulfill the above rule;
  - space for word line decoding large enough for high voltage transistors;
- Cell efficiency of 50 to 65 % is targeted according to the die size dependency introduced in Chap. 5:
  - Peripheral circuit overhead is added derived from high voltage requirements, layout consideration for sense amplifier layout solutions grouped in bit line pitch;
- Program cycle time is set to a value linked to a certain memory array and cell architecture:
  - SLC NAND has a typical average page program time of 200 μs.
    The model setup is for example 250 μs.
- Data throughput is based on the effective available page size per die:
  - A page size doubling is assumed for each new NAND **A**rchitecture **T**echnology **N**ode;

The NAND Performance Indicator Model is based on the rule that every NAND **A**rchitecture **T**echnology **N**ode doubles the memory density and doubles the program throughput.

The Performance Indicator Model trend line for SLC NAND flash is set as most aggressive model option assuming the doubling of the effective NAND page size per die ensured by design innovation, by array segmentation and by CMOS shrink capability shown in Fig. 6.5.

The **NAND Array Model** is an iterative calculation with the technology node as input value:

1. The achievable memory density is derived from technology node, cell efficiency and allowed die size (depending on package size and package type e.g. MCP).

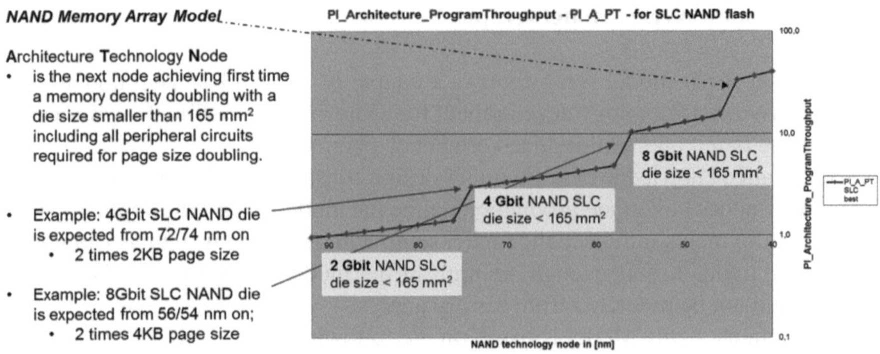

Fig. 6.5 NAND Performance indicator model: definition of architecture technology node

2. The array efficiency is adapted accordingly.
3. The effective page size is increased to fulfill the PIM rule.
4. The final page size is calculated and the achieved cell efficiency has to fulfill the targets.
5. If the targets are fulfilled the performance indicator calculation is finished, if not the process is started again changing at least one parameter.

Two independent and one dependent parameter are used to calculate the NAND architecture nodes:

- Doubling the memory density in terms of cells per die.
- Doubling the program throughput in terms of MB/s.
- Fulfilling both above parameter targets first time and achieving a competitive design size in mm^2 fitting in the standard package defines the architecture technology node.

The Performance Indicator Model for NAND is used to generate trend charts for performance and cost indicators to analyze the flash memory trend of interest. This will be done in the next chapter.

6.3.2.2 Model-Based Performance Indicator Trend charts

The defined and introduced PIM is used to calculate the expected performance indicator value PI_A_PT over all technology nodes from 90 nm on. A model-based performance indicator trend chart is generated for the parameter PI_A_PT shown in Fig. 6.6 and named "best". It is characterizing the density and performance roadmap of SLC NAND designs along the CMOS shrink roadmap.

The iterative process to define the memory architecture model is described.

A graphical representation is introduced to represent the development of the performance indicator with a trend chart. The graphical performance indicator trend

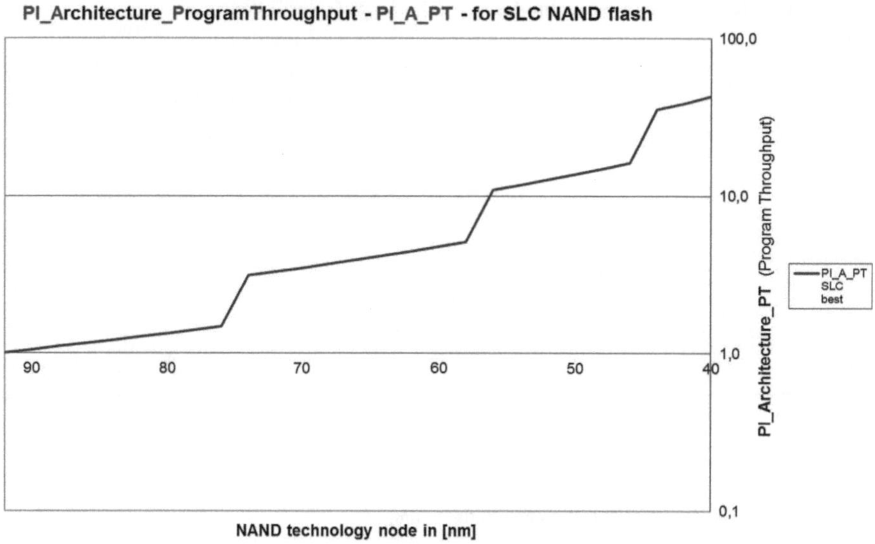

Fig. 6.6 Trend chart of PIM for PI_A_PT for SLC NAND over technology nodes

charts are the way to apply the methodology and to evaluate design and technology trends and innovations and derive conclusions.

6.3.2.3 Confidence Interval of Model-Based PI Trend Charts

The PI trend chart describing the development roadmap of the NAND memory architecture is used to assess available design data and verify the prediction quality of the Memory Architecture Model.

The model is enlarged by a second conservative calculation rule, not taking into account innovations required to double the program performance. This "conservative" trend chart is defined as the bottom line for the NAND technology roadmap. A third calculation rule generates a trend chart called "expected", which includes all design and technology specific dependencies. These trend charts are shown in Fig. 6.7. The cell efficiency is based on the die size dependent efficiency calculation model (see Fig. 5.8) applied with a 2 nm granularity for the following graphical trend charts.

A confidence interval is defined on both sides of the expected trend chart for SLC NAND designs in volume production. This target design and technology space—the area between best and conservative trend charts—is the level of confidence given by the model-based PI trend analysis.

The rules applied in the model setup for NAND has to be validated with a significant large number of high volume non-volatile memory designs to prove the

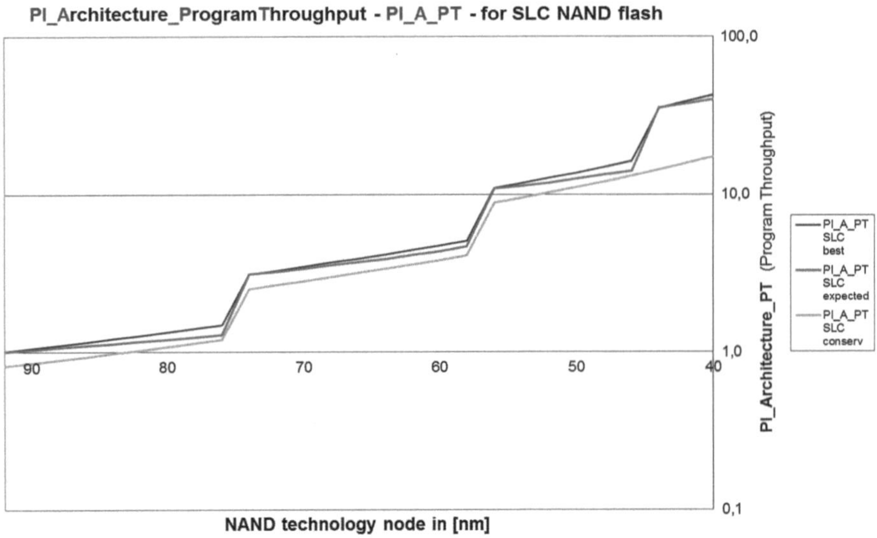

Fig. 6.7 Trend chart with confidence interval of PIM for SLC NAND over technology nodes

concept. This validation procedure of the model is the first learning step to apply the performance indicator methodology.

The validation procedure starts with the calculation of PI_A_PT values for each investigated NAND design and add these values to the graphical representation. SLC NAND flash designs should be selected from three technology nodes as shown in Fig. 6.8. The marked values for the investigated SLC NAND designs are on or below the model-based "expected" PI trend chart for SLC NAND designs.

In this book the default chart technique uses only one trend line per performance indicator. The target is to illustrate the performance indicator methodology clear enough within printed figures.

The "best" trend line has the most simple building rules and is selected as the preferred trend chart for the next chapters. The default graphical PI representation for SLC NAND is shown in Fig. 6.9.

The PIM model visualize the known memory design principles in a proper way especially combining density, cell efficiency and chip size together with cost and performance parameters in a single performance indicator value. A design and vendor specific calculated PI_A_PT value represents the quality of this design with respect to the model-based PIM trend line for the performance indicator characterizing the memory architecture targeting the program throughput.

The option to include a confidence level surrounding the expected model-based trend line enables use of the methodology as concept proof for memory design innovations.

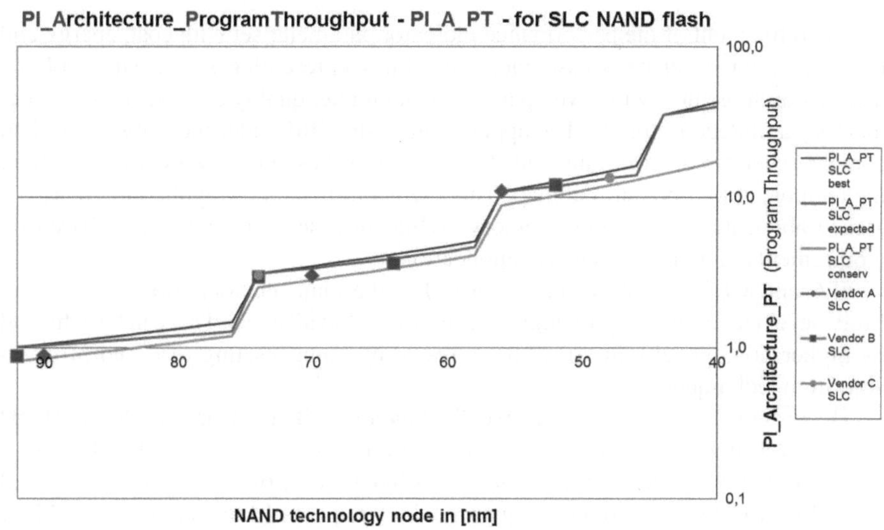

Fig. 6.8 Trend chart of PIM for PI_A_PT including SLC NAND design data of three vendors

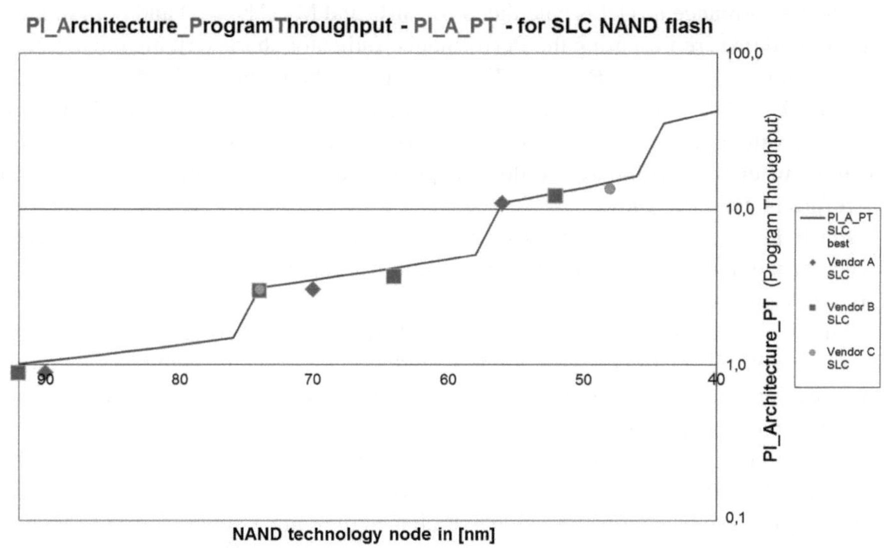

Fig. 6.9 Trend chart of PIM including SLC design data: one trend line only

6.3.3 PIM: Application for SLC and MLC NAND Assessment

The introduced Performance Indicator Model and the graphical representation based on the trend lines can be simply judged as another form to represent the non-volatile semiconductor memory development based on Moore's Law.

The refinement of the performance indicator parameter set with cost, energy and reliability figures and the assessment of trend lines representing different non-volatile memory architectures will develop a significant higher quality to present and predict memory architecture trends. The application of the PIM under the above described context generates precise statements forecasting the best memory architecture along the memory development roadmap. This methodology is now applied to create evidence about the non-volatile memory architecture development targets based on requirements of a memory centric application.

SLC and MLC NAND designs are based on the same memory array architecture. Multi-level techniques require higher accuracy within all areas. This can be achieved by higher homogeneity in array structures or by compensating read and program algorithm techniques.

The program performance is an excellent parameter to compare Single-Level and Multi-Level memories. The number of programming pulses required to finalize a SLC NAND flash page program sequence is in the range of 3 to 5 pulses compared to a MLC NAND page program sequence which requires 14 to 20 pulses. The MLC program time is roughly four times longer and therefore the MLC program throughput is expected to be four times lower than for SLC NAND.

The performance model is calculated for SLC and MLC NAND and both model-based trend lines representing the **P**erformance **I**ndicator characterizing the memory **A**rchitecture targeting the **P**rogram **T**hroughput are shown in Fig. 6.10.

The **PI_A_PT** values from two flash manufacturers are added. The performance indicator values are based on volume products. The high volume production NAND designs differ often compared to designs presented on conferences in die size and therefore in cell efficiency.

Flash product design decisions can be supported applying the performance indicator methodology only with a very detailed competitor database regarding cell efficiency and performance parameters.

The two major NAND flash supplier (marked with A and B) are always close to aggressive PI_A_PT trend charts for SLC and MLC NAND independent of significant differences within their technology roadmaps especially metallization concepts and technology nodes.

The difference in program performance is translated into a distance between the SLC and the MLC trend line. The next array architecture node for MLC NAND creates a performance indicator value, which is in the same range as the SLC NAND values the nodes before. The following conclusions are derived based on the graphical representation of model-based PI_A_PT shown in Fig. 6.10:

- A memory design exactly on the array architecture node—jump of trend line—has a significant performance and density benefit.
- A MLC NAND design from the next array architecture node—like developed from vendor A in Fig. 6.10—can nearly replace SLC NAND designs from the previous array architecture node.

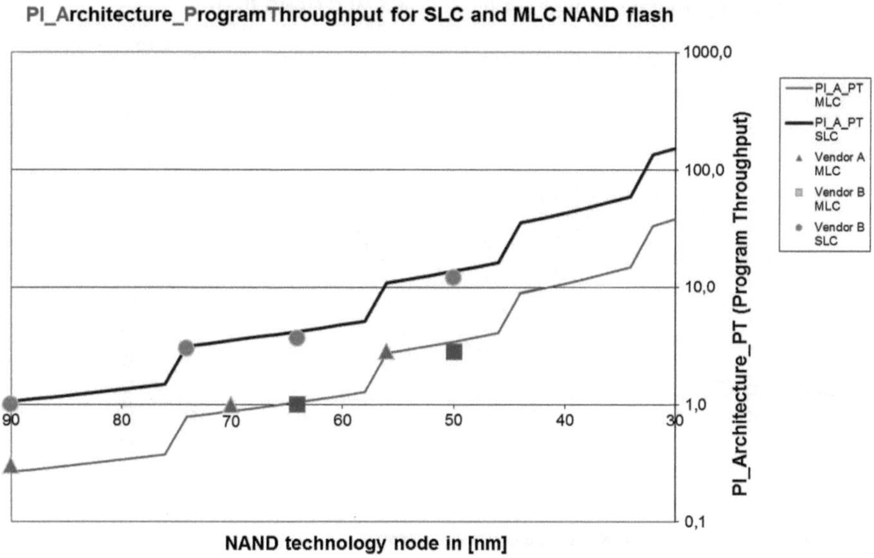

Fig. 6.10 Trend chart of PIM for PI_A_PT for SLC and MLC NAND including NAND design data

Business and cost figures have to be based on bit efficiency. The PI_AC_PT values compare the more efficiency based figures in which 2-bit per cell are counted twice. The customer pays for twice the number of bits compared to 1-bit SLC NAND.

The distance between SLC and MLC NAND becomes even smaller for the model-based trend chart based on the key performance parameter PI_AC_PT. The application of the PIM using different PI-values and comparing the relative differences between SLC and MLC NAND architectures is supporting the preparation for memory centric application development decisions.

The difference in terms of technology nodes and the corresponding timeline when the MLC architecture can replace the SLC architecture are shown for PI_A_PT and PI_AC_PT in Figs. 6.11 and 6.12.

The distance between SLC and MLC NAND becomes even smaller for the model-based trend chart based on the key performance parameter PI_AC_PT shown in Fig. 6.12.

MLC NAND was replacing the SLC NAND architecture based on performance and cost in most of the mobile application. The **P**erformance **I**ndicator characterizing the memory **A**rchitecture and **C**ost targeting the **P**rogram **T**hroughput is characterizing the NAND flash market in an excellent way. The performance benefit in absolute values measured with **PI_AC_PT** is lower compared to **PI_A_PT** values. The performance increase of SLC NAND does not outperform the cost benefit of MLC NAND.

The difference in program performance and cost is translated into a distance between the SLC and the MLC trend line. Again the next array architecture node for

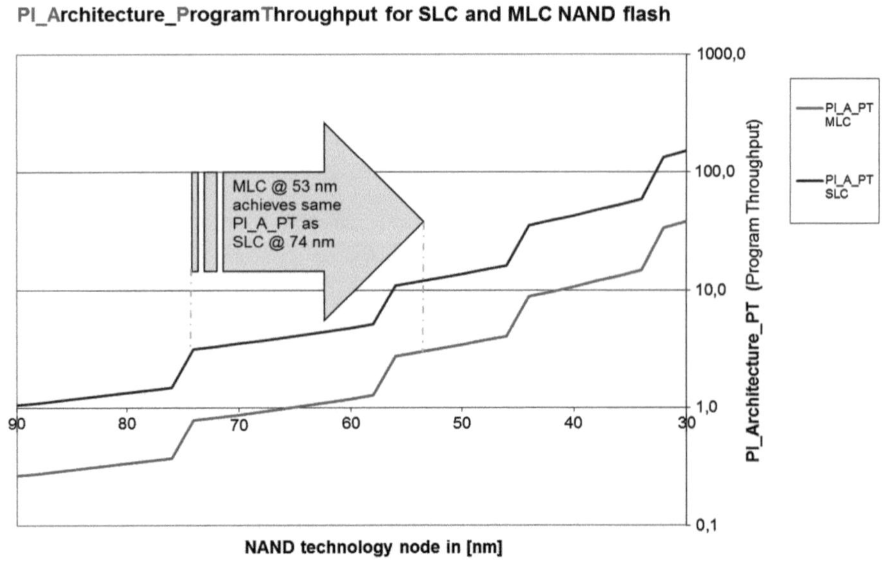

Fig. 6.11 PI_A_PT trend charts with distance between same PI values for SLC and MLC NAND

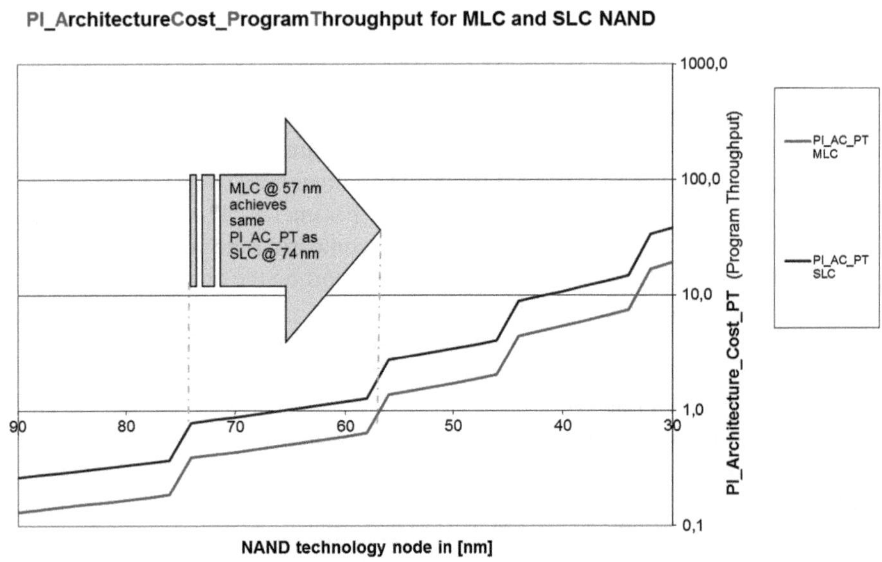

Fig. 6.12 PI_AC_PT trend charts with distance between same PI values for SLC and MLC NAND

MLC NAND creates a performance indicator value, which is higher as SLC NAND values from the nodes before. The conclusion derived from the PI_AC_PT trend charts comparing SLC and MLC NAND is the same:

- A memory design on the array architecture node—jump of trend line—has a performance and density benefit.
- MLC NAND can replace SLC NAND always with the next array architecture node.

The combination out of doubling density and doubling program performance is the major driver for the fast replacement based on the performance indicator value assessment.

6.3.4 Performance Indicator Model Enhancement with Durability

The refinement of the performance indicators with reliability parameters and the assessment of the corresponding trend lines for SLC and MLC NAND flash is the next step. The methodology is applied to create evidence about the impact of reliability degradation on flash memory development targets.

The PIM methodology targets reduction of complexity. The flash memory reliability analysis has introduced **R**eliability **F**actors described in Table 4.3. A robust and simple rule is defined to generate a first normalized durability parameter.

The average access to a storage memory is split into write and read operation. For the following performance indicator the write durability is selected and normalized for a bit density of 1 bit per F^2.

The **P**erformance **I**ndicator characterizing memory **A**rchitecture, **C**ost and **D**urability targeting the Program throughput—**PI_ACD_PT**—is defined:

$$PI\_ACD\_PT_{AC\_Durability} = \frac{PGM\_Throughput * Bit\_Efficiency_{Die} * Durability_{WR}}{Technology\_Node^2}$$

The model assumes a durability degradation of approximately 10 % from node to node along the shrink road.

The distance between SLC and MLC is significantly increased due to the difference in endurance values between SLC and MLC NAND. The PIM trend lines are calculated for SLC, MLC and XLC NAND and are shown in Fig. 6.13.

The difference in program performance, cost and durability is translated into a distance between the trend lines which incorporates durability of NAND flash. Density and performance increase have to over compensate the significant reduced durability of MLC compared to SLC NAND.

A different conclusion is derived from the PI_ACD_PT trend charts comparing SLC and MLC NAND:

- Reliability can be compensated by increased density and increased performance.

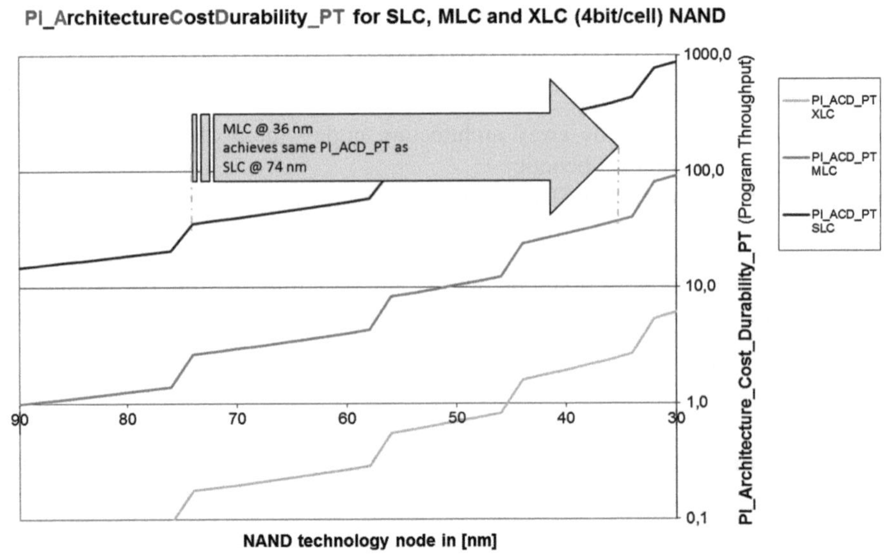

Fig. 6.13 PI_ACD_PT trend charts with distance between same PI values for SLC and MLC NAND

- Durability parameter definition is an application specific key process.

The performance indicator value of PI_ACD_PT incorporates the specified reliability parameter endurance for flash memories into an overall cost, performance and durability quantification. In contrast to this assessment most applications adapt MLC NAND based on market data faster.

The average access to a storage memory is split into write and read operation. For the following performance indicator a weighted and combined read and write durability parameter is used. The different reliability factors are summarized in Table 4.3. The data throughput is based on a split into 25 % write and 75 % read operations. This combination improves the flash durability parameter and includes the read disturbance robustness in parallel. The switch from program throughput to data throughput (combination of read and write) is marked with DT at the end.

The **P**erformance **I**ndicator characterizing memory **A**rchitecture, **C**ost and **D**urability targeting the **D**ata **T**hroughput—**PI_ACD_DT**—is defined:

$$PI\_ACD\_PT_{AC\_Durability} = \frac{PGM\_Throughput * Bit\_Efficiency_{Die} * Durability_{WR\_3RD}}{Technology\_Node^2}$$

The second durability based performance indicator PI_ACD_DT generates a unique value to characterize a flash memory. The analysis of the PI_ACD_DT trend charts shown in Fig. 6.14 allows the following conclusions:

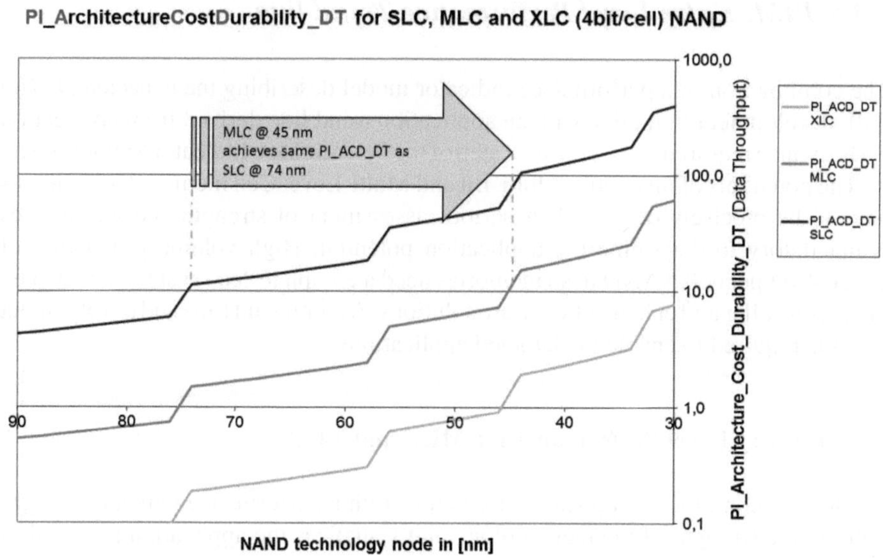

Fig. 6.14 PI_ACD_DT trend charts with distance between same PI values for SLC and MLC NAND

- The MLC NAND array architecture node (n+2) achieves the same program performance as the SLC NAND array architecture node (n) and enables eight times (8x) the density. The density increase can be accepted as a solution to compensate the reduced durability for a predefined application space.
- The MLC NAND flash can replace SLC NAND with array architecture node (n+2) based on PI_ACD_DT. This holds true under the condition that every design on a new NAND Architecture Technology Node doubles performance and density.

The performance indicator based on PI_ACD_**DT** is the preferred PIM (Performance Indicator Model) to be applied if durability degradation can be compensated by memory density.

The performance indicator based on PI_ACD_**PT** is the preferred PIM to be used if durability degradation can only be compensated by increased performance combined with increased memory density.

The two **PI_ACD** performance indicators cover the typical read/write durability and disturbance requirements. Additional application specific reliability issues can be included based on the above described methodology.

This work will now focus on two key topics:

- How long does the density and performance doubling trend have to be maintained?
- The interdependency between application requirements and memory architecture solutions.

6.3.5 PIM: Entry Level Performance Trend line

The combination of a performance indicator model describing the expected NAND flash development trend line with an application trend line derived from application performance requirements is the next step of the Performance Indicator Methodology.

The potentials of innovative Multi-Bit and Multi-Level Cell memory technologies have to be precisely described. A serious assessment of strength and weaknesses is mandatory to determine the application potential. High volume production of 3- and 4-bit per cell NAND flash memories need a complete "eco system" of adapted microcontroller and enhanced software solutions. A significant time and development effort is required to enable the targeted applications.

6.3.5.1 Entry Level Performance for MLC and XLC

The idea to combine an application trend line with a performance indicator is highlighting the strength of the methodology and visualizes the application potential of multi-level cell NAND flash.

The combination of both trend lines extracts the capability of the memory architecture benchmarked against the targeted application performance parameter. The application trend line is the entry level for a single NAND flash performance indicator value. Both trend lines reflect the expected increase over technology node (translated into generations on application level). The performance indicator PI_A_PT is selected for this assessment shown in Fig. 6.15.

Figure 6.15 illustrates the development of an application over time. The SLC NAND performance is high enough from 90 nm on to enable the market entry. MLC and XLC performance indicator trend lines cross the application trend line a certain number of NAND **A**rchitecture **T**echnology **N**odes later (56 and 32 nm). A performance indicator above the application trend line indicates the entry level to enter the market with MLC and XLC NAND flash based application solutions.

The **P**erformance **I**ndicator **M**odel based trend chart assessment indicates the demand for MLC and XLC NAND technology and design solutions. The performance improvement achieved with two NAND **A**rchitecture **T**echnology **N**odes enable the shift of a complete application market from MLC to XLC NAND for Audio application products and two nodes later for Video applications, shown in Fig. 6.16.

A NAND design development team can derive the amount of innovations required to establish a certain MLC or XLC technology as the default memory solution. The calculated design specific performance indicator value has to be higher than the enabling application performance trend line.

All conclusions based on **P**erformance **I**ndicator **M**odel are made for a single die performance assessment. The known principles to increase the system performance operating flash dies in parallel in a MCP package are not considered. An energy optimized solution is assumed in the mobile application market on a single flash die capable to deliver the required performance level.

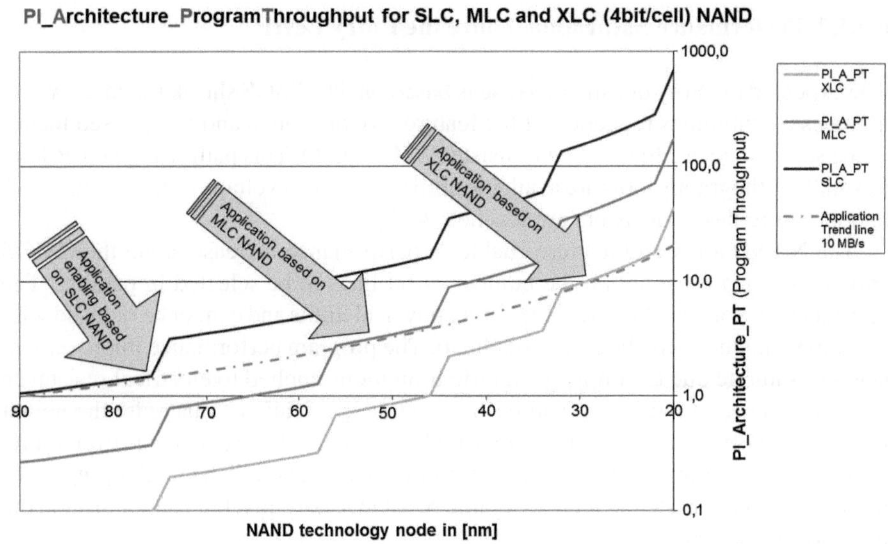

Fig. 6.15 PI_A_PT trend chart with application trend line example for SLC, MLC and XLC NAND

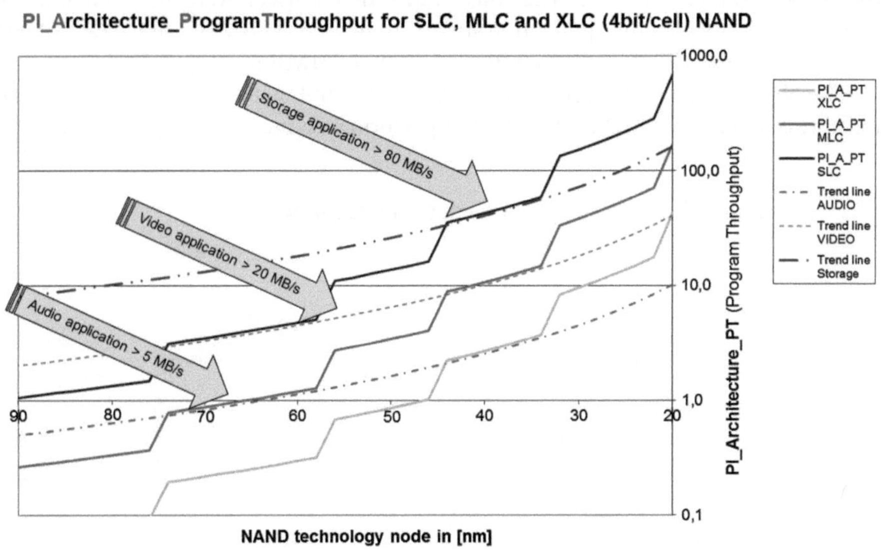

Fig. 6.16 PI_A_PT trend chart for SLC, MLC and XLC NAND with enabling performance trend lines for three applications

6.3.5.2 Performance Saturation Above the Entry Level

The expected memory density increase is based on the CMOS shrink roadmap which assumes a continuous reduction of the feature size of Silicon and SiO_2 based metal-oxide semiconductor field-effect transistors (MOSFET). This path will go on as long as smaller lithography tools are available until the physical voltage limits of MOSFET and therefore flash transistors are reached [4, 5].

The NAND array architecture enables a performance increase along the CMOS shrink roadmap by doubling the number of bit lines to be selected in parallel. The resistance of word and bit lines is reaching physical limits and cannot be reduced with the same amount along the shrink roadmap. The program performance improvement starts to saturate due to simple geometric limitations applied to electrical resistance and capacitance of long word and bit lines. The aspect ratio can increase the height of the lines keeping the resistance constant but increases the capacitance even more. Advanced algorithmic solutions, low current sensing and adiabatic charging techniques can compensate the increasing line resistance partially, but cannot inverse the trend to slow down.

The conservative model-based trend line incorporates the increase of the line resistance with a certain percentage and can be applied to investigate the bottom line for the performance increase. The conservative doted SLC NAND model line down to 20 nm does not pass the application storage trend line of 80 MB/s and MLC NAND does not pass the video application trend line with 20 MB/s in Fig. 6.17.

The NAND flash array architecture is a unique combination which delivers along the shrink roadmap a higher density and performance increase keeping the energy

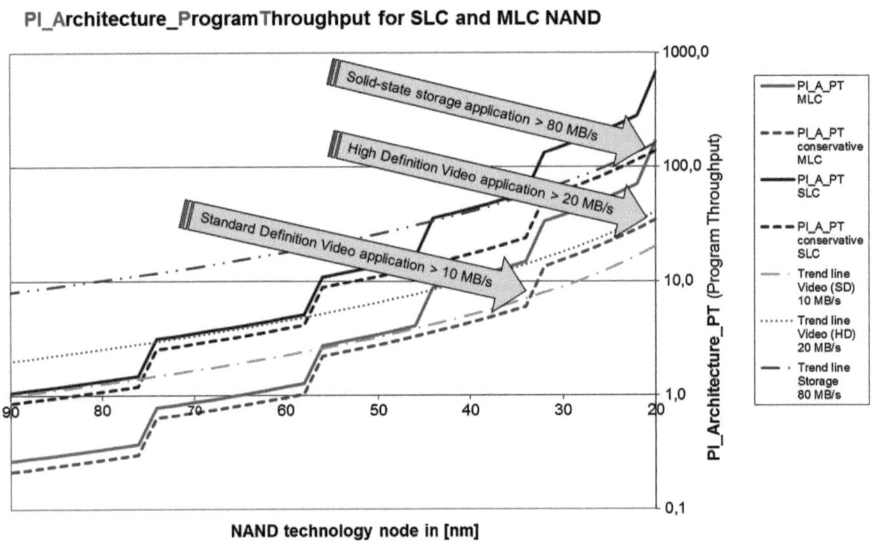

Fig. 6.17 PI_A_PT trend chart for SLC and MLC NAND with conservative model conditions

consumption constant. The high voltage operation is limited on a small die size area—0.025 %—of the large NAND array and the energy to charge and discharge this large plate capacitor—Erase Block—is in first order technology node independent.

Every NAND array architecture node requires design and technology innovations to follow the Performance Indicator Model trend line and enable new application for NAND flash based memory-centric systems.

The system performance specification values can be increased by operating more NAND flash devices in parallel and adding more independent flash access channels. The performance indicator model focuses on the performance of a single flash device, because this is the most energy efficient solution. Additional NAND devices can be applied to hide erase operation times. The control of the write amplification impact enforced by the asymmetric segment sizes between program and erase data package sizes of flash memories becomes more complex utilizing more flash devices in parallel. The performance impact calculation was made in Chap. 3.2.5.2 to demonstrate that the worst case data throughput can be kept up to the single die NAND flash data transfer rate values.

6.3.6 Performance Indicator Model: Summary

The performance indicator is calculated with one memory performance and one efficiency parameter normalized over the technology node (square of the minimal feature size). The predicted memory performance and efficiency values are generated by a simplified memory array model. This approach focuses on cell and array performance. A structured memory array model is defined based on pre-defined efficiency parameters. The read and program performance values are based on charging and discharging sequences of bit- and word-lines.

The performance indicator is based on a subset of parameters and reduces the non-volatile memory complexity triangle into a single parameter—the **P**erformance **I**ndicator (**PI**) value. A limited number of performance indicators are introduced targeting array and performance—PI_A_xx, array, cost and performance—PI_AC_xx and array, cost, durability and performance PI_ACD_xx.

The model-based Performance Indicator Methodology is the graphical representation of the performance indicator values along the shrink roadmap. The performance indicator trend lines combined with the values for flash memory designs enable a judgement of modification and innovation obtained by different memory vendors.

The Performance Indicator Methodology answers the questions: "What will be the impact of an innovation on the competitiveness of a new flash design?" with a quantitative measure. The difference of the performance indicator value calculated for the investigated design once with and second without the innovation is a clear and unique measure of effectiveness of product innovations.

An excellent matching of the memory architecture with system application requirements is an important point for memory centric applications. The performance indicator trend chart of the selected array architecture along the shrink roadmap defines the competitiveness of the target system. The development of the model and the usage of the model-based Performance Indicator Methodology ensure a combination of market understanding and foreseeable innovation potential. The development of the model improves automatically the understanding of the details to match best non-volatile memories—like the introduced array architecture nodes—with the application requirements—like application performance trend lines.

The strength of the method is the translation of memory "improvements" into quantitative "Performance Indicator values" which are applicable for judgement of innovations and evolutionary architecture improvements.

The "weakness" of the methodology is the depth of knowledge required to apply this approach. The selection of performance parameters and assumptions made to develop the model-based trend lines has to incorporate detailed knowledge on cell, array, algorithm and technology topics.

The refinement of the performance indicator parameter set with cost, energy and reliability figures and the assessment of trend lines representing different non-volatile memory architectures enables a higher quality to forecast memory architecture trends.

6.4 Application of Performance Indicator Methodology

The model-based Performance Indicator Methodology is applied to investigate flash memory development options of an established technology. The NAND performance indicator model predicts a doubling of bit density and program throughput for every array architecture node.

Two cost effective ways are predicted to achieve performance doubling.

- Array segmentation is cutting the bit lines by half and doubles the number of array segments which doubles program and read performance. This approach is based on maximizing the reuse of known concepts like the shielded bit line sensing concept.
- Maximizing the parallelism is utilizing all bit lines during program and read operation which doubles the program and read performance. This approach requires a couple of innovations. A significantly improved sensing and new algorithm solutions have to compensate the effect of shielded bit lines during read and program.

The methodology is applied to analyze NAND flash product innovations and conclusions are derived based on the corresponding performance indicator set.

6.4.1 NAND Performance: Array Segmentation and Interface Data Rate

A cell efficiency optimized NAND product design is developed based on two array segments which corresponds to two planes. A further array segmentation can offer two benefits reducing the bit line length and enlarging the available effective page size as shown in Fig. 6.18.

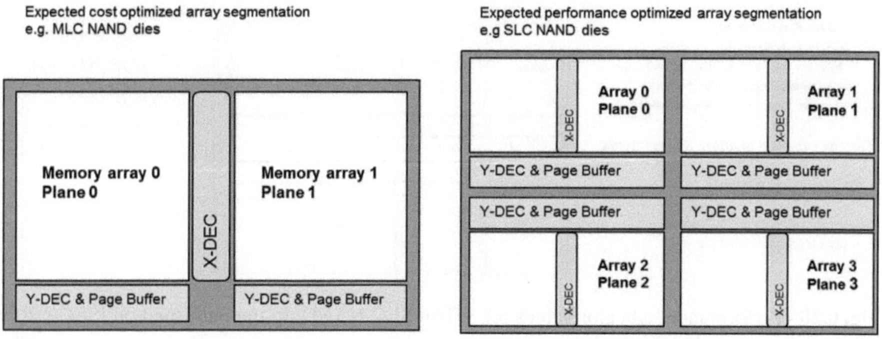

Fig. 6.18 Memory array segmentation boosts NAND program and read performance

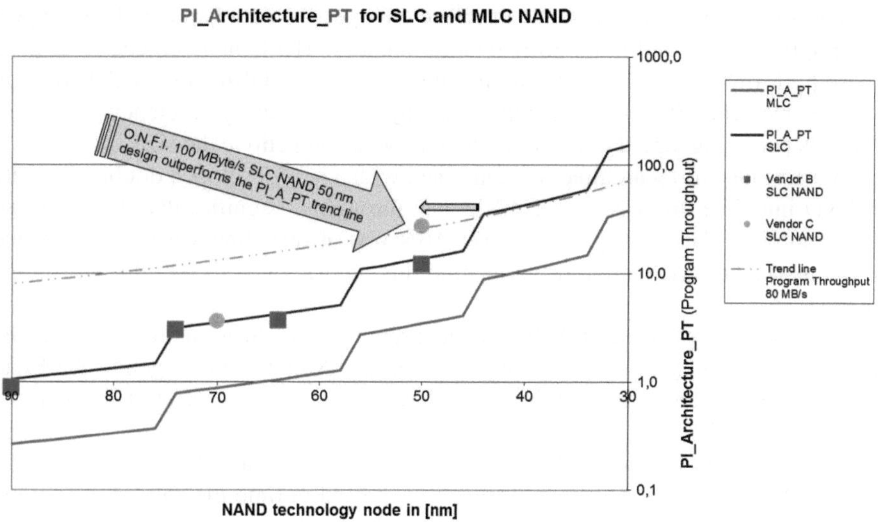

Fig. 6.19 PI_A_PT SLC NAND trend line outperformed by O.N.F.I. SLC NAND design in 2008

Both design changes boost the read and the program performance values.

The logic overhead for a design with four planes is significant as shown in Fig. 6.18 and has to be compensated by an additionally improved array efficiency within the four array segments. NAND flash offers a straightforward way to increase the array efficiency selecting the next longer NAND string combination (from 32 to 64, 64 to 128). The first O.N.F.I SLC NAND design [6] presented at ISSCC 2008 is now selected to add to the calculated PI_A_PT values to the PI trend chart analysis shown in Fig. 6.19.

This SLC NAND design [6] outperforms the read and program throughput expected in a 50 nm node. The model-based trend line has expected a similar PI_A_PT value at the 44 nm NAND **A**rchitecture **T**echnology **N**ode. The achieved program performance improvement shifts the design above the 80 MB/s application program throughput trend line.

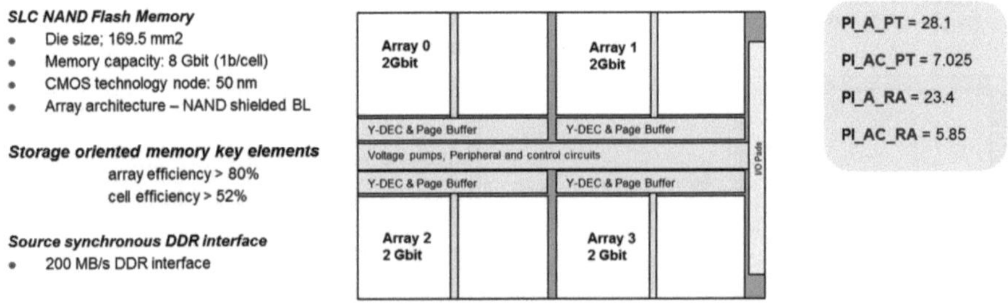

Fig. 6.20 Performance Indicator values for 8 GBit SLC NAND design published on ISSCC 2008 [6]

The source synchronous DDR interface enlarges the data bandwidth to support the improved read and write array data throughput. The logic overhead to achieve these performance values is visible in Fig. 6.20 for the 8GBit SLC NAND design. The reduced cell efficiency is compensated by increased array efficiency—usage of 64 cells per string already in 50 nm—to achieve a competitive die size.

A four plane array architecture combined with a new developed double data rate NAND interface improves the NAND data throughput significantly. The reported I/O read/write data throughput of 200 MB/s was at that time a new performance benchmark. The application driver behind this innovation is the solid-state storage market.

Adding a normalized 160 MB/s data throughput application trend line to the performance indicator characterizing the memory **A**rchitecture targeting the **R**ead **A**ccess shows the perfect fit of the above described 8 GBit SLC NAND design illustrated in Fig. 6.21.

Three innovations are developed and applied to outperform the SLC NAND performance trend line and to enter earlier as expected in time the solid-state storage application market:

- Array segmentation with four planes to double the read and program data throughput.
- Double data rate NAND interface to support the available higher memory array data rate.
- Increase number of cells per NAND string—from 32 to 64—to improve the array efficiency.

The listed innovations are predicted by the Performance Indicator Model for NAND technology nodes from 44 nm on to achieve the expected Performance Indicator trend line.

Two major conclusions can be derived based on the **P**erformance **I**ndicator **M**ethodology from this SLC NAND application example:

- Sub-45 nm SLC NAND designs have to fulfill the solid-state storage application requirements to enable a high performance class of solid-state disc—SSD—devices.

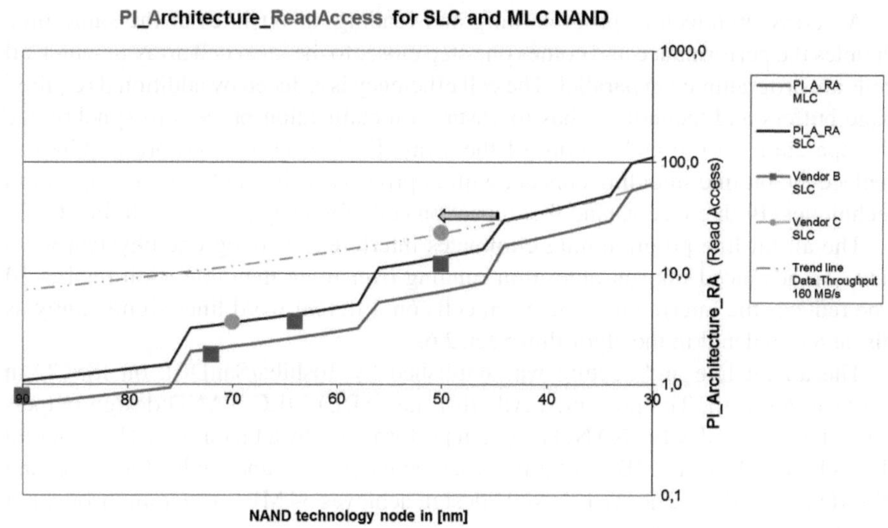

Fig. 6.21 PI_A_RA SLC NAND trend line compared with the solid state application trend line

- Sub 34 nm MLC NAND designs show the capability to enter the same solid-state storage applications developed with SLC NAND devices based on the performance indicator analysis.

The NAND interface performance has to follow the memory array bandwidth and is analyzed in Chap. 7.1. Package technologies like Through Silicon Via (TSV) for chip to chip to controller interconnects are necessary to ensure access to the increasing NAND array data throughput as forecasted in [7].

6.4.2 NAND Performance and Reliability: All Bit Line Architecture

The aggressive NAND performance indicator based trend line predicts a doubling of bit density and program throughput around the 42 nm technology node. The classical two plane architecture would not be able to deliver the expected program and read data throughput. Array segmentation was discussed in the last chapter as one of the solution. The development target for MLC and XLC NAND is to improve performance and reliability together.

The success of the NAND array architecture was founded on the Shielded Bit Line sensing method introduced already in 1994 [8]. The neighbor bit lines—for example the odd lines—are grounded to shield the program verify and read operation on the even bit lines. This shielding effect was the basis for the optimized single sided layout and design optimized Y-mux and sense amplifier architecture.

An array architecture programming and reading all bit lines at the same time doubles the performance and comes one step closer to the ideal cell array in which all cells are programmed in parallel. The cell efficiency is reduced by additional required page buffers and technology has to ensure a metallization process to synchronize the operation on top and bottom of the array. Design innovations are required to replace the bit line shielding concept with improved sensing and program algorithm techniques. Both concepts and the impact on cell efficiency are shown in Fig. 6.22.

The all bit line programming eliminates interferences along one physical word line. An advanced imprint algorithm running over more than one additional word line reduces the interferences between cells on different word lines significantly as discussed in detail in the algorithm Sect. 2.6.

The all bit line architecture was published by Toshiba/SanDisk on ISSCC in the year 2008 [9]. The presented All-Bit-Line (ABL) MLC NAND design outperforms the classical MLC NAND program performance by a factor of 3. The program throughput is 10 to 12 MB/s at 56 nm based on the performance indicator model and NAND device data. The ABL NAND design achieves 34 MB/s program throughput based on published data.

MLC NAND designs based on the ABL architecture compete with SLC NAND designs. The performance indicator trend analysis comparing shared with all bit line architectures is shown in Fig. 6.23. The ABL architecture combined with array segmentation enables the use of MLC NAND technology for the solid-state storage application market.

The design and algorithm innovations [9] outperform the MLC NAND performance trend line for the referenced 56 nm node. This innovative concept is mandatory to follow the XLC NAND performance trend line for smaller technology nodes with

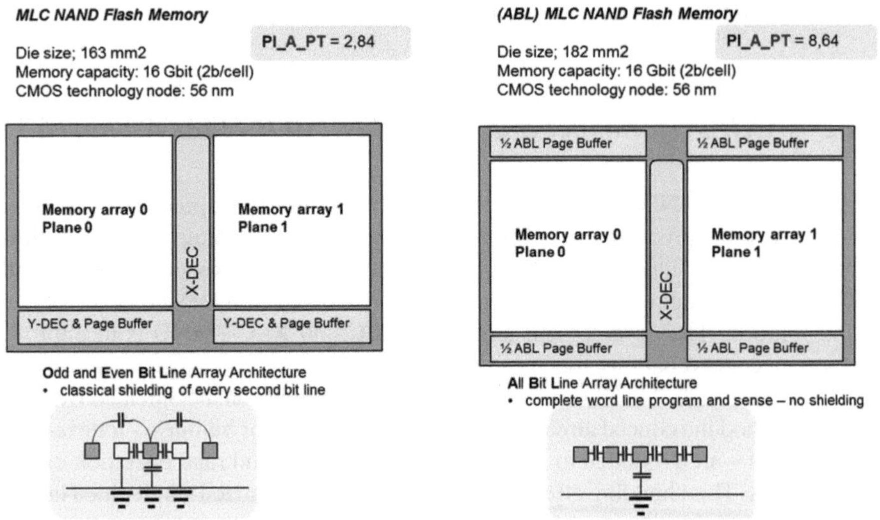

Fig. 6.22 NAND array architecture comparison between shielded bit line and all bit line

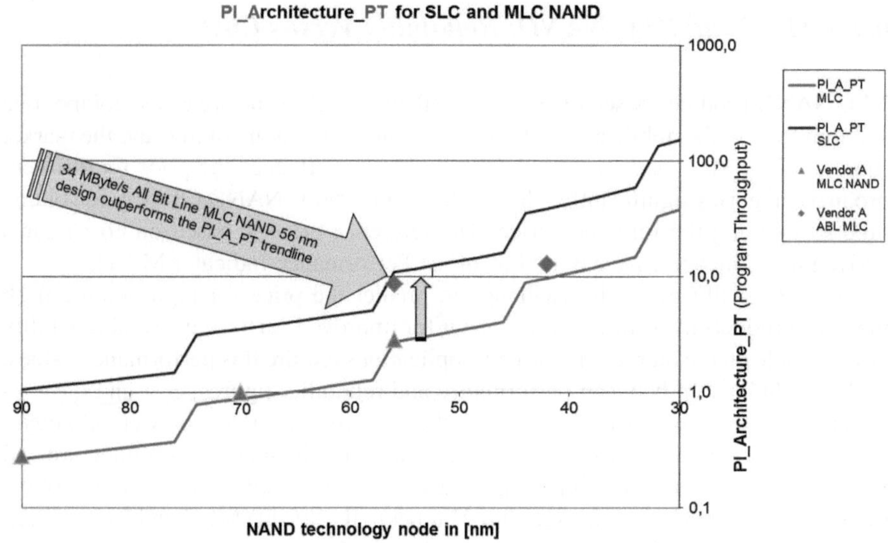

Fig. 6.23 PI_A_PT MLC trend line was outperformed by the All-Bit-Line MLC NAND architecture

higher interferences between cells and bit lines. Improving performance and reliability in parallel is based on two important improvements:

- Sensing technique reading all cells along a word line without discharging bit line voltages.
- Imprint program algorithm incorporating a pre-defined number of logical pages within a multiple phase program procedure.

Three conclusions can be derived based on the **P**erformance **I**ndicator **M**ethodology:

- The program performance increase for 40 and 30 nm technology nodes can be supported by all bit line NAND architecture.
- The increase of cell to cell interferences can be compensated by advanced imprint algorithms including next two neighbor word lines based on the all bit line NAND architecture.
- The distance between logical page sizes (e.g. 4 kByte) and physical word line programming scheme is becoming bigger, physically 8 to 32 kByte data segments are required to start an advanced MLC imprint programming sequence.

The interaction between performance, cost and reliability based on margin consideration and counter measures for random bit failures are targeted in the next chapter.

6.4.3 MLC and XLC NAND: Reliability Versus Cost

XLC NAND products based on 3-bit per cell and All Bit Line are a cost competitive solution for a wide mobile application range. The production volume and the market share of SLC, MLC and XLC NAND flash devices influence the price for different product categories significantly. This work assumes MLC NAND (2b/cell) as volume product defining the reference price. The reference price is based on cost figures derived from the aggressive trend line of the Performance Indicator Model.

Figure 6.24 illustrates the challenge to predict the price for high volume flash memory products over time. A price adder for improved performance and reliability is achievable on the market in case new applications require this performance values.

This relationship between performance and reliability figures, cost and flash architecture type is now investigated using the Performance Indicator Methodology.

The NAND **P**erformance **I**ndicator characterizing the memory **A**rchitecture and **C**ost targeting the **P**rogram **T**hroughput based trend line predicts a penetration of the mobile solid-state storage market with 2 bit per cell (2b/cell) MLC and 3 bit per cell (3b/cell) XLC NAND designs as shown in Fig. 6.25.

MLC and XLC flash product development requires an application specific solution for the issues highlighted in the flash reliability optimization triangle in Fig. 5.16. V_{th} distribution window margin definition and correct assumptions of distribution widening are key points illustrated in Fig. 6.26 to overcome reliability and durability weaknesses of MLC and XLC NAND flash based products.

The classical memory margin definition cannot deliver the expected performance, cost and reliability parameter guaranteed for each cell. The stand-alone development focus on reliability margin per die decreases performance and increases cost of bits. Additional spare area is required for ECC on every page and for durability optimization—write amplification—on system level a second time. The reliability chapter has introduced **R**eliability **F**actors to incorporate automatically the larger bit

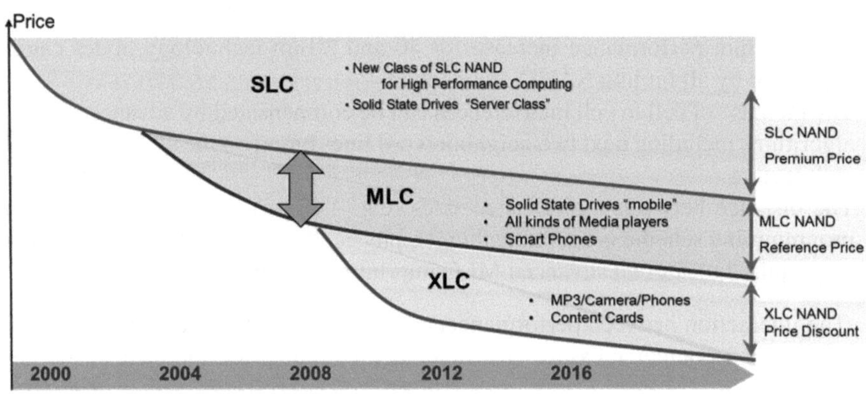

Fig. 6.24 NAND price flexibility for SLC, MLC and XLC (3b/cell and 4b/cell)

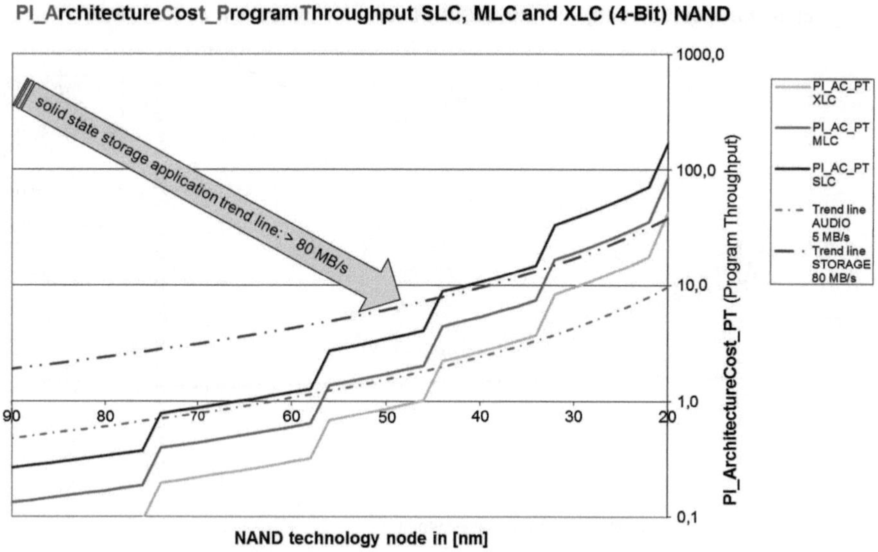

Fig. 6.25 PI_AC_PT trend chart for SLC, MLC and XLC NAND with two enabling performance trend lines

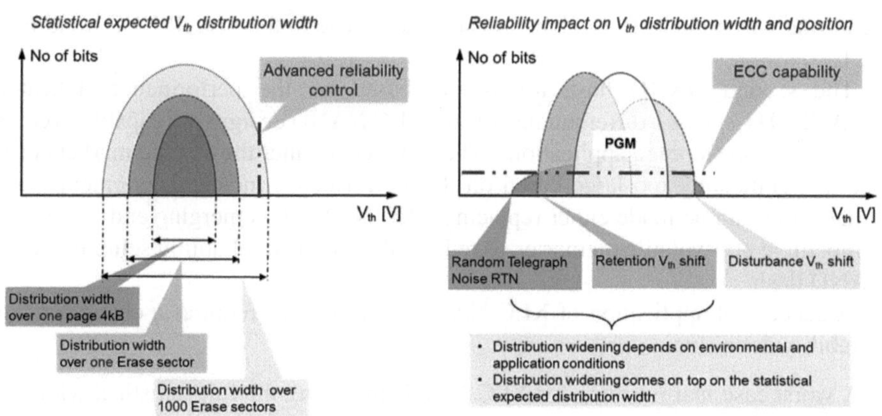

Fig. 6.26 Reliability margin consideration: distribution widening impacts XLC NAND

density and the lower program performance into the system durability calculation. Therefore the **P**erformance **I**ndicator characterizing memory **A**rchitecture, **C**ost and **D**urability targeting the **D**ata **T**hroughput (25 % PGM and 75 % RD operations) is selected to assess latest generations MLC and XLC NAND designs.

The All Bit Line NAND architecture improves the MLC and XLC performance and reliability values significantly as shown in Fig. 6.27 and over achieves the PIM trend line. The 3b/cell XLC NAND designs [10] [11] based on the shielded bit line concept are not able to achieve the aggressive target lines of the PIM (a 3b/cell XLC

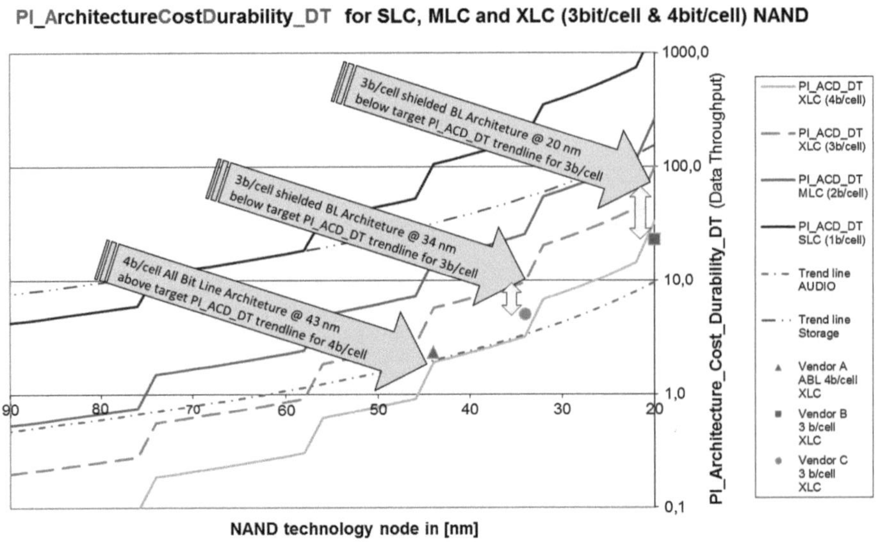

Fig. 6.27 Assessment of 4b/cell All Bit Line versus 3b/cell shielded bit line NAND architecture

trend line was added). Both 3b/cell designs are close to the 4b/cell XLC trend line [12].

The second NAND flash assessment based on the performance indicator **PI_ACD_DT** shows a differentiated result. XLC NAND designs are capable to enter audio or mobile storage applications. The distance to enter the storage market entry line is significantly too large based on this performance indicator judgment. Two conclusions can be made either replacing XLC NAND by emerging and more cost competitive non-volatile memories or rethink the classical reliability setup for XLC NAND flash.

A successful application of XLC NAND flash devices requires a change of the reliability definition:

- A worst case margin setup for XLC NAND including the full statistical width of 128 GBit flash cells reduces the achievable reliability parameter to less meaningful values.
- The independent definition of reliability parameter has to be replaced by a combined system reliability figure—for example a combined performance, density and durability factor representing the behavior of 99.999 % of all cells on system level.

Two conclusions are derived from the reliability and cost assessment of MLC and XLC NAND designs based on the **P**erformance **I**ndicator **M**ethodology:

- On-die reliability diagnostic of a pre-defined quantity of cells has to be the default operation of XLC NAND products during lifetime. A default and a recovery operation mode are mandatory requirements for the usage of XLC NAND flash devices.

Table 6.4 Most common performance characteristics: sequential and random operations

Benchmark/measurement	IOPS description	Remark regarding PIM
Total IOPS	Total number of I/O operations per second (mix of read and write tests)	
Random Read IOPS	Average number of random read I/O operations per second	Associated with small data transfer sizes e.g. 4 Kbyte
Random Write IOPS	Average number of random write I/O operations per second	Associated with small data transfer sizes e.g. 4 Kbyte
Sequential Read IOPS	Average number of sequential read I/O operations per second	Associated with large data transfer sizes e.g. 128 KByte
Sequential Write IOPS	Average number of sequential write I/O operations per second	Associated with large data transfer sizes e.g. 128 KByte

- MLC and XLC NAND flash designs require a high data throughput to system level to operate a reliability assessment control unit. A dedicated controller has to be connected with highest parallelism to ensure the data throughput between memory array and reliability control unit.

Failure tolerant system architectures with read recovery levels are preferred for solid-state storage devices. 3D interconnect techniques are required for fast SLC and for MLC/XLC NAND application to ensure a cost competitive high data throughput between array and on-line reliability control unit.

6.4.4 Performance Versus Energy Balancing

The parameter IOPS—**I**nput/**O**utput operations **P**er **S**econd—is known to calculate the throughput of randomly distributed read and write operations specified with a certain block size. This parameter is a performance measurement used to benchmark memory-based devices like hard disk drives (HDD), solid state drives (SSD), and storage area networks (SAN).

The performance measurement based on the parameter IOPS is organized in a procedure shown in Table 6.4 and has to be combined with the Performance Indicator Model assessment methodology.

The achieved number of IOPS can be combined with the current/energy consumption of each test sequence and results into an excellent performance and energy benchmark. Random read and write operations of small data packages can reduce the memory bandwidth of a flash based storage system significantly up to more than an order of magnitude.

The IOPS parameter can be improved on system level by an increased number of independent flash dies and/or planes and banks, a fast data interface to all flash memories (DDR interface >100 MHz) and by suspend commands for program and erase operation to prioritize read.

Application of an IOPS performance benchmark and a Performance Indicator assessment of non-volatile memories results into two conclusions:

- A strict differentiation into access oriented and storage oriented memories is a must to support a focused system energy and performance optimization.
- Array energy consumption has to be nearly independent from CMOS shrink roadmap for all array operation parameters—like the voltage driven capacitive coupling of NAND—to ensure a memory array data throughput increase without increase of required energy.

6.5 Performance Indicator Methodology Summary

The complexity of non-volatile memory development is described for flash memories and summarized into an initial flash complexity figure shown in Fig. 6.1. A straightforward approach is introduced to reduce the number of parameters impacting non-volatile memory products and to calculate instead Performance Indicator values based on a memory array model approach.

A simplified memory array model is defined and successfully applied for NAND flash memories. The main performance parameter for non-volatile memories is the program throughput—**PT**—because programming incorporates the complete physics of every non-volatile storage element. Different classes of performance indicators—**PI**—are developed to characterize and quantify the memory design and technology development options.

- **PI_A_PT: PI** characterizing memory **A**rchitecture targeting the **PT**
- **PI_AC_PT: PI** characterizing memory **A**rchitecture and **C**ost targeting the **PT**
- **PI_ACD_PT: PI** characterizing memory **A**rchitecture, **C**ost and **D**urability targeting the PT

Two examples are given widening the scope of the **P**erformance **I**ndicator **M**ethodology:

- **PI_A_RA: PI** characterizing memory **A**rchitecture targeting the **R**ead **A**ccess
- **PI_ACD_DT: PI** characterizing memory **A**rchitecture, **C**ost and **D**urability targeting the Data Throughput–a combined read and write data transfer rate

The benefit of applying the Performance Indicator Methodology is worked out for three application examples. The strength of the method is based on the unique quantitative judgment of different architectures and array cell combinations. Memory density and program performance can compensate the durability performance for the selected non-volatile memory application and the amount of compensation can be derived from the PI values.

The combination out of model-based performance indicator trend lines for a specific flash architecture, application performance entry level trend lines and the placement of real silicon data from different vendors ensures an excellent decision process regarding the number of required innovations for a specific memory development project.

The Performance Indicator Methodology was applied in Sect. 6.4 to predict and evaluate innovations required to follow the aggressive model-based performance trend lines:

- Aggressive performance doubling—additional array segmentation combined with interface improvement based on design improvements—like double data rate interfaces—and physical technology innovation—like highly parallel TSV chip interconnects.
- Reliability and performance improvement—ensured with new algorithms which program or move most of the cells at the same time to reduce the impact of all interferences—like the described all bit line array architecture.
- Array efficiency has to over compensate the expected reduction of the cell efficiency—like the introduction of 64 cells per string earlier than expected.

Published SLC and MLC NAND data have been used to verify the model-based approach. Years after the first usage of the performance trend lines [13] new publications are fitting to the predicted behavior.

XLC NAND flash technology enters step by step most of the applications enabled by SLC and 2bit/cell MLC NAND flash products. Product parameter weaknesses of 3bit and 4bit/cell XLC NAND flash can be compensated on performance indicator level, but not on single product parameter level.

The combination of density, performance and algorithm solutions on flash die and on system level ensures a system behavior which fulfills the application requirements on system level. Techniques how to handle the statistical behavior of sub 40 nm memory cells are part of the quantitative measure of the PIM to judge a non-volatile solid-state storage sub-system.

The performance indicator subset PI_ACD_XX summarizes this complete assessment of a non-volatile memory. On application level this approach is already used. The reliability figures for an SSD are specified as a maximum number of operations per system density.

The Performance Indicator Methodology can be continuously evolved step by step and can accomplish a clear differentiation in the quality of decisions on system level:

- Straightforward focus on application trends and on the analysis of various drivers behind them.
- Detailed Investigation of all key system components and their roadmaps.

  - Exclude the influence of common causes by using simplified memory models.
  - Compare only the differences between competing memories/components.

- Investigate the differences in detail and start a search process for product innovation.

The non-volatile solid-state memory technology will become a driver of real 3D non-volatile memory integration. The Performance Indicator Methodology is a powerful tool to quantify the analysis and to support the decision making along this disruptive innovation for the semiconductor industry:

- Convert the performance potential of 3D memory array technology developments into performance indicator values to quantify the additional gain in durability and performance of 3D over 2D memory array architectures.
- Focus on interface challenges supporting the increased 3D memory data throughput.

The third dimension is the additional missing parameter to compensate the 2D issues and will ensure the execution of the application based trend lines predicted by the model-based performance indicator trend charts.

References

1. T. Hara, K. Fukuda, K. Kanazawa, N. Shibata, K. Hosono, H. Maejima, M. Nakagawa, T. Abe, M. Kojima, M. Fujiu, Y. Takeuchi, K. Amemiya, M. Morooka, T. Kamei, H. Nasu, C.-M. Wang, K. Sakurai, N. Tokiwa, H. Waki, T. Maruyama, S. Yoshikawa, A 146-mm2 8-Gb Multi-level NAND flashmemory with 70-nm CMOS technology. IEEE J. Solid-State Circuits **41**, 161–169 (2006)
2. D. Byeon, S.-S. Lee, Y.-H. Lim, J.-S. Park, W. Han, P.-S. Kwak, D.-H. Kim, D.-H. Chae, S.-H. Moon, S. Lee, H.-C. Cho, J.-W. Lee, M.-S. Kim, J.-S. Yang, Y.-W. Park, D.-W. Bae, J.-D. Choi, An 8Gb Multi-Level NAND Flash Memory with 63nm STI CMOS Process Technology, in *ISSCC, Digest of Technical Papers*, pp. 46–47, San Francisco, 2005
3. K. Takeuchi, Y. Kameda, S. Fujimura, H. Otake, K. Hosono, H. Shiga, Y. Watanabe, T. Futat suyama, Y. Shindo, M. Kojima, M. Iwai, M. Shirakawa, M. Ichige, K. Hatakeyama, S. Tanaka, T. Kamei, J. Fu, A. Cernea, Y. Li, M. Higashitani, A 56nm CMOS 99mm2 8Gb Multi-level NAND Flash Memory with 10MB/s Program Throughput, in *ISSCC, Digest of Technical Papers*, pp. 507–516, San Francisco, 2006
4. G.M. Borsuk, T. Coffey, Moore's Law: A Department of Defense Perspective, in *Defense Horizons*, pp. 1–8, July 2003
5. J. Meindl, Special Issue on Limits of Semiconductor Technology, in *Proceedings of IEEE*, **89** (3), March 2001
6. D. Nobunaga, E. Abedifard, F. Roohparvar, J. Lee, E. Yu, A. Vahidimowlavi, M. Abraham, S. Talreja, R. Sundaram, R. Rozman, L. Vu, C.L. Chen, U. Chandrasekhar, R. Bains, V. Viajedor, W. Mak, M. Choi, D. Udeshi, M. Luo, S. Qureshi, A 50nm 8Gb NAND Flash Memory with 100MB/s Program Throughput and 200MB/s DDR Interface, in *ISSCC Digest of Technical Papers*, pp. 426–428, San Francisco, 2008
7. International Technology Roadmap For Semiconductors—Interconnect, *ITRS*, 2009
8. T. Tanaka, A quick intelligent page-programming architecture and a shielded bitline sensing-method for 3 V-Only NAND flash memory. IEEE JSSC Bd. **29**(11), 1366 (1994)

9. R. Cernea, Pham, F. Moogat, S. Chan, B. Le, Y. Li, S. Tsao, T.-Y. Tseng, K. Nguyen, J. Li, J. Hu, J. Park, C. Hsu, F. Zhang, T. Kamei, H. Nasu, P. Kliza, K. Htoo, J. Lutze, & Y. Dong, A 34MB/s-program-throughput 16Gb MLC NAND with all-bitline architecture in 56nm, in *ISSCC, Digest of technical Papers*, pp. 420–424, San Francisco, 2008
10. G. Marotta, A. Macerola, A. D'Alessandro, A. Torsi, C. Cerafogli, C. Lattaro, C. Musilli,D. Rivers, E. Sirizotti, F. Paolini, G. Imondi, G. Naso, G. Santin, L. Botticchio, L. De Santis, L. Pilolli, M. Gallese, M. Incarnati, M. Tiburzi, A 3bit/Cell 32Gb NAND Flash Memory at 34 nm with 6 MB/s Program Throughput and with Dynamic 2b/cell Blocks Configuration Mode for a ProgramThroughput Increase up to 13MB/s, in *IEEE ISSCC, Techn. Digest*, pp. 444–445, San Francisco, 2010
11. K.-T. Park, O. Kwon, S. Yoon, M.-H. Choi, I.-M. Kim, B.-G. Kim, M.-S. Kim, Y.-H. Choi, S.-H. Shin, Y. Song, J.-Y. Park, J.-E. Lee, C.-G. Eun, H.-C. Lee, H.-J. Kim, J.-H. Lee, A 7MB/s 64Gb 3-Bit/Cell DDR NAND Flash Memory in 20nm-Node Technology, in *IEEE ISSCC, Digest of Technical Papers*, pp. 212–213, San Francisco, 2011
12. C. Trinh, N. Shibata, T. Nakano, M. Ogawa, J. Sato, Y. Takeyama, K. Isobe, B. Le, F. Moogat, N. Mokhlesi, K. Kozakai, P. Hong, T. Kamei, K. Iwasa, J. Nakai, T. Shimizu, M. Honma, S. Sakai, T. Kawaai, S. Hoshi, J. Yuh, A 5.6MB/s 64Gb 4b/Cell NAND Flash memory in 43nm CMOS, in *ISSCC, Digest of Technical Papers*, pp. 246–247, San Francisco, 2009
13. D. Richter, Vorlesung Halbleiter Bauelemente—Nichtfluchtige Speicher Titel: NVM Shrink Roadmap 3–4bit/cell NAND - Key Performance Indicator, Munchen: TUM, Lehrstuhl fur Technische Elektronik, Fakultat fur Elektro- und Informationstechnik, 16. December 2008

Chapter 7
System Optimization Based on Performance Indicator Models

The industry is adapting multi-core microprocessor hardware in various application areas. The available additional calculation power accelerates the change from hardware into software based solutions. The system architecture has to handle the random bit failure rates occurring in all memory sub-systems in a predictable way. Software on application level combined with dedicated low level embedded software algorithms can ensure the acceptance of multi-bit and multi-level (2, 3 and 4 bit per cell) non-volatile memories.

The performance indicator methodology is now applied on system level. Performance parameters on system level are linked to application requirements based on typical use cases. A quantification of use cases is hard to make 2–4 years in front of the system introduction in the market. The memory array model based trends are applied to quantify memory and system architectures.

7.1 Economic Principles of Memory-Centric System Development

Complex systems combining hardware and software can be only developed in time applying requirement based system development processes. A flash memory based system can be classified as a complex system. The system architecture has to be derived from application cases and from application requirements.

The requirement based development defines rules how to partitioning a system and how to start the development of sub-systems. Two different ways are in use to develop a memory today:

- A memory can be developed as a memory device following the standardization and specification process linked to the memory development roadmap as described in Chaps. 2–5.
- A memory can be seen in parallel as a system component fulfilling the application and system requirements as introduced in this work in Chaps. 5 and 6.

D. Richter, *Flash Memories*, Springer Series in Advanced Microelectronics 40,
DOI: 10.1007/978-94-007-6082-0_7,

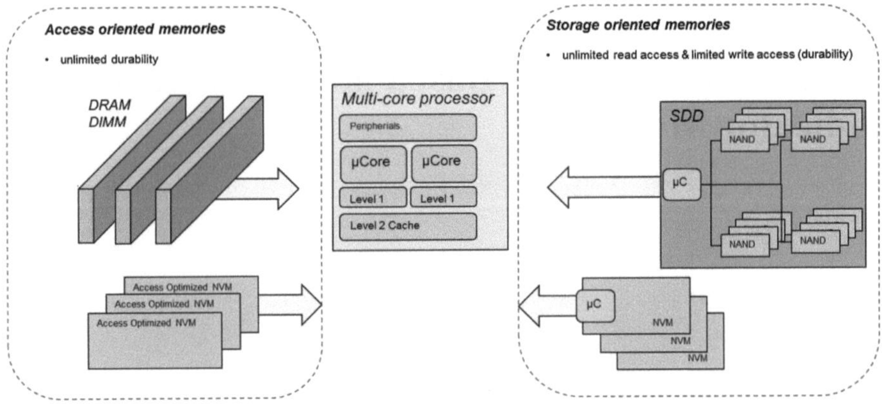

Fig. 7.1 Access and storage oriented memories

The performance indicator methodology supports the transfer of contradicting memory performance parameters into one target memory architecture. Figure 7.1 illustrates generic access and storage requirements to memories in the system context which are mapped to application trend lines.

This development is supported by a memory array model and the corresponding performance indicator trend lines to predict the array performance, cost and durability figures. We put the focus on the memory access model to guide the decision process for the system development.

The standard development process has to answer the question:

- Which memories are available and fit best to the system architecture?

A memory centric requirement based system development process generates a different question:

- Which memory solutions are required to ensure a cost competitive system?

The performance indicator characterizing memory **A**rchitecture, **C**ost and **D**urability targeting the **D**ata **T**hroughput includes non-volatile memories and is based on a combined read and write data transfer rate. All reliability parameter of a non-volatile memory are compressed into the durability part of the performance indicator. Therefore the required durability values on application level will become the important reliability parameter for the memory selection process instead of specified endurance values for a corner case combined with a dedicated JEDEC test pattern. Table 7.1 compares access and storage oriented memories and shows the preferred performance indicators.

For access oriented memories the durability parameter—endurance—is set to 10^{12} for volatile memories which are competing with access oriented non-volatile memories.

For storage oriented memories every reliability parameter has to be part of a total system performance calculation. The principle of the Reliability Factors introduced

Table 7.1 Comparison between access oriented and storage oriented memories

Access oriented memories	Storage oriented memories
• Read cycle = write cycle	• Read cycle different than write cycle
• Short random read access time	• Highly parallel read and write array access
• Energy optimized page access	• Energy optimized non-volatile write access
• Durability—continuously fault free operation on a small address size – 1 to 4 bit ECC solution for Random bit failure rate	• Durability—continuously fault free operation within a large address size – Random bit failure rate is covered with 10 to 24 bit ECC solutions
• Package optimized for data interface performance	• Package optimized for die stacking and cost
• Performance Indicators preferred: – PI_ACD_PT—durability – PI_A_RA—read access – PI_AC_RA—access and cost	• Performance Indicators preferred: – PI_A_PT—array performance – PI_AC_PT—array and cost – PI_ACD_DT—performance, cost and durability

in Sect. 4.3 is recommended to be applied to balance the system architecture between cost and reliability.

Additional rules and restriction for memory-centric system decisions have to be considered during the packaging development process. Storage oriented flash memories would gain from larger die sizes, because the high voltage circuit overhead is constant but the cell and array area is increased. The left side of Fig. 7.2 shows this storage oriented single die optimization. For high density storage application multiple-die stacking is the preferred solutions as shown on the right side of Fig. 7.2, which has different requirements to the allowed maximum die size.

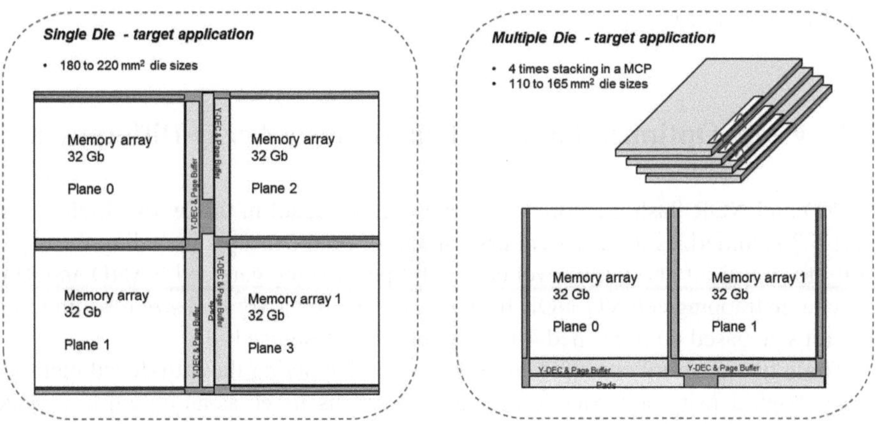

Fig. 7.2 Optimization for single die and multiple die target application—multi-chip packages "MCP"

Table 7.2 NAND interface bandwidth follows read data throughput based on page size

Plane number	Page size (kByte)	Random read cycle	Read throughput	NAND interface bandwidth 8 bit interface width
2	4	30 μs (SLC)	133 MB/s	SDR; 133 MHz; 7,5 ns cycle time DDR ; 66 MHz; 15,0 ns cycle time
2	4	50 μs (MLC)	80 MB/s	SDR; 80 MHz; 12,5 ns cycle time
2	8	30 μs (SLC)	266 MB/s	DDR ; 133 MHz; 7,5 ns cycle time
2	8	50 μs (MLC)	160 MB/s	DDR; 80 MHz; 12,5 ns cycle time
4	8	30 μs (SLC)	533 MB/s	DDR ; 266 MHz; 3,75 ns cycle time
4	8	50 μs (MLC)	320 MB/s	DDR ; 160 MHz; 6,25 ns cycle time

The performance indicator methodology developed in Chap. 6 is based on a NAND Architecture Technology Node which assumes 165 mm^2 as the target die size to double density and performance.

The data transfer rate of the specified interface has to follow the available physical page size on the NAND flash dies. Table 7.2 summaries the required interface data bandwidth for the read operations:

Reduced interface voltage values, on-die termination and other solutions are well known from the DDR-SDRAM market to integrate the required data bandwidth for non-volatile memories.

The error detection and error correction concept and the derived system architecture and the partitioning have an influence on the required data transfer rates per single NAND.

A serious assessment of competitive memory architectures and emerging memory concepts is continuously done in the industry. All required data are available to make these assessments, but the methodology to compare different architectures can be improved. The next chapter applies the Performance Indicator Methodology to compare different memory architectures based on quantitative facts derived from performance indicator trends.

7.2 System Optimization Based on Memory Array Differences

NAND and NOR flash memories are compared in detail in this work. Table 5.3 in Sect. 5.3 summarizes the classical assessment of both concepts including the capability for multi-bit and multi-level cells. 4 bit per floating gate cell NAND and 4 bit per charge trapping cell VG-NOR flash memories offer a $4F^2$ cell size resulting into $1F^2$ bit size based on published 4-bit per cell flash designs [1–3].

These two memory concepts are now assessed applying the introduced methodology. High density non-volatile memories were discussed as an option to replace large DRAM DIMM banks to save energy on system level. Multiple bits per cell and the absence of continuous refresh due to the non-volatile behaviour are two technical reasons to discuss flash as a serious DRAM replacement.

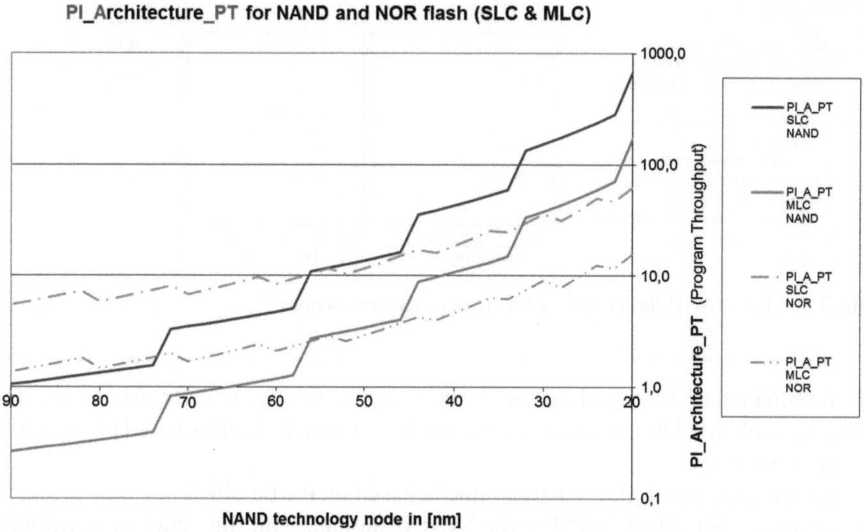

Fig. 7.3 PI_A_PT—FG SLC and MLC NAND and NOR architecture trend chart

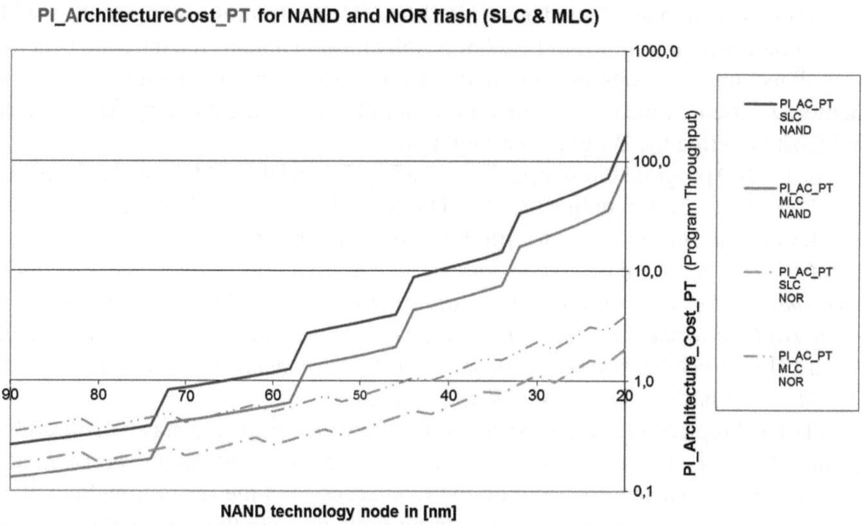

Fig. 7.4 PI_AC_PT—FG SLC and MLC NAND and NOR architecture trend chart

The performance indicator methodology is used to quantify the required factor of improvement for this expected transition. We start the analysis with performance indicators focusing on the program throughput as show in Figs. 7.3 and 7.4.

The PI_A_PT trend lines indicate a competition between both architectures and a slow transition from NOR flash dominated into a NAND flash dominated non-volatile market. The PI_A_PT does not reflect the transition from NOR to NAND as

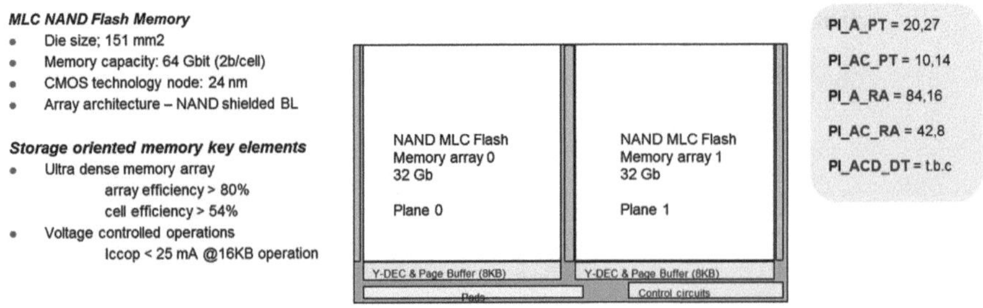

Fig. 7.5 MLC NAND flash example for storage oriented memory

the dominant flash array architecture in the market correctly. A performance indicator based on cell efficiency is an excellent tool to compare memories based on the same array architecture.

The cost driven PI_AC_PT trend line is based on the bit efficiency. The program performance benefit of NAND over NOR combined with the better bit efficiency results into a significant better performance indicator for NAND from the 70 nm technology node on as shown in Fig. 7.4.

The class of storage oriented non-volatile memories requires a memory array with highest bit efficiency shown in Fig. 7.5. NAND flash maintains the intrinsic benefits over all technology nodes and continuously outperforms the competing non-volatile memories based on the performance indicator characterizing memory **A**rchitecture and **C**ost targeting the **P**rogram **T**hroughput.

The NAND program throughput based on Fig. 7.4 and the NAND read throughput based on Table 7.2 are enabling the NAND flash technology based on achievable data transfer rates to start to compete with DRAM data transfer rates.

NAND can enter the computer domain as a non-volatile main memory to boost performance, but most of these applications are still linked to a storage oriented usage.High performance computing (HPC) is an application area in which NAND based [4] *and VG-NOR based flash solution* [5] *have chances to substitute DRAM DIMM's even in the access oriented usage domain.*

A DRAM replacement can be successful in case the competing non-volatile architecture offers a competitive read access performance. We continue the performance indicator based analysis focusing on the read access and the read throughput. The access oriented performance indicator trend lines indicate the expected behaviour that NOR outperforms NAND as shown in Fig. 7.6.

All access oriented performance indicators are dominated by the difference in first random access time between NOR and NAND. If the target application is using DRAM as an access oriented memory only a NOR based architecture can be a potential candidate to substitute DRAM based on Fig. 7.6.

The third analysis step is based on the architecture, cost and durability performance indicator targeting the **D**ata **T**hroughput.

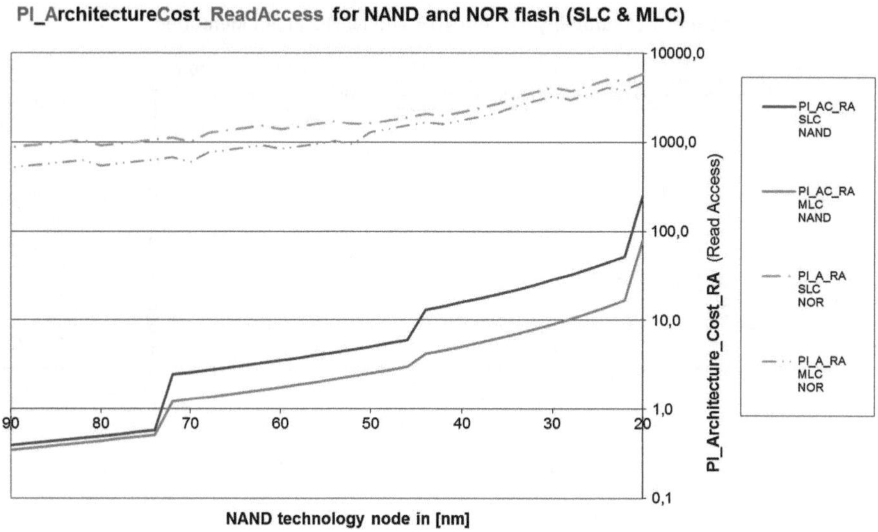

Fig. 7.6 PI_AC_RA—FG SLC and MLC NAND and NOR architecture trend chart

The model-based performance indicator analyses for storage oriented memories (PI_AC_PT) indicate a quantitative benefit of the NAND flash architecture of one order of magnitude latest from 45 nm on shown in Fig. 7.4.

The access oriented memory domain shows a differentiated result in this analysis. The PI trend chart analysis indicates the same direction. Low latency optimized SLC NAND flash can compete based on the superior cost and density position. The read cycle time for low latency NAND flash is assumed between 5 and 10 μs. The large page sizes and the corresponding number of page buffers can be judged as cache on system level and compensate the large read cycle times. The performance indicator analysis (PI_ACD_DT) based on the combined read and write data throughput indicates a serious option for SLC NAND flash to enter the market even for access oriented applications as shown in Fig. 7.7. Low latency SLC NAND flash designs have to outperform the aggressive performance indicator trend line to enter this memory market for sub 20 nm technology nodes.

A latency tolerant application can be combined with a system algorithm which predicts address request and executes speculative NAND read operations in advance. This combination has the capability to hide with a certain probability the physical random read latency of NAND. A smart system algorithm can additionally utilize the speculative reads for reliability analysis of the corresponding NAND block and supports a hidden durability improvement of NAND flash memories.

The excellent cost position and the achieved performance level required for different applications are the key points for the strong position of NAND flash memories. In case the deterministic durability and reliability behavior can be fully incorporated

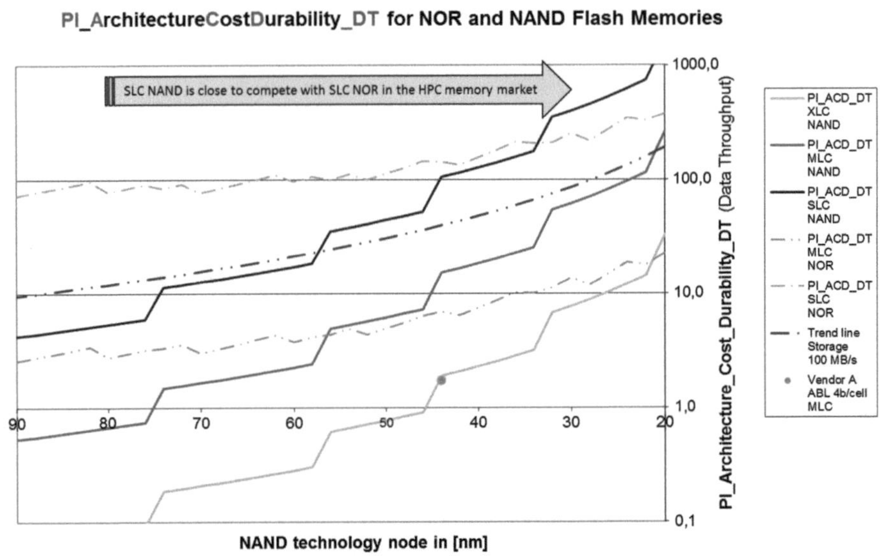

Fig. 7.7 PI_ACD_DT—FG SLC and MLC NAND and NOR architecture trend chart

by an adaptive Design for Durability NAND flash could be a serious candidate to replace portions of volatile main memories.

A very economic memory solution for access oriented memory systems is the combination of volatile and non-volatile memories within a Multi-Chip-Package (e.g. LPDRAM and PCM [6]).

7.3 Integral Memory-centric System Optimization

A memory-centric system development has to activate the available optimization potential on different integration levels. The model-based performance indicator methodology can be applied as a verification measure to assess the achieved efficiency of various system optimization strategies.

The performance indicator trend lines for SLC and MLC NAND—this time with confidence level—and the application performance trend lines define the expected design space for next generation NAND designs. Published data of SLC and MLC NAND designs are added to Fig. 7.8. MLC NAND design examples for 32, 24 and 21 nm indicate a slowdown of the performance doubling due to the effort to compensate all parasitic effects and statistical variation [7] along the 2D shrink roadmap.

Program algorithm innovations are required to compensate local process variation—bit- and word line dimensions—and statistical variation of each transistor. The predicted program throughput increase from technology node to node slows down especially for SLC and MLC. The classical 2D-NAND development effort has

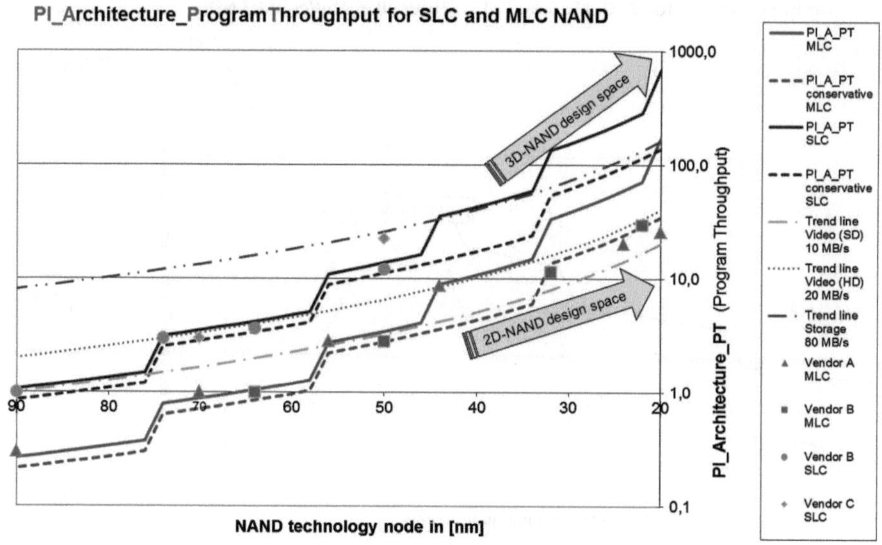

Fig. 7.8 PI_A_PT trend charts with confidence level for SLC and MLC NAND

to be spent to enable 3-bit/cell and 4 bit/cell XLC NAND designs to ensure the density increase year by year. XLC NAND flash requires V_{th} stability. The advanced time consuming XLC NAND program algorithm can compensate statistical variations.

The aggressive performance indicator trend lines based on the program throughput are not achieved by published design data as shown in Fig. 7.8. The solid-state storage based application trend will enforce and enable the required non-volatile memory architecture which combines the requested performance and density. The number of 3D-NAND publications [8, 9] have increased over the last years. 3D NAND is the most promising candidate to fulfil again the application performance trend line for non-volatile storage memories indicated in Fig. 7.8.

Figure 7.9 illustrates the growing difference between achieved MLC and XLC NAND program throughput performance values and the expected performance indicator model trend lines. The "Audio" and "Video I" application trend lines are fulfilled, which confirms the expected saturation above a certain application performance as already introduced and discussed in Fig. 6.3.

All elements defining successful XLC NAND application are part of an integral system optimization.

- Flash cell V_{th} window margin challenges are addressed by algorithm and design techniques on lowest level. V_{th} stability has to be ensured by floating gate like cell construction and by strong cell to cell interferences of all neighbour cells (as stronger the interaction as smaller the impact of the individual cell onto the V_{th} stability).
- V_{th} distribution widening has to be limited to a predictable margin reduction enforced by array operations over a specified lifetime. Durability figures and reli-

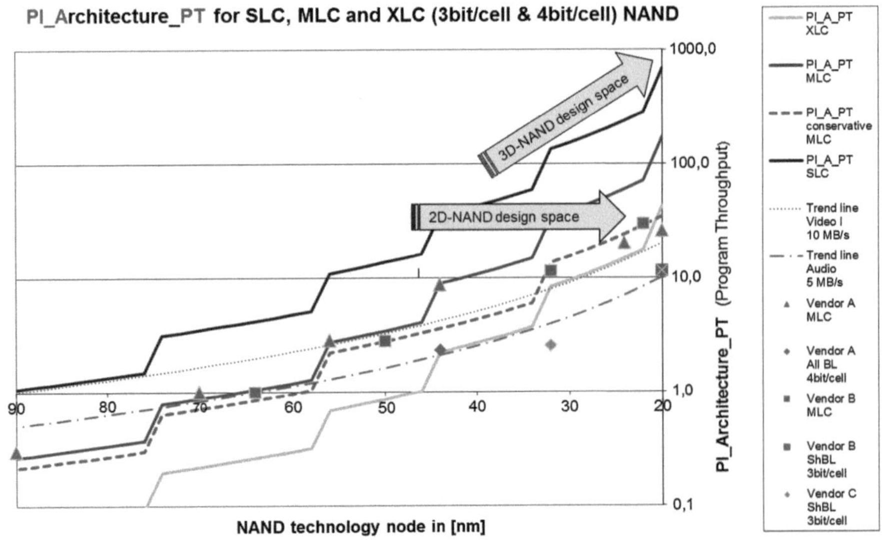

Fig. 7.9 PI_A_PT trend chart with MLC and XLC 2D-NAND and SLC 3D-NAND development trends

ability results are strongly application specific and have to be detected on lowest design level and improved with counter measures on next higher level by software solutions.

- Access optimization, remapping of addresses, elimination of worst case data pattern and repetitive refreshes are executed on flash controller level to keep and maintain the specified narrow distribution width over lifetime.

Extending the 2D MLC and XLC NAND roadmap we can conclude that a single software based optimization strategy cannot overcome the memory array and cell based effects. These weaknesses in design and technology can limit the mathematical proven potential of error correction codes. Whether the next technology generation of lithography can improve this situation cannot be assessed seriously in this work.

The other **P**erformance **I**ndicator characterizing the memory **A**rchitecture targeting the **R**ead **A**ccess is the PI_A_RA. SLC and MLC NAND performance indicator values are determined and this trend is following the predicted trend lines as shown in Fig. 7.10. The excellent read data throughput values of 400 MB/s reported in [10, 11] are not fully visible in the performance indicator values, because the read access cycle time is increased to values between 40 to 80 μs and the cell efficiency is falling below 40–50 % for the analysed designs.

Floating gate and charge trapping FinFET 3D flash cell transistors have demonstrated an increase of durability and a higher stability of retention values compared to the same cell principle in 2D. The next development applying all benefits of 3D flash cell transistors is a 3D NAND array architecture.

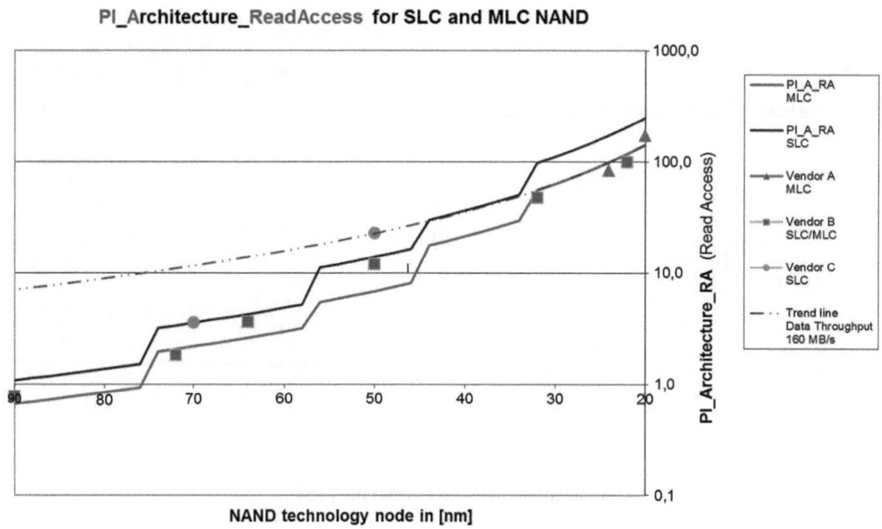

Fig. 7.10 PI_A_RA trend chart with SLC and MLC NAND design in 2012

The 3D NAND architecture publications [8, 9, 12–15] converge into a common principle architecture shown on the right side of Fig. 7.11. The third dimension offers the required flexibility to increase the bit density per F^2 and in parallel recover the program performance due to shorter global lines.

The 3D NAND array architecture has to overcome major challenges for high volume production, but 3D NAND offers intrinsic benefits for the future non-volatile solid-state based memory architecture:

- Utilization of real 3-D transistors;
- Gate length defined by the height and independent of lithography variations;
- Multiple bits are placed in different location—different cells along the Z-dimension;

The above assumed 3D NAND string architecture fulfils requirements defined for an ideal non-volatile cell concept in this work. A 3D cell improves the charge trapping cell behaviour regarding FN tunnelling operations and can enable charge trapping cells within a NAND array. The cell decision is seen as one of the critical design innovations for the 3D NAND string architecture due to the already discussed weaknesses of the charge trapping cell.

The Performance Indicator Methodology compares the concepts based on cell or bit efficiency, performance and durability data. Based on first estimations only 3D non-volatile memories can outperform the SLC performance indicator trend line and support the performance and density requirements created by multi-core microprocessor based mobile applications.

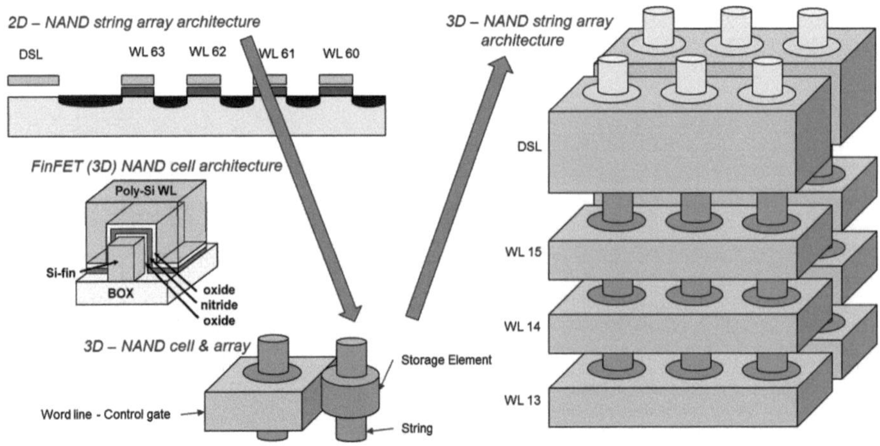

Fig. 7.11 2D versus 3D NAND array architecture concepts [16]

7.4 Failure Tolerant Systems: Design for Durability

An excellent cost position of memory-centric systems is based on an overall cost optimized non-volatile memory and on a guaranteed data integrity within a specified time frame. The mapping between the strengths of the memory array architecture and the key application requirements is supported by the introduced performance indicator methodology.

The performance parameters are linked to application requirements on typical use cases. Application knowledge is the basis to set up relevant use cases. A quantification of use cases is hard to make, therefore the model-based application trend lines are introduced to represent the application requirements. Average values of read and write data throughput are selected as the main benchmark criterion for storage oriented non-volatile memories. The memory array data rate—the program and read data throughput—is the basis for the developed set of Performance Indicators.

An access oriented non-volatile memory was the target of the Performance Indicator assessment in Sect. 7.2 comparing the two memory array architectures summarized in Sect. 2.8. NAND flash has to follow the aggressive performance indicator trend lines to be the competitive NVM in this application segment too. Memory innovations are required and expected on the storage element and on rise time reduction for all array lines by 3D memory array architectures. Durability and reliability are focused on memory and cell level. The memory architecture has to adapt failure tolerant design and architecture concepts known from safety or security systems on lowest level for access oriented memories.

The storage oriented non-volatile memories have a long history applying advanced methods for error detection and error correction. The storage oriented NAND flash memory was the target of the Performance Indicator assessment in Sect. 7.3 comparing SLC, MLC and XLC NAND flash concepts. The aggressive performance indicator trend line targeting the read access and the read throughput (PI_A_RA) is fulfilled for SLC and MLC NAND flash memories along the complete development roadmap from 90 down to 19 nm [11].

The conservative performance indicator trend line targeting the program throughput is in average still fulfilled, but not outperformed. The performance gain by the continuous doubling of the page size is partially consumed by advanced programming algorithms to compensate all technology and cell deviations and by the increase of the bit and word line rise time values. The required additional amount of spare area for EDC, ECC and adaptive read techniques reduces the achieved cell efficiency.

The predicted program performance increase is fully consumed by Design for Durability measures. The effort and the complexity to ensure the data integrity of MLC and XLC flash memories under all conditions was discussed in this work in detail and the introduced techniques are all applied today.

An additional indicator quantifying a failure tolerant memory system could support the Design for Durability during the system concept phase. The approach increasing EDC and ECC will not automatically result into a failure tolerant system. Each physical root cause for a random bit failure on application level has to be analysed. A checklist is proposed to prove the effectiveness of design and software measures to guarantee an increased failure tolerance.

The assessment of the memory system has to be complete including the counter measures for the expected random failure rate and the unexpected statistical relevant noise effects. The focus of this assessment has to be set on the identification of weaknesses, because the strength of each memory concept is well documented. Only known and accepted weaknesses of a memory sub-system can be compensated. The development effort in software to compensate weaknesses is high. Therefore the long term potential of the selected memory architecture has to be analyzed under different aspects.

A system architecture decision has to select the memory technology offering the highest potential for performance and reliability as well as reusability. The reusability is especially very important for the surrounding system software package.

Classification of data to be stored on flash based systems can improve the reliability and durability behaviour significantly. Every data package can have a property describing the reliability requirements specifically for each package or for groups of packages. A failure tolerant hardware and software memory sub-system automatically selects the suitable counter measure along the data pipeline.

References

1. A. Shappir, E. Lusky, G. Cohen, B. Eitan, NROM window sensing for 2 and 4-bits per cell products, in *NVSMWS*, Monterey, CA, (2006)
2. N. Shibata, H. Maejima, K. Isobe, K. Iwasa, M. Nakagawa, M. Fujiu, T. Shimizu, M. Honma, S. Hoshi, T. Kawaai, K. Kanebako, S. Yoshikawa, H. Tabata, A. Inoue, T. Takahashi, T. Shano,Y. Komatsu, K. Nagaba, M. Kosakai, N.Motohashi, A 70 nm 16 Gb 16-level-cell NAND flash-memory. IEEE J. Solid-State Circuits **43**, 929–937 (2008)

3. C. Trinh, N. Shibata, T. Nakano, M. Ogawa, J. Sato, Y. Takeyama, K. Isobe, B. Le, F. Moogat, N. Mokhlesi, K. Kozakai, P. Hong, T. Kamei, K. Iwasa, J. Nakai, T. Shimizu, M. Honma, S. Sakai, T. Kawaai, S. Hoshi, J. Yuh, A 5.6MB/s 64Gb 4b/Cell NAND Flash memory in 43nm CMOS, in *ISSCC, Digest of Technical Papers*, pp. 246–247, San Francisco, 2009
4. M. Deutscher, Fusion-io's flash NAS software makes your servers run 25X faster. SiliconAngle. 03 Aug 2012
5. M. Feldman, HPCWire. 22 Jan 2010. http://www.hpcwire.com/hpcwire/2010-01-22/govt_brews_national_cloud_for_science.html. Zugriff am 17 Juni 2011
6. Micron Technology, Inc., Product Brief LPDDR2-PCM and Mobile LPDDR2 121-Ball MCP. Micron Product Specification, Boise, 2012
7. M. Miranda, When every atom counts—as transistors shrink, the problem of chip variability grows. IEEE Spectrum, 30–35 (July 2012)
8. H. Tanaka, M. Kido, K. Yahashi, M. Oomura, R. Katsumata, M. Kito, Y. Fukuzumi, M. Sato, Y. Nagata, Y. Matsuoka, Y. Iwata, H. Aochi, A. Nitayama, Bit cost scalable technology with punch and plug process for ultra high density flash memory, in *IEEE Symposium on VLSI Technology, 2007*, pp. 14–15, Kyoto (2007)
9. A. Nitayama, H. Atochi, Bit cost scalable (BiCS) flash technology for future ultra high density storage devices, in *VLSI Technology Systems and Applications (VLSI-TSA)*, pp. 130–131, Hsinchu (2010)
10. C. Kim, J. Ryu, T. Lee, H. Kim, J. Lim, J. Jeong, S. Seo, H. Jeon, B. Kim, I. Lee, D. Lee, P. Kwak, S. Cho, Y. Yim, C. Cho, W. Jeong, K. Park, J.-M. Han, D. Song, K. Kyung, A 21nm High Performance 64 GbMLC NAND flash memory With 400MB/s asynchronous toggle DDR interface. IEEE J. Solid-State Circuits 47, 981–989 (2012)
11. K. Kanda, N. Shibata, T. Hisada, K. Isobe, M. Sato, Y. Shimizu, T. Shimizu, T. Sugimoto, T. Kobayashi, N. Kanagawa, Y. Kajitani, T. Ogawa, K. Iwasa, M. Kojima, T. Suzuki, Y. Suzuki, S. Sakai, T. Fujimura, Y. Utsunomiya, T. Hashimoto, A 19 nm 112.8 mm2 64 Gb multi-level flash memory with 400 Mbit/sec/pin 1.8 V toggle mode interface. IEEE J. Solid-State Circ. **48**, 1–9 (2013)
12. T. Maeda, K. Itagaki, T. Hishida, R. Katsumata, M. Kito, Y. Fukuzumi, M. Kido, H. Tanaka, Y. Komori, M. Ishiduki, J. Matsunami, T. Fujiwara, H. Aochi, Y. Iwata, Multi-stacked 1G cell/layer Pipe-shaped BiCS flash memory, in *Symposium on VLSI Circuits, 2009*, pp. 22–23, Kyoto (2009)
13. R. Katsumata, M. Kito, Y. Fukuzumi, M. Kido, H. Tanaka, Y. Komori, M. Ishiduki, J. Matsunami, T. Fujiwara, Y. Nagata, L. Zhang, Y. Iwata, R. Kirisawa, H. Aochi, A. Nitayama, Pipe-shaped BiCS flash memory with 16 stacked layers and multi-level-cell operation for ultra high density storage devices, in *Symposium on VLSI Technology, 2009*, pp. 136–137, Honolulu (2009)
14. J. Jang, H.-S. Kim, W. Cho, H. Cho, J. Kim, S. I. Shim, Y. Jang, J.-H. Jeong, B.-K. Son, D. W. Kim, Kihyun, J.-J. Shim, J. Lim, K.-H. Kim, S. Y. Yi, J.-Y. Lim, D. Chung, H.-C. Moon, Vertical cell array using TCAT (Terabit Cell Array Transistor) technology for ultra high density NAND flash memory, in *Symposium on VLSI Technology, 2009*, pp. 192–193, Honolulu (2009)
15. Y. Yanagihara, K. Miyaji, K. Takeuchi, Control gate length, spacing and stacked layer number design for 3D-stackable NAND flash memory, in *4th IEEE International Memory Workshop (IMW)*, 2012, Milan (2012)
16. C. Friederich, M. Specht, T. Lutz, F. Hofinann, L. Dreeskornfeld, W. Weber, J. Kretz, T. Melde, W. Rosner, E. Landgraf, J. Hartwich, M. Stadele, L. Risch, D. Richter, Multi-level p+ tri-gate SONOS NAND string arrays, in *IEDM Technical Digest*, pp. 1–4, Washington, 2006

Chapter 8
Conclusion and Outlook

Non-volatile memories were selected to introduce a model-based quantitative performance indicator methodology. The complex example of flash memories is used to introduce and apply the methodology to quantify product innovation during the memory development process. Non-volatile solid-state storage systems are a key enabler technology for mostly all mobile devices and for multi-core based systems in general. The development requirements for flash memories are extremely high managing the non-volatile design and technology complexity, achieving the cost targets and incorporating disruptive innovation at the right moment in time.

Identification of application trends and a matching process between these application requirements and necessary design and technology innovation are the key elements to be competitive in the high volume "commodity" non-volatile memory market.

8.1 Summary

Cell, array, performance and reliability effects of flash memories are introduced and analyzed. Key performance parameters are derived to handle the flash complexity. A performance and array memory model is developed and a set of performance indicators characterizing architecture, cost and durability is defined. The dependencies between performance, reliability and design and algorithm innovations are explained applying the V_{th} window margin analysis for flash memories.

The goal of this work was to establish a robust methodology based on technical and economic measures to quantify design and technology decisions during the flash roadmap and development process.

The complexity of hundreds of design and technology parameters is reduced into a structured set of Performance Indicators. A model-based approach is introduced to identify and predict trends of Performance Indicator values along the shrink roadmap. The graphical representation based on trend lines over technology nodes is

D. Richter, *Flash Memories*, Springer Series in Advanced Microelectronics 40,
DOI: 10.1007/978-94-007-6082-0_8, © Springer Science+Business Media Dordrecht 2014

the basis for the analysis of competitor products, innovations and system optimization strategies.

Floating gate NAND and charge trapping VG-NOR flash product experience out of five technology generations and published non-volatile memory designs are used to make a proof of concept of the model-based Performance Indicator Methodology. Single aspects of the methodology are successfully used to guide MLC NAND product specification, design and technology innovations. The complete Performance Indicator Methodology was developed over the last 4 years verifying the predicted trend lines with high volume MLC NAND product data.

The Performance Indicator Methodology is applied on SLC, MLC and 4-bit per cell XLC flash memories to demonstrate the importance of hidden memory parameters for a commercially successful product and system development roadmap. The potential and the clarity of quantitative assessments based on graphical performance indicator trend lines for performance, cost and reliability parameters of flash memories are demonstrated for different application cases. The iterative development of the memory model ensures the knowledge and visualizes the understanding for the important technology parameters. The selective usage of different efficiency parameters for the performance indicator calculation makes the difference compared to standard parameter assessments and identifies the potential of successful product innovation.

The combination of model-based performance indicator trend charts, specific application trend lines and competitor products represents a powerful toolset to develop and optimize memories and system architecture for non-volatile storage technologies.

The development of the memory models is based on an intensive characterization program of real silicon data based on two technology nodes.

A successful product development strategy for non-volatile memories applies product innovations for two technology nodes in advance and combines this design concept with increased performance or density targets to ensure a proof of concept. The array noise and reliability effect learning cycle on silicon has to be combined with stochastic modelling of intrinsic noise effects impacting the V_{th} operation window [1].

8.2 Outlook for Flash Memories

A successful memory development process has to become application and system driven.ssss

Economic principles are the key for the commercial success of semiconductor memory roadmaps.

The identification of the key application driver and the conversion into performance indicators including the complete complexity out of design, technology and embedded software are major steps for a memory sub-system concept development process.

Memory architectures fitting "100 %" to the system requirements are becoming automatically the driver for the complete market. We have to reference again to Herbert Kroemer: "*The principal applications of any sufficiently new and innovative technology always have been—and will continue to be—applications created by that technology*" [2].

Application case knowledge transferred into system requirements defines the reliability space of a memory sub-system. Intrinsic fault tolerant memory architectures are becoming the default for the commercial success of future memory array architectures. Design for Durability on system level and adaptive design techniques on memory level are mandatory for non-volatile semiconductor memories.

Error detection, error correction and software based data handling has to be a part of the memory sub-system architecture and requires a deep understanding of all statistical relevant effects and case by case dedicated and different counter measures at the optimum design level.

The introduced methodology—"How to apply model-based key performance indicators"—supports a faster decision process to develop the required innovations for cost and performance optimized products, and to execute the aggressive memory performance and architecture trend line over the next decade.

References

1. C. Friederich, J. Hayek, A. Kux, T. Muller, N. Chan, G. Kobernik, M. Specht, D. Richter, D. Schmitt-Landsiedel, Novel model for cell—system interaction (MCSI) in NAND Flash, in *IEDM Technical Digest*, (Washington, 2008), pp. 1–4
2. H. Kroemer, Nobel Physica Laureate, in *Lex Prix Nobel*, (2000)

Index

D. Richter, *Flash Memories*, Springer Series in Advanced Microelectronics 40,
DOI: 10.1007/978-94-007-6082-0,

MIX
Papier aus verantwortungsvollen Quellen
Paper from responsible sources
FSC® C105338
FSC
www.fsc.org

If you have any concerns about our products, you can contact us on
ProductSafety@springernature.com

In case Publisher is established outside the EU, the EU authorized representative is:
Springer Nature Customer Service Center GmbH
Europaplatz 3, 69115 Heidelberg, Germany

Printed by Libri Plureos GmbH
in Hamburg, Germany